Calculus

Instruction Manual

By Lisa Angle

1·888·854·MATH(6284) - mathusee.com
sales@mathusee.com

Calculus Instruction Manual

©2011 Lisa Angle
Published and distributed by Demme Learning

All rights reserved. No part of this book may be reproduced, stored in a retrieval system, or transmitted in any form by any means—electronic, mechanical, photocopying, recording, or otherwise—without prior written permission from Demme Learning.

mathusee.com

1-888-854-6284 or +1 717-283-1448 | demmelearning.com
Lancaster, Pennsylvania USA

ISBN 978-1-60826-047-8
Revision Code 0312-B

Printed in the United States of America by Bindery Associates LLC
 2 3 4 5 6 7 8 9 10

For information regarding CPSIA on this printed material call: 1-888-854-6284 and provide reference #0312-112718

Calculus

SCOPE & SEQUENCE

HOW TO USE MATH-U-SEE

LESSON 1	Terminology and Graphing
LESSON 2	Parabola, Circle, Ellipse
LESSON 3	Hyperbolas and Systems of Equations
LESSON 4	Functions
LESSON 5	Trigonometry
LESSON 6	Exponential and Logarithmic Functions
LESSON 7	Limits
LESSON 8	Limits and Continuity
LESSON 9	Definition of a Derivative
LESSON 10	Derivative Rules
LESSON 11	Chain Rule
LESSON 12	Derivatives of Trig Functions
LESSON 13	Derivative of e^x and $\ln(x)$
LESSON 14	Implicit Differentiation
LESSON 15	Graphing with the 1st Derivative
LESSON 16	Graphing with the 2nd Derivative
LESSON 17	Mean Value Theorem; L'Hôpital's Rule
LESSON 18	Physics Applications
LESSON 19	Economics Applications
LESSON 20	Optimization
LESSON 21	Related Rates
LESSON 22	Antiderivatives
LESSON 23	Integration Formulas
LESSON 24	Area Under a Curve
LESSON 25	Definite Integrals
LESSON 26	Area Between Two Curves
LESSON 27	Inverse Trigonometric Functions
LESSON 28	Integration Using an Integral Table
LESSON 29	Differential Equations
LESSON 30	Integral Application: Differential Equations

STUDENT SOLUTIONS
TEST SOLUTIONS

SYMBOLS & TABLES
GLOSSARY OF TERMS
SECONDARY LEVELS MASTER INDEX
CALCULUS INDEX

Curriculum Sequence

∫	**Calculus**
cos	**PreCalculus** with Trigonometry
xy	**Algebra 2**
Δ	**Geometry**
x^2	**Algebra 1**
x	**Pre-Algebra**
ζ	**Zeta** Decimals and Percents
ε	**Epsilon** Fractions
δ	**Delta** Division
γ	**Gamma** Multiplication
β	**Beta** Multiple-Digit Addition and Subtraction
α	**Alpha** Single-Digit Addition and Subtraction
P	**Primer** Introducing Math

Math-U-See is a complete, K-12 math curriculum that uses manipulatives to illustrate and teach math concepts. We strive toward "Building Understanding" by using a mastery-based approach suitable for all levels and learning preferences. While each book concentrates on a specific theme, other math topics are introduced where appropriate. Subsequent books continuously review and integrate topics and concepts presented in previous levels.

Where to Start
Because Math-U-See is mastery-based, students may start at any level. We use the Greek alphabet to show the sequence of concepts taught rather than the grade level. Go to MathUSee.com for more placement help.

Each level builds on previously learned skills to prepare a solid foundation so the student is then ready to apply these concepts to algebra and other upper-level courses.

Major concepts and skills for Calculus:

- Graphs of conic sections
- Exponential and logarithmic functions
- Limits and continuity
- Derivatives
- First and second derivatives
- Mean value theorem and L'Hôpital's Rule
- Applications of calculus to physics and economics
- Optimization
- Antiderivatives
- Integration formulas
- Area under a curve and between curves
- Definite integrals
- Differential equations

Find more information and products at MathUSee.com

HOW TO USE MATH·U·SEE

Welcome to *Calculus*. I believe you will have a positive experience with the unique Math·U·See approach to teaching math. These first few pages explain the essence of this methodology which has worked for thousands of students and teachers. I hope you will take five minutes and read through these steps carefully.

If you are using the program properly and still need additional help, you may contact your authorized representative or visit Math·U·See online at MathUSee.com/support.html.

THE SUGGESTED MATH-U-SEE APPROACH

In order to train students to be confident problem solvers, here are two steps that I suggest you use to get the most from the Math·U·See curriculum at this level:

Step 1. Preparation for the lesson.
Step 2. Progression after mastery.

Step 1. Preparation for the lesson.

This course assumes a knowledge of geometry, algebra 1 and 2, and precalculus (including trigonometry). The first few lessons review some important concepts from previous levels and apply them to the study of calculus. Watch the DVD to learn the concepts. Study the written explanations and examples in the instruction manual. Many students watch the DVD along with their instructor. Students at this level who have taken responsibility to study this course themselves will do well to watch the DVD and read through the instruction manual.

Step 2. Progression after mastery.

Once understanding of the new concept is demonstrated, begin doing the pages in the student text for that lesson. Mastery can be demonstrated by having each student teach the new material back to you. The goal is not to fill in worksheets, but to be able to teach back what has been learned.

Proceed to the lesson tests. They were designed to be an assessment tool to help determine mastery, but they may also be used as extra worksheets. Your students will be ready for the next lesson only after demonstrating mastery of the new concept.

Confucius is reputed to have said, "Tell me, I forget; show me, I understand; let me do it, I will remember." To which we add, **"Let me teach it and I will have achieved mastery!"**

Length of a Lesson

So how long should a lesson take? This will vary from student to student and from topic to topic. You may spend a day on a new topic, or you may spend several days. There are so many factors that influence this process that it is impossible to predict the length of time from one lesson to another. I have spent three days on a lesson, and I have also invested three weeks in a lesson. This occurred in the same book with the same student. If you move from lesson to lesson too quickly without the student demonstrating mastery, he will become overwhelmed and discouraged as he is exposed to more new material without having learned the previous topics. But if you move too slowly, your student may become bored and lose interest in math. I believe that as you regularly spend time working along with your student, you will sense when is the right time to take the test and progress through the book.

By following the two steps outlined above, you will have a much greater opportunity to succeed. Math must be taught sequentially, as it builds line upon line and precept upon precept on previously learned material. I hope you will try this methodology and move at your student's pace. As you do, I think you will be helping to create a confident problem solver who enjoys the study of math.

LESSON 1

Terminology and Graphing

Variable
  A quantity designated by a letter to which any value can be assigned.

Constant
  A quantity whose value is fixed.

Coefficient
  A constant that is multiplied by a variable. (Example: 3 is the coefficient of x in 3x.)

Function
  When two variables are so related that the value of the first variable is determined uniquely by the value of the second variable, the first variable is said to be a *function* of the second. The second variable, to which you may assign any value, is called the *independent* variable; the first variable is determined by the choice of the second variable and is called the *dependent* variable. For example, in the function y = x + 2, x is the independent variable and y is the dependent variable. When x = 3, y is forced to be 5, and when x = –1, y is forced to be 1.

  Functions exist everywhere in life. For instance, the gross dollar amount of my paycheck depends on the number of hours I worked. If I make $10 per hour, then P = 10h, where P = dollars paid and h = hours worked. The distance a car can travel at a constant speed is dependent upon the time traveled. If the car's constant speed is 40 mph, then d = 40t, where d = distance in miles and t = time in hours.

Figure 1

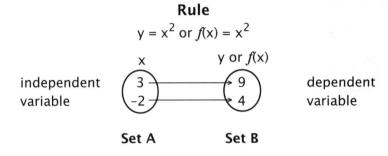

domain: all real numbers   range: numbers ≥ 0

Mathematical Function

A function is a rule which maps each x-value in set A to a unique y-value in set B.

Set A = **domain** of the function f or "values that x may have"
Set B = **range** of the function f or "values that y may have"

Absolute value of a constant a is represented by $|a|$. Thus, $|-3| = 3 = |3|$. The symbol $|a|$ is read "the absolute value of a."

Interval

We will often restrict variables to a portion of the number line. If a and b are endpoints of the *interval*, with a < b, we can include or exclude either or both of them. We use parentheses to exclude them and brackets to include them.

Here are graphical representations of each possible interval:

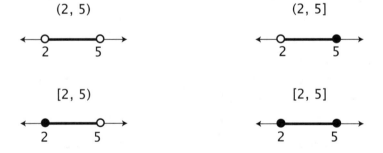

Tangent Lines

These lines play a prominent role in the study of calculus. We learned in geometry that a *tangent line* touches a curve in exactly one point.

Consider $y = 3$ and $y = 3 - x^2$. Their graphs are given below. Notice that these curves meet in exactly one point, namely $(0, 3)$. Therefore, $y = 3$ is a tangent line to $y = 3 - x^2$ at $(0, 3)$.

Figure 2

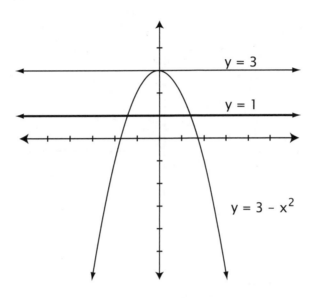

Secant Lines

Lines that touch the curve in exactly two points are called *secant lines*. The darker line $y = 1$ represents a secant line to $y = 3 - x^2$.

GRAPHING ABSOLUTE VALUE PROBLEMS IN ONE DIMENSION

The absolute value function is defined as:

$$y = \begin{cases} x \text{ for } x \geq 0 \\ -x \text{ for } x < 0 \end{cases}$$

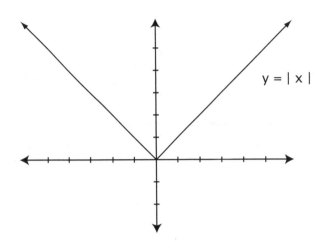

$y = |x|$

Example 1

Find the solutions to $|x + 2| = 3$.

Because there are two definitions, we have two problems to solve:

$x + 2 = 3$ and $-x - 2 = 3$

This yields two answers, $x = 1$ and $x = -5$.

Example 2

Find the solutions to $|2x + 1| = 2$.

$2x + 1 = 2$ and $-2x - 1 = 2$

This yields two answers: $x = 1/2$ and $x = -3/2$.

These next examples involve inequalities. Recall that when solving inequalities, you must reverse the direction of the inequality when you multiply or divide by a negative number.

Example 3

Graph the solutions to $|x + 2| \leq 2$.

$x + 2 \leq 2$ and $-x - 2 \leq 2$
$x \leq 0$ and $-x \leq 4$ This is the same as $x \geq -4$.

```
←——●————————●——→
   -4        0
```

Checking values in the region we shaded:
   when $x = -1$, we get $1 \leq 2$, which is true.

Checking values in the non-shaded region:
   when $x = 3$, we get $5 \leq 2$, which is false.

Example 4

$|3 - x| > 4$
$3 - x > 4$ and $-3 + x > 4$
$-x > 1$ and $x > 7$
$x < -1$

```
←——○————————————○——→
   -1           7
```

Checking values in the shaded region:
   when $x = 8$, we get $5 > 4$, which is true.

Checking values in the non-shaded region:
   when $x = 0$, we get $3 > 4$, which is false.

GRAPHING TWO DIMENSIONAL PROBLEMS

It will be a tremendous advantage to the calculus student to be able to sketch graphs accurately and quickly. Plotting points can be used for unfamiliar graphs. Familiar graphs like lines, conic sections, trig, natural log, and e^x should be easily obtained or memorized. It would also be wise to memorize the following graphs.

1. $y = \dfrac{1}{x}$

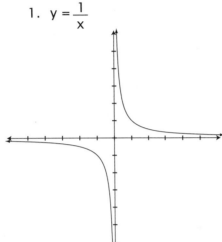

2. $y = x$

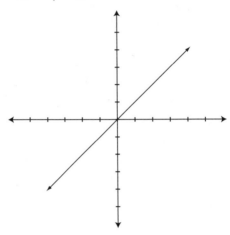

3. $y = x^2$

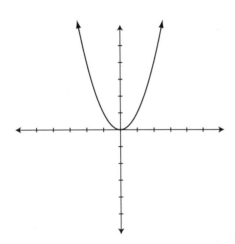

4. $y = x^3$

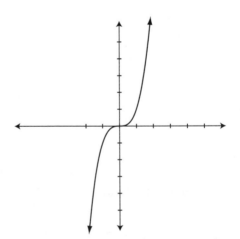

5. $y = \sqrt{x}$

6. $y = |x|$

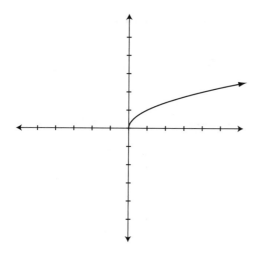
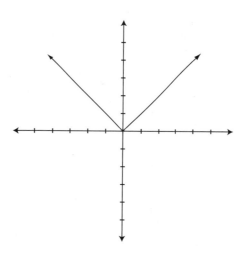

Example 5

Draw the graph of $\dfrac{1}{(x-2)}$.

x	y
0	$-\frac{1}{2}$
1	-1
1.5	-2
-2	$-\frac{1}{4}$
2.5	2
3	1
4	$\frac{1}{2}$

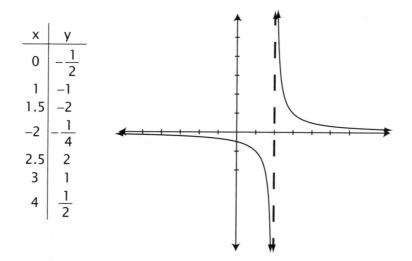

This graph is similar to y = 1/x. It is helpful to note where the graph does not exist. These points will provide vertical asymptotes. An *asymptote* is a straight line that is closely approached but never met by the curve. In this case, there is a vertical asymptote at x = 2.

Example 6

$y = 2x^2 + 3$

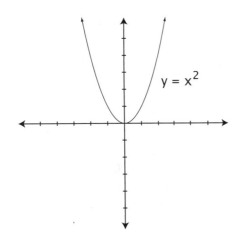

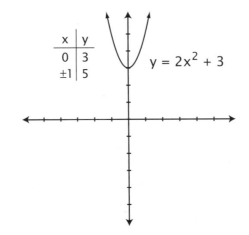

Remembering our work with translations and dilations from *Geometry*, we see that this graph is similar to $y = x^2$, but it is translated up three units. It is also narrower than the graph of $y = x^2$ because of the factor of two which makes y-values larger more quickly.

Example 7

$y = \sqrt{x - 3}$

Begin with $y = \sqrt{x}$. Shift the graph to the right 3 units.

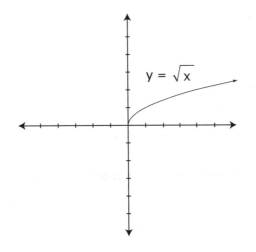

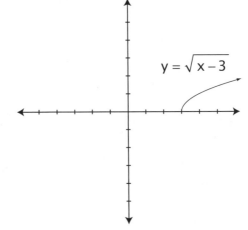

Example 8

y = | x + 2 | − 1

Begin with y = | x |.

This shifts the graph 2 units to the left and down 1 unit.

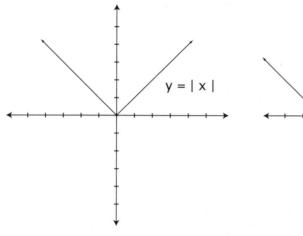
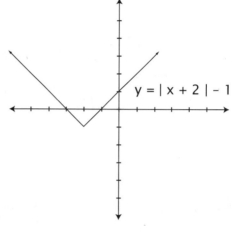

Example 9

$y = \frac{1}{2}(x - 4)^2 - 8$

Begin with y = x².

Enlarge the "U" shape by a factor of 2.

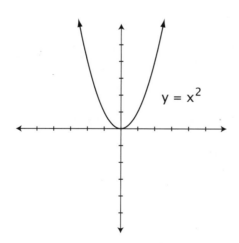
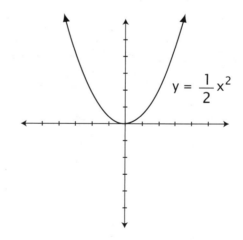

Shift 4 units right and 8 units down.

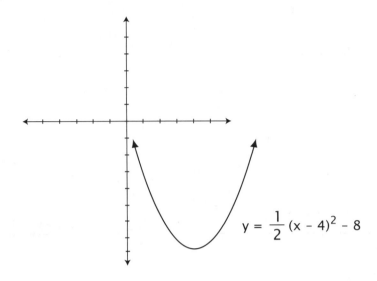

$y = \frac{1}{2}(x-4)^2 - 8$

GRAPHS WITH MULTIPLE DEFINITIONS

Graphs can be defined differently depending upon the interval. We saw earlier that the absolute value has two definitions. Here are two more examples:

Example 10

$$y = \begin{cases} x+2 \text{ if } x > 3 \\ x \text{ if } x \leq 3 \end{cases}$$

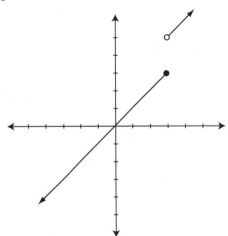

Example 11

$$y = \begin{cases} x \text{ if } x = 2 \\ -x \text{ if } x \neq 2 \end{cases}$$

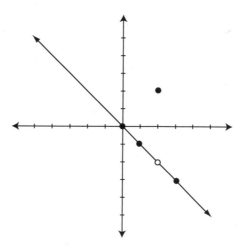

CONTINUITY

A graph which is *continuous* has no hole or break in it. If you were to draw the graph of a continuous graph from the smallest x-value to the largest x-value, you would never need to lift your pencil off the page. All lines and parabolas are continuous graphs. If there is a place where the graph is undefined, there is a discontinuity there.

Consider the following graphs:

A. $y = |x|$

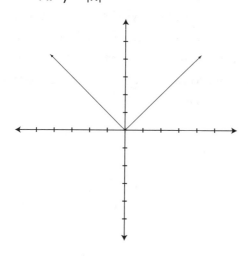

B. $y = x^2$

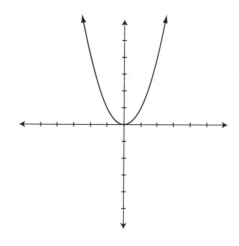

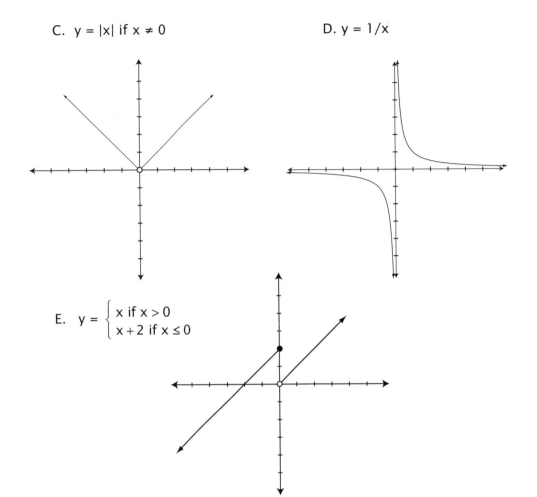

In graphs A and B, the graph "continues." In graphs C, D, and E, there is a break or hole. A and B are continuous graphs. C, D, and E have discontinuities at $x = 0$. Graphs C, D, and E illustrate the three common types of discontinuities. The discontinuity in C is the only one of those three which is considered "removable." The discontinuities in D and E are unable to be repaired. We'll see more about this concept of a removable discontinuity in later chapters.

Example 12

Does y have a discontinuity? Explain your answer. Graph it.

$$y = \begin{cases} \dfrac{1}{x} & \text{for } x \text{ in } (0, 1) \\ \sqrt{x} & \text{for } x \geq 1 \end{cases}$$

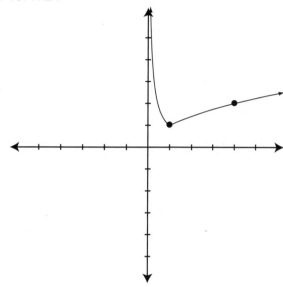

No, the graph is continuous for all x > 0.

Example 13

Does y have a discontinuity? Explain your answer. Graph it.

$$y = \begin{cases} 4 - x^2 \text{ for } x < 2 \\ x^2 \text{ for } x \geq 2 \end{cases}$$

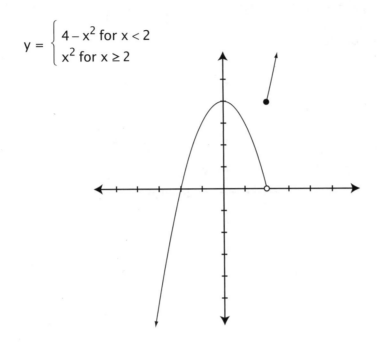

There is discontinuity at x = 2. It is not removable.

LESSON 2

Parabola, Circle, Ellipse

Quadratic comes from the Latin root QUADRATUS, which means square. A ***quadratic equation*** has no variable with an exponential power higher than 2.

Lines

Recall the slope-intercept form of a line (y = mx + b) and the standard form of a line (Ax + By = C). There is a third form for the equation of a line, $\frac{x}{a} + \frac{y}{b} = 1$, where x and y are the variable coordinates and a and b are the x- and y-intercepts respectively.

Example 1

x + y + 3 = 0.

In slope-intercept form, we have y = -x - 3.

The slope is -1 and the y-intercept is -3.

If we choose to put this equation in the new form above, we have $\frac{x}{-3} + \frac{x}{-3} = 1$.

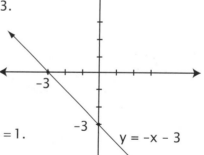

The x- and y-intercepts are both -3. See the graph at right.

Below are two more examples of graphing lines using both methods.

Example 2

$$2x + 3y = 6; \text{ dividing by } 6$$

$$\frac{x}{3} + \frac{y}{2} = 1$$

x-intercept = 3, y-intercept = 2

Using slope-intercept: $3y = -2x + 6$

$$y = -\frac{2}{3}x + 2$$

y-intercept = 2, slope = $-\frac{2}{3}$

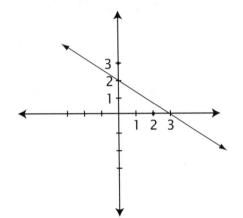

Example 3

$$-2x + 3y = 12; \text{ dividing by } 12$$

$$\frac{x}{-6} + \frac{y}{4} = 1$$

x-intercept = –6, y-intercept = 4

Using slope-intercept: $3y = 2x + 12$

$$y = \frac{2}{3}x + 4$$

$$m = \frac{2}{3}, b = 4$$

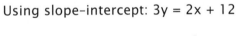

Parabolas

An example of an up/down parabola is $x^2 - y = 0$. All parabolas are identified by having one squared term. In this case, we have $y = x^2$. The standard form for these parabolas is $y = a(x - h)^2 + k$ where the vertex is (h, k). If *a* is positive, then it is a "smile" and if 'a' is negative, it is a "frown." The ***axis of symmetry*** is $x = h$.

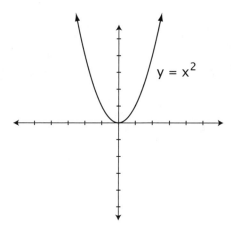

Axis of symmetry is $x = 0$.

$y = a(x - h)^2 + k$ $a = 1, h = 0, k = 0$
$y = 1(x - 0)^2 + 0 \Rightarrow y = x^2$

vertex = $(0, 0)$

Example 4

Draw a sketch of $x^2 - 4x - 8y - 20 = 0$.

$x^2 - 4x + 4 - 8y = 20 + 4$ Complete the square.
$(x - 2)^2 - 8y = 24$

$-8y = -(x - 2)^2 + 24$
$8y = (x - 2)^2 - 24$
$y = \frac{1}{8}(x - 2)^2 - 3$

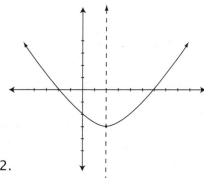

The vertex is $(2, -3)$.

The axis of symmetry is $x = 2$.

axis of symmetry

The 1/8 makes it very wide.

An example of a right/left parabola is $x - y^2 = 0$. The standard form for these parabolas is $x = a(y - k)^2 + h$. where the vertex is (h, k). If 'a' is positive, it has the shape of a *c* and if *a* is negative, a "backwards *c*." The axis of symmetry is y = k.

Example 5

$$x = 3y^2 - 6y + 7$$
$$x = 3(y^2 - 2y) + 7$$
$$x = 3(y^2 - 2y + 1) + 7 - 3 \quad \text{Complete the square.}$$
$$x = 3(y - 1)^2 + 4$$

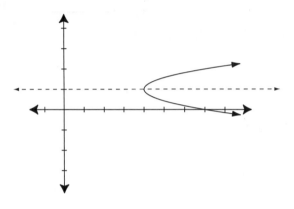

The vertex is (4, 1) and the axis of symmetry is y = 1.

Circles

The formula for a circle is $(x - h)^2 + (y - k)^2 = r^2$ with center (h, k) and radius *r*. If the center is at the origin, the formula reduces to $x^2 + y^2 = r^2$.

Example 6

Find the center and radius and draw a graph.

$$x^2 + y^2 - 8y + 4x - 5 = 0$$

Solution — Let's put this equation into the correct form by completing the square.

$$(x^2 + 4x +) + (y^2 - 8y) = 5$$

Notice that we have grouped the terms together and made room for the number to complete each square. Don't forget to add the same amount to both sides of the equation.

$$(x^2 + 4x + 4) + (y^2 - 8y + 16) = 5 + 4 + 16$$

$$(x + 2)^2 + (y - 4)^2 = 25 \text{ or } 5^2$$

center (−2, 4)
radius 5

In calculus we are concerned with functions, so we will be working with semi-circles. Recall that a function is a rule which maps each x to a unique y-value. Because we are dealing with positive square root values only, we have a semi-circle with all y-values positive.

Example 7

$$y = \sqrt{4 - x^2}$$

If you square both sides, you have $y^2 = 4 - x^2$ or $x^2 + y^2 = 4$.

Ellipses

Ellipses have a more complicated definition, but simply stated, an ellipse is similar to a circle that has been stretched in one direction. The standard formula for an ellipse is as follows:

$$\frac{(x-h)^2}{a^2} + \frac{(y-k)^2}{b^2} = 1$$

with center (h, k) and extreme points (h ± a, k) (h, k ± b).

When the center is at the origin, we have $\frac{x^2}{a^2} + \frac{y^2}{b^2} = 1$.

The formula for an ellipse is easily identified. There must be two squared terms with unequal coefficients with the same sign.

Example 8

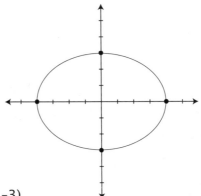

$9x^2 + 16y^2 = 144$

$\frac{x^2}{16} + \frac{y^2}{9} = 1$

$\frac{(x-0)^2}{4^2} + \frac{(y-0)^2}{3^2} = 1$

a = 4, b = 3, h = 0, k = 0

center (0, 0)
extreme points (4, 0) (−4, 0) (0, 3) (0, −3)

Example 9

$$4x^2 + 25y^2 = 100$$

Divide by 100 in order to make the equation fit the standard form.

$$\frac{4x^2}{100} + \frac{25y^2}{100} = 1$$

$$\frac{x^2}{25} + \frac{y^2}{4} = 1$$

$$\frac{(x-0)^2}{5^2} + \frac{(y-0)^2}{2^2} = 1$$

a = 5, b = 2, h = 0, k = 0

center (0, 0)
extreme points (±5, 0) (0, ±2)

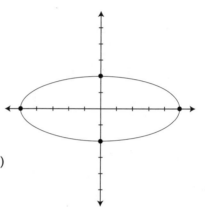

Example 10

$$x^2 + 9y^2 - 4x + 18y + 4 = 0$$

$$x^2 - 4x + 9y^2 + 18y = -4$$

Completing the square for the x term:

$$x^2 - 4x + 4 + 9(y + 2y)^2 = -4 + 4$$

Completing the square for the y term:

$$(x - 2)^2 + 9(y + 2y + 1) = 0 + 9$$

$$\frac{(x-2)^2}{9} + \frac{(y+1)^2}{1} = 1$$

$$\frac{(x-2)^2}{3^2} + \frac{(y+1)^2}{1^2} = 1$$

$a = 3, b = 1, h = 2, k = -1$

center $(2, -1)$
extreme points $(5, -1)\ (-1, -1)\ (2, 0)\ (2, -2)$

Completing the square helps us to rearrange and rewrite the unfamiliar information into a standard equation that we can recognize more readily.

LESSON 3

Hyperbolas and Systems of Equations

We will begin our discussion of hyperbolas with left-right hyperbolas. The x^2 and y^2 terms of circles and ellipses are both positive. What distinguishes a hyperbola is that one component is negative. The equation $x^2 + y^2 = 5^2$ is a circle and $x^2 + 2y^2 = 5^2$ is an ellipse, but $x^2 - y^2 = 5$ or $y^2 - x^2 = 2$ is a hyperbola. The standard form of a hyperbola centered at the origin is:

$$\frac{x^2}{a^2} - \frac{y^2}{b^2} = 1$$

Vertices are $(\pm a, 0)$. Asymptotes are $y = \pm \frac{b}{a} x$.

Example 1

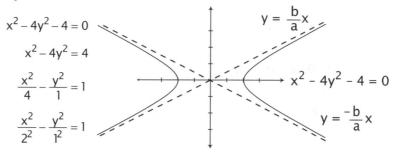

$x^2 - 4y^2 - 4 = 0$

$x^2 - 4y^2 = 4$

$\dfrac{x^2}{4} - \dfrac{y^2}{1} = 1$

$\dfrac{x^2}{2^2} - \dfrac{y^2}{1^2} = 1$

The vertices are (2, 0) and (–2, 0). The asymptotes are $y = \pm \dfrac{1}{2} x$.

The up-down or north-south hyperbolas are quite similar.

The standard form centered at the origin is:

$$\frac{y^2}{a^2} - \frac{x^2}{b^2} = 1$$

Vertices are (0, ± a). The asymptotes are $y = \pm \frac{a}{b} x$.

The *a* in the denominator is always paired with the positive component. In the standard form of a hyperbola on the previous page, a^2 is under x^2. In this example, a^2 is under y^2. When finding the asymptote using a and b, the letter that accompanies y is always the numerator, since slope is computed by the rise (y-component) over the run (x-component)!

Example 2

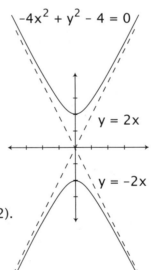

$-4x^2 + y^2 - 4 = 0$

$y^2 - 4x^2 = 4$

$\frac{y^2}{4} - \frac{x^2}{1} = 1$

$\frac{y^2}{4} - \frac{x^2}{1} = 1$

a = 2, b = 1, h = 0, k = 0

The vertices are (0, 2) and (0, -2).

The asymptotes are y = ±2x.

You may have noticed that rearranging the equation will give you more information, which helps us to draw a graph.

Lastly, there are hyperbolas which have the equation xy = +c or xy = −c.

Example 3

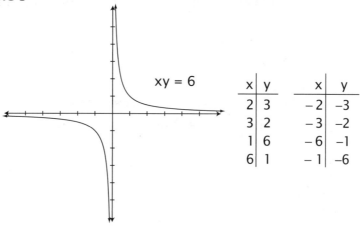

Plotting points will quickly yield a graph in quadrants I and III.

Example 4

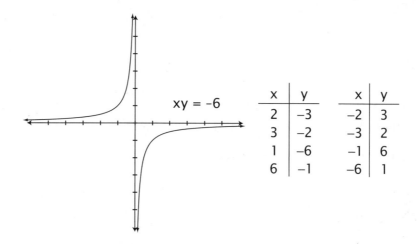

The graph of xy = −6 yields a graph in quadrants II and IV.

SYSTEMS OF EQUATIONS

Especially in the second semester of calculus, we will be learning about areas between curves, and we will need to solve systems of equations and graph them.

Example 5

Find the intersection of $y = 1 - x^2$ and $y = x^2 - 1$.

These are parabolas (one squared term). They both have a vertical axis of symmetry ($x = 0$). The first is a "frown" and the second is a "smile."

$$1 - x^2 = x^2 - 1$$
$$2 = 2x^2$$
$$1 = x^2$$
$$x = \pm 1$$

Plugging into either of the original equations, we get $y = 1 - 1 = 0$.

Intersection is $(\pm 1, 0)$

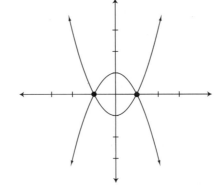

Example 6

Find the intersection of $y = x^2$ and $x - y + 2 = 0$.

$$x^2 = x + 2$$
$$x^2 - x - 2 = 0$$
$$(x + 1)(x - 2) = 0$$
$$x = -1, 2$$

Subsituting into $y = x^2$

$$y = (-1)^2 = 1$$
$$y = 2^2 = 4$$

Intersection $(-1, 1)$ $(2, 4)$

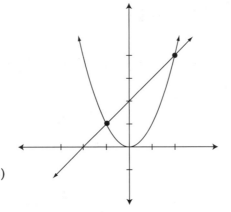

INEQUALITIES

Let's revisit example 5. Where is $y \geq 1 - x^2$? First, we graph $y = 1 - x^2$. Then we plug in arbitrary values for x and y and determine whether or not those values are above or below the curve.

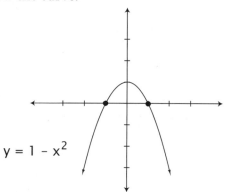

Let's try (0, 0). Plugging into $y \geq 1 - x^2$ we get:

$0 > 1 - 0^2$, or
$0 > 1$, which is false, so (0, 0) is not a solution.

Now try (1, 4). Plugging into $y \geq 1 - x^2$ we get $4 \geq 1 - 1^2$, or
$4 \geq 0$, which is true. We can now with confidence shade
our solutions, which include (1, 4) and exclude (0, 0).

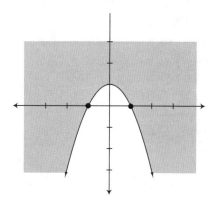

Example 7

Shade the region which represents all solutions to $y < x^2$.

Drawing $y = x^2$ with dotted lines we have:

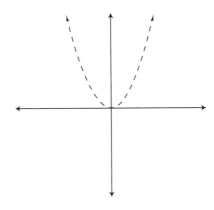

Picking (0, 2), we get $2 < 0^2$, which is false, so (0, 2) is not a solution.
Picking (0, −2) we get $-2 < 0^2$, so $-2 < 0$, which is true.

Our answer is:

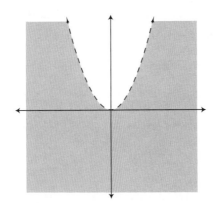

All quadratic equations follow the form:

$$Ax^2 + Bxy + Cy^2 + Dx + Ey + F = 0$$

Curve	A	B	C	D	E	F	Comments
Line	0	0	0	RN	RN	RN	D = 0 yields a horizontal line
							E = 0 yields a vertical line
Parabola (up/down)	RN	0	0	R	RN	R	A, E same signs = "frown"
							A, E different signs = "smile"
Parabola (left/right)	0	0	RN	RN	R	R	C, D same sign = "c" on left
							C, D different sign = "c" on right
Circle	RN	0	RN	R	R	RN	A must equal C
Ellipse	RN	0	RN	R	R	RN	A, C must be the same sign
							x > y = football, y > x = egg
Hyperbola (left/right)	RN	0	RN	R	R	RN	+A = symmetry about y-axis
							A & C must be opposite signs
Hyperbola (up/down)	RN	0	RN	R	R	RN	−A = symmetry about y-axis
							A & C must be opposite signs
Hyperbola (quadrants)	0	RN	0	0	0	RN	+ F = graph in quadrants II, IV
							− F = graph in quadrants I, III

0 - means that the value of the coefficient must be zero.

R - the value of the coefficient may be any positive or negative real number, including zero.

RN - the value of the coefficient may be any positive or negative real number, not including zero.

LESSON 4

Functions

The definition of a *function* is that for every x-value there is one and only one y-value that corresponds to it. Functions adhere to the vertical line test. A vertical line passes through the function and will intersect the function in only one value. A vertical line has one value for x. The vertical line x = −2 intersects the x-axis at the point −2 and extends north and south indefinitely through an infinite number of y-values. Therefore, all lines are functions with one exception—all vertical lines! Parabolas that are centered on the y-axis are functions, but those centered on the x-axis are not.

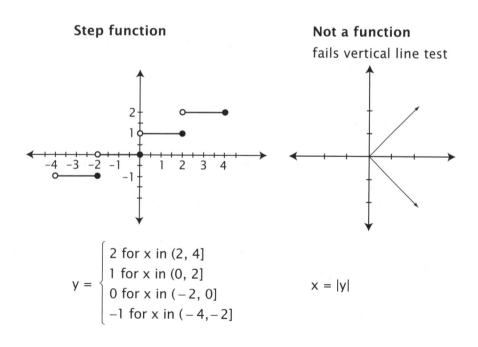

Step function

$$y = \begin{cases} 2 \text{ for x in } (2, 4] \\ 1 \text{ for x in } (0, 2] \\ 0 \text{ for x in } (-2, 0] \\ -1 \text{ for x in } (-4, -2] \end{cases}$$

Not a function
fails vertical line test

$$x = |y|$$

Natural log function

Not a function
fails vertical line test

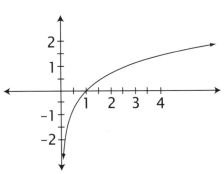

$f(x) = \ln(x)$

$x^2 + y^2 = 1$

EVALUATION

The symbol $f(x)$ is used to denote a function of x, and is read "f of x." In order to distinguish between different functions, the prefixed letter is changed, as $F(x)$, $G(x)$, $f(x)$, etc.

Evaluating functions involves substituting the value of the variable in each occurrence of x in the equation. To evaluate $f(x) = x^2 - 9x + 14$ when x = 0, we have $f(0) = 0^2 - 9(0) + 14 = 14$.

Here are two more examples of evaluating the function when x = –1 and when x = 3:

$$f(-1) = (-1)^2 - 9(-1) + 14 = 24.$$
$$f(3) = 3^2 - 9(3) + 14 = -4.$$

Evaluating the function creates points on the function. The points created above are (0, 14), (–1, 24), and (3, –4).

We can also substitute letters for the independent variable:

$$f(b) = (b)^2 - 9(b) + 14 = b^2 - 9b + 14$$
$$f(b + 1) = (b + 1)^2 - 9(b + 1) + 14 = b^2 + 2b + 1 - 9b - 9 + 14$$
$$= b^2 - 7b + 6$$
$$f(x + h) = (x + h)^2 - 9(x + h) + 14 = x^2 + 2Xh + h^2 - 9x - 9h + 14$$

FUNCTIONS: DEPENDENT AND INDEPENDENT VARIABLES

Consider the line $y = x + 1$. To graph this line by plotting points, we choose values for x and place them in the equation to determine the value for y. If x was chosen to be 5, then y would be determined to be 6. Therefore, y is a dependent variable. Its value depends on what is chosen for x.

Simple functions have both a dependent variable and an independent variable (y and x respectively in the above example). We would say that y is a function of x. Function notation replaces the dependent variable with f(independent variable). So, our example would read $f(x) = x + 1$. A graph is given below.

Figure 1

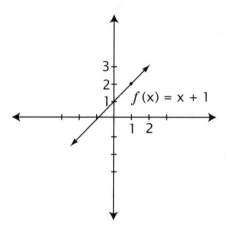

Let's try another: $y = x^2 - 1$. Again, x is the independent variable (the one we choose), and y is dependent upon the value of x chosen. If $x = 2$, then $y = 3$, creating the point (2, 3) on the graph. If $x = -1$, then $y = 0$. Written in function notation, we have $f(x) = x^2 - 1$.

Figure 2

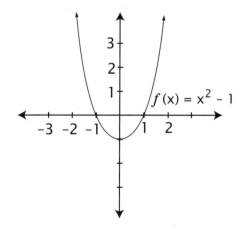

Example 1

$$f(x) = x - 3$$

Find $f(4)$, $f(-3)$, $f(a)$, $f(a + b)$.

$f(4) = 4 - 3 = 1$ $\qquad$ $f(a) = a - 3$
$f(-3) = -3 - 3 = -6$ $\qquad$ $f(a + b) = a + b - 3$

Moreover, we use the name *domain* to mean the values that the independent variable can assume and *range* to mean the values that the dependent variable may assume. Domain is all the permissible x-values and range is the permissible y-values. If $y = \sqrt{x}$ we see that the values of x are limited to all positive values and zero. The domain can be expressed as $x \geq 0$. (We will not be working with complex numbers, so the negative values are excluded.) The range can be expressed as $y \geq 0$ also. (We are assuming principal square roots in this course, so the answer to the square root of a number will be positive.) Infinity is not a specific value that exists. Positive infinity communicates that the values are getting larger and larger positively while negative infinity conveys that values are getting smaller and smaller negatively. Using interval notation we have: domain $[0, \infty)$ and range $[0, \infty)$.

Example 2
Find the domain and range for each of the following:

1. $f(x) = x - 3$ domain and range are both $(-\infty, +\infty)$
2. $f(x) = x^2 + 2$ domain $(-\infty, +\infty)$; range $[2, \infty)$
3. $f(x) = e^x$ domain $(-\infty, +\infty)$; range $(0, +\infty)$
4. $f(x) = |x| - 1$ domain $(-\infty, +\infty)$; range $[-1, +\infty)$

NOTATION

In practical applications we see that the perimeter of a square is dependent upon the length of a side. Written mathematically, we have $p = 4s$. If the side has a length of 2 inches, then the perimeter is 8 inches. Written in function notation, we have $p(s) = 4s$.

Example 3

Rewrite $y = \sqrt{x}$ in function notation. Graph this function.

$y = \sqrt{x}$ in function notation is $f(x) = \sqrt{x}$

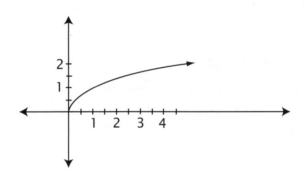

Example 4

Rewrite $y = \sin(x)$ in function notation. Graph this function.

$y = \sin(x)$ in function notation is $f(x) = \sin(x)$.

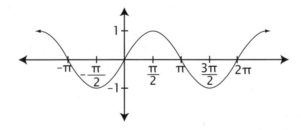

Example 5
Write a function that shows that the area of a square depends on the length of a side.

$$A(s) = s^2$$

Example 6
Write a function that shows the area of a circle depends on the length of the radius.

$$A(r) = \pi r^2$$

COMPOSITE FUNCTIONS

Composite functions are defined as a function of a function. If $f(x) = x - 3$ and $g(x) = x + 2$, then we can evaluate $f(g(x))$. Begin in the middle and work out. Put $g(x)$ into the function f for the value of x. We get $f(g(x)) = (x + 2) - 3$ which becomes $x - 1$.

Example 7
If $f(x) = 2x + 3$ and $g(x) = x^2$, then find $f(g(x))$ and $g(f(x))$.

$$f(g(x)) = f(x^2) = 2x^2 + 3$$
$$g(f(x)) = g(2x + 3) = (2x + 3)^2 = 4x^2 + 12x + 9$$

Example 8
If $r(x) = 2 - x$ and $s(x) = 2x + 1$, then find $r(s(x))$ and $s(r(x))$.

$$\begin{aligned}r(s(x)) &= r(2x + 1)\\ &= 2 - (2x + 1)\\ &= 2 - 2x - 1\\ &= 1 - 2x\end{aligned}$$

$$\begin{aligned}s(r(x)) &= s(2 - x)\\ &= 2(2 - x) + 1\\ &= 4 - 2x + 1\\ &= 5 - 2x\end{aligned}$$

INVERSE FUNCTIONS

Consider $f(x) = 3x$.

Figure 3

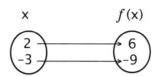

We know that functions of x map each x-value to a specific $f(x)$, or y-value.

The inverse of $f(x)$, denoted as $f^{-1}(x)$, is the rule that maps $f(x)$ back to x. In this case, $f^{-1}(x) = x/3$. To check this, determine $f^{-1}(f(x))$. Since f^{-1} and f are inverse functions, you should have $f^{-1}(f(x)) = x$ every time. In our case $f^{-1}(f(x)) = \frac{1}{3}(3x) = x$. It checks.

Let's look at the graphs of inverse functions. Notice that each line is the same as the other line reflected about y = x.

Figure 4

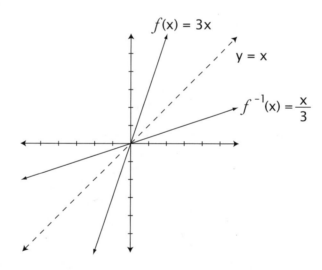

Here's how to find the inverse of a function using the function $f(x) = 3x$.

1. Replace $f(x)$ with y. $y = 3x$
2. Interchange x and y. $x = 3y$
3. Solve for y. $y = \frac{x}{3}$ or $y = \frac{1}{3}x$
4. Rename y as $f^{-1}(x)$. $f^{-1}(x) = \frac{x}{3}$

Does every function have an inverse function? No. Consider $f(x) = x^2$.

1. Replace $f(x)$ with y. $\qquad$ $y = x^2$
2. Interchange x and y. $\qquad$ $x = y^2$
3. Solve for y. $\qquad$ $y = \pm\sqrt{x}$
4. Rename y as $f^{-1}(x)$. $\qquad$ $f^{-1}(x) = \pm\sqrt{x}$

You can see by the vertical line test that $f^{-1}(x)$ is not a function. A quick way to ascertain if the original function has an inverse function is to use the horizontal line test. (Check to see if all horizontal lines pass through only one x-coordinate.) If the original function passes the horizontal line test, then the inverse function will exist and will pass the vertical line test. In our example the horizontal line test fails and thus there is not an inverse function. There is an inverse to $f(x) = x^2$, but it is not a function.

Figure 5

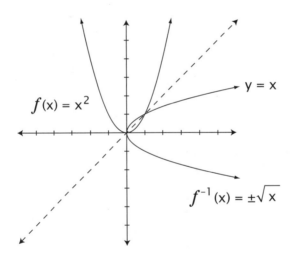

Notice that the reflection of $y = x^2$ in $y = x$ is not a function. In order for the inverse function to exist, the point $(-2, 4)$ which belongs to $f(x) = x^2$ should have a companion point in the inverse function, namely $(4, -2)$. This cannot happen because the square root of 4 is 2. Basically a reflection does exist, but it is not a function, so there is no inverse function.

Example 9

Find the inverse to the following function: $f(x) = -x + 2$.

1. $y = -x + 2$
2. $x = -y + 2$
3. $y = -x + 2$
4. $f^{-1}(x) = -x + 2$

Example 10

Find the inverse to the following function: $f(x) = 4x + 3$.

1. $y = 4x + 3$
2. $x = 4y + 3$
3. $y = \frac{1}{4}x - \frac{3}{4}$
4. $f^{-1}(x) = \frac{1}{4}x - \frac{3}{4}$

LESSON 5

Trigonometry

In calculus, we generally use radian measure. Therefore, the serious student should be familiar with the values of all of the trigonometric functions at $0, \frac{\pi}{6}, \frac{\pi}{4}, \frac{\pi}{3}, \frac{\pi}{2}$, and their multiples. This is a review of what you learned in *PreCalculus*.

Recall that y = sin(x) has a period of 2π and looks like:

Figure 1

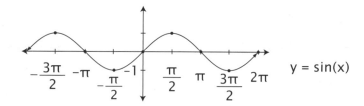

And y = cos(x) has a period of 2π and looks like:

Figure 2

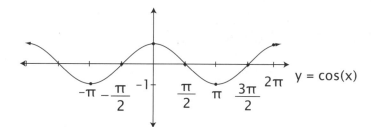

And y = tan(x) has a period of π and looks like the graph below.

Figure 3

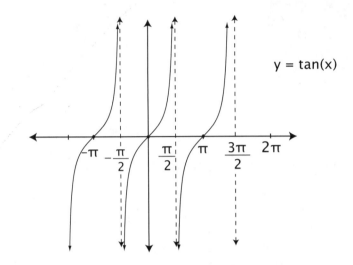

$$\tan(x) = \frac{\sin(x)}{\cos(x)}$$

Where cos(x) becomes 0, we have a vertical asymptote. Memorize these graphs!

You will recall that evaluating a trig expression involving a 30, 45, or 60 degree angle involves making a triangle and using our knowledge of special triangles.

Example 1

$$\tan\left(\frac{\pi}{4}\right) = \tan 45° = \frac{\text{opposite}}{\text{adjacent}} = \frac{1}{1} = 1$$

Example 2

$$\tan\left(\frac{\pi}{6}\right) = \tan 30° = \frac{\text{opposite}}{\text{adjacent}} = \frac{1}{\sqrt{3}} = \frac{\sqrt{3}}{3}$$

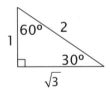

You should be able to quickly obtain all the values in this table:

θ	radians	sin(θ)	cos(θ)	tan(θ)
0°	0	0	1	0
30°	$\frac{\pi}{6}$	$\frac{1}{2}$	$\frac{\sqrt{3}}{2}$	$\frac{\sqrt{3}}{3}$
45°	$\frac{\pi}{4}$	$\frac{\sqrt{2}}{2}$	$\frac{\sqrt{2}}{2}$	1
60°	$\frac{\pi}{3}$	$\frac{\sqrt{3}}{2}$	$\frac{1}{2}$	$\sqrt{3}$
90°	$\frac{\pi}{2}$	1	0	undefined

From this table, csc(θ), sec(θ), and cot(θ) would also be easily obtained with the appropriate reciprocals.

Figure 4

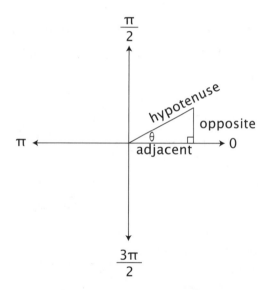

From the drawing above, you see that the opposite side, which is the numerator for sin(θ), corresponds to a y-value and is therefore positive in quadrants I and II. Likewise, the adjacent side, which is the numerator for cos(θ), corresponds to an x-value and is positive in quadrants I and IV. This information is found in the chart in figure 5.

Figure 5

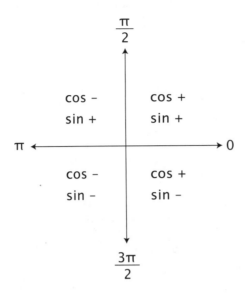

In quadrants I and III, $\tan(\theta)$ will be positive because $\tan(\theta) = \sin(\theta)/\cos(\theta)$. You can also determine the signs of all the other trigonometric values. In quadrants I and II both $\sin(\theta)$ and $\csc(\theta)$ are positive and in quadrants I and IV both $\cos(\theta)$ and $\sec(\theta)$ are positive.

Example 3

Find $\sin\left(\dfrac{5\pi}{3}\right)$ (without a calculator).

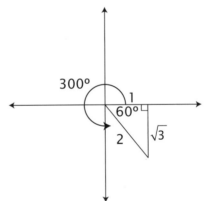

What quadrant is it in?

Quadrant IV, because $\dfrac{5\pi}{3}$ is 300°. Therefore, the answer is negative.

Draw a diagram. The reference angle is 60° or $\dfrac{\pi}{3}$.

Conclusion: $\sin\left(\dfrac{\pi}{3}\right) = \dfrac{\sqrt{3}}{2}$, so $\sin\left(\dfrac{5\pi}{3}\right) = -\dfrac{\sqrt{3}}{2}$.

Example 4

Find $\tan\left(\dfrac{5\pi}{4}\right)$.

What quadrant is it in?

Quadrant III, because $\dfrac{5\pi}{4} = 225°$.

Therefore, the answer is positive.

Draw a diagram. The reference angle is 45° or $\dfrac{\pi}{4}$.

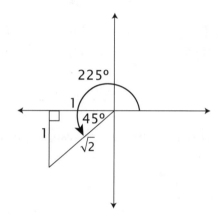

Conclusion: $\tan\left(\dfrac{\pi}{4}\right) = 1$, so $\tan\left(\dfrac{5\pi}{4}\right) = 1$.

The general forms of the sine and cosine functions are:

$y = d + a \sin [b(x - c)]$ and $y = d + a \cos [b(x - c)]$ where:

d = vertical shift, $|a|$ = amplitude, period = $\frac{2\pi}{b}$, and c = phase shift.

Amplitude ($|a|$) – one-half of the distance between the maximum and minimum values of the periodic function.

Vertical shift (d) – if d is positive, the graph is shifted d units upwards; if it is negative, it is shifted d units downwards.

Phase shift (horizontal shift) (c) – if c is positive, the shift is c units to the right; if c is negative, the shift is c units to the left.

Period = $\frac{2\pi}{b}$ – horizontal length of one complete cycle

Frequency = b – number of cycles the function completes in a period of 2π.

Note: Frequency = 1/Period.

If $f(x) = 1 + 3 \sin \left[2\left(x - \frac{\pi}{2} \right) \right]$, then the amplitude = 3, the period = $\frac{2\pi}{2} = \pi$, the phase shift is $\frac{\pi}{2}$ (to the right) and the vertical shift is 1.

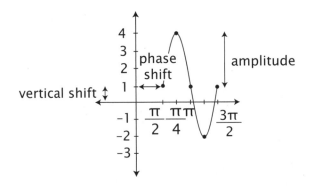

Example 5

What is the frequency of $y = \sin\left(\frac{1}{2}x\right)$?

The graph $y = \sin\left(\frac{1}{2}x\right)$ does not complete an entire cycle.

It only completes 1/2 of its cycle. The 1/2 in front of the x term gives the frequency.

Example 6

Find the amplitude, period, vertical shift, and phase shift. Draw a sketch.

$y = 1 + 2\sin\left[\frac{1}{2}(x+\pi)\right]$ amplitude = 2, vertical shift = 1, period = 4π, phase shift = -π

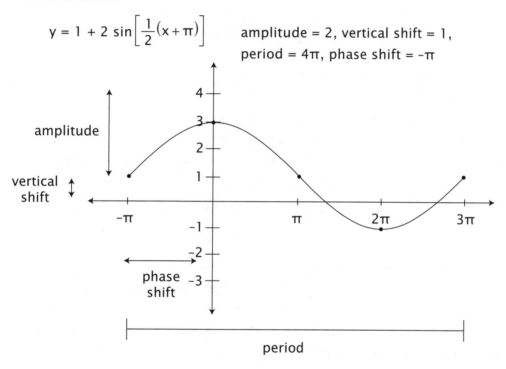

Example 7

Solve for x: sin(2x + π) = 1 at [0, π].

$$\sin(\theta) = 1, \text{ when } \theta = \frac{\pi}{2}, \frac{5\pi}{2}, \frac{9\pi}{2} \ldots$$

$$2x + \pi = \frac{\pi}{2} \qquad 2x + \pi = \frac{5\pi}{2} \qquad 2x + \pi = \frac{9\pi}{2}$$

$$2x = -\frac{\pi}{2} \qquad 2x = \frac{3\pi}{2} \qquad 2x = \frac{5\pi}{2}$$

$$x = -\frac{\pi}{4} \qquad x = \frac{3\pi}{4} \qquad x = \frac{5\pi}{4}$$

$x = -\frac{\pi}{4}$ and $x = \frac{5\pi}{4}$ are outside [0, π].

$x = \frac{3\pi}{4}$ falls within [0, π], so it is the solution.

When solving these problems go to the left and right of the interval until you find solutions that do not satisfy the requirements.

To sum up this lesson, we'll use the variables T, A, P, and S to represent the four factors that influence our graphing. Different texts use different variables to represent these translations. It makes it easier if we use the initials for the words themselves.

T = translation (vertical shift)   P = used to determine period.
A = amplitude   S = shift (phase shift)

Here is a little compilation of the salient features of the graph of the sine and cosine function. I hope it helps.

y = A sin [P (θ - S)] + T			
↕ A > 1	P > 1 squishes	S > 0 →	T > 0 ↑
↕ A < 1	0 < P < 1 expands P < 0 inverts the graph	S < 0 →	T < 0 ↓

LESSON 6

Exponential and Logarithmic Functions

Exponential functions are of the form $y = a^x$ where *a* is a constant greater than zero and not equal to one and x is a variable. Both $y = 2^x$ and $y = e^x$ are exponential functions. The function, e^x, is extensively used in calculus. You should memorize its approximate value when x = 1. ($e^1 \approx 2.718$)

You should also be able to quickly graph $y = e^x$ without the aid of a calculator. A simple graph of $y = e^x$ is shown below.

Figure 1

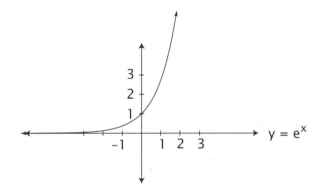

LOGARITHMIC FUNCTIONS

The equation $y = \log_a x$ is the same as $a^y = x$. The inverse of the exponential function is $y = a^x$. In this course we will restrict our study of logarithms to log base e which will be written as ln(x). The equation $y = \ln(x)$ is the inverse function of $y = e^x$. Notice that the graph of ln(x) is a reflection of graph of e^x around the line y = x. You should be able to quickly sketch from memory $y = \ln(x)$. It will also be important to remember the basic logarithm rules listed on the next page.

Figure 2

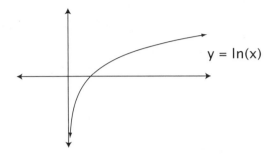

I. $\ln(1) = 0$

II. $\ln(e) = 1$

III. $\ln(e^x) = x$

IV. $e^{\ln(x)} = x$

V. Product: $\ln(xy) = \ln(x) + \ln(y)$

VI. Quotient: $\ln(x/y) = \ln(x) - \ln(y)$

VII. Power: $\ln(x^a) = a\ln(x)$

Remember that the natural log of a negative number is undefined. Some books specify ln(x) as ln |x|. We will use ln(x) for this book. Be careful to use only positive, non-zero values for x when employing the natural log function.

The natural log function can be used to free variable exponents from their exponential functions. Conversely, the exponential function can do the same for the natural log functions.

Example 1

Solve for x.

$$e^{2x} = 1$$

Taking ln of both sides:

$$\ln(e^{2x}) = \ln(1)$$
$$2x = 0$$
$$x = 0$$

checking $e^{2(0)} = e^0 = 1$

Example 2

Solve for x.

$$\ln(x + 5) = 0$$

Use each side of the equation as the exponent for e.

$$e^{\ln(x + 5)} = e^0$$
$$x + 5 = 1; \text{ so } x = -4$$

Sometimes the equations are complex and we need to use substitution to solve them. See example 3 on the next page.

Example 3
Solve for x.

$$e^{2x} - 4e^x + 3 = 0$$

Substituting $u = e^x$, $u^2 - 4u + 3 = 0$.

Factoring, we get $(u - 3)(u - 1) = 0$.

Replacing u with e^x, we get $(e^x - 3)(e^x - 1) = 0$.

Solving each factor, we get: $e^x = 3$; $e^x = 1$.

Taking the ln of both sides:

$$\ln(e^x) = \ln(3) \qquad \ln(e^x) = \ln(1)$$
$$x = \ln(3) \qquad\qquad x = 0$$

Example 4
Draw the graph of $y = 2e^x$ and its inverse.

$y = 2e^x$

$x = 2e^y$ switch variables

$\frac{1}{2}x = e^y$

$\ln\left(\frac{1}{2}x\right) = \ln(e^y)$

$\ln\left(\frac{1}{2}x\right) = y$

$f^{-1}(x) = \ln\left(\frac{1}{2}x\right)$

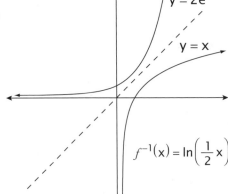

LESSON 7

Limits

You will recall from *PreCalculus* that a limit as x approaches a value 'a' is found by considering the value of the function for points very close to 'a'.

Figure 1

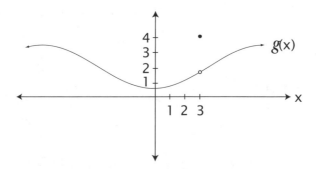

If a = 3 in g(x) above, the value of g(a) = g(3) = 4. (See figure 1.) **The value of the function might be the limit, but it is not required to be the limit.** To evaluate the limit of g(x) as x approaches 3, we must look at the values of the function both from the left and the right. Coming from the left, the function appears to be approaching a value of 2 as x gets closer to 3. Coming from the right, the function appears to be approaching a value of 2 again. Therefore, because the right-hand and left-hand limits agree, the value of the limit is 2. This is written in mathematical notation like this:

$$\lim_{x \to 3} g(x) = 2.$$

Consider the following series: $3 + 1, 3 + 1/3, 3 + 1/9, \ldots 3 + 1/3^n$. What is the limit of this series? It should be obvious that the limit is 3. Will this series ever reach 3? No, but it will get very close!

Some limits are the same as the value of the function. $\lim_{x \to 2} (x + 3)$ is read as "the limit as x approaches 2 of x + 3." The function $f(x) = x + 3$ is a line. There are no holes or discontinuities, therefore the value of the limit as x approaches 2 will be the value of the function at x = 2. Therefore, $\lim_{x \to 2} (x + 3) = 2 + 3 = 5$.

To calculate the limit of a function, the following theorems apply:

Given that u, v, and w are functions of a variable x and $\lim_{x \to a} u = A$, $\lim_{x \to a} v = B$, and $\lim_{x \to a} w = C$.

1. $\lim_{x \to a} C = C$.

   The limit of a constant is the constant.

2. $\lim_{x \to a} (u + v - w) = \lim_{x \to a} u + \lim_{x \to a} v - \lim_{x \to a} w = A + B - C$

   The limit of the sum is the sum of the limits.

3. $\lim_{x \to a} \dfrac{u}{v} = \dfrac{A}{B}$ if B is not zero.

   The limit of the quotient is the quotient of the limits as long as the denominator ≠ 0.

4. $\lim_{x \to a} Cu = C \lim_{x \to a} u$

   The limit of a constant times a function is the constant times the limit of the function

5. $\lim_{x \to a} (uvw) = ABC$

   The limit of the product is the product of the limits.

Example 1
$$\lim_{x \to 2} 4 = 4$$

Rule 1 applies. The limit of a constant is the constant. $f(x) = 4$ is the function we are evaluating. The graph is a horizontal line.

For all values of x, $f(x) = 4$, so the $\lim_{x \to 2} 4 = 4$.

Example 2
$$\lim_{x \to 2}(x^2 + 4x) =$$
$$\lim_{x \to 2} x^2 + \lim_{x \to 2} 4x = \lim_{x \to 2}(x)(x) + 4x = (2)(2) + 4(2) = 12$$

Rule 2 applies. The limit of the sum is the sum of the limits.

Example 3
$$\lim_{x \to 2}\frac{(x^2 - 9)}{(x + 2)} = \frac{\lim_{x \to 2}(x^2 - 9)}{\lim_{x \to 2}(x + 2)} = -\frac{5}{4}$$

The limit of the quotient is the quotient of the limits as long as the denominator $\neq 0$.

Example 4
$$\lim_{\theta \to \frac{\pi}{6}} 3\sin(\theta) = 3\lim_{\theta \to \frac{\pi}{6}} \sin(\theta) = 3\left(\frac{1}{2}\right) = \frac{3}{2}$$

The limit of a constant times a function is the constant times the limit of the function.

Example 5

$$\lim_{x \to 0} x\cos(x) = \lim_{x \to 0} x \cdot \lim_{x \to 0} \cos(x) = 0 \cdot 1 = 0$$

The limit of the product is the product of the limits.

Example 6

$$\lim_{x \to 1} \frac{(x^2-1)}{(x-1)}$$

If we attempt to evaluate $f(1)$, we get an undefined answer. (division by zero)

This does NOT necessarily mean that the limit does not exist. We can factor out the discontinuity at x = 1.

When you get 0/0 for the answer when evaluating, you can be reasonably sure that your rational fraction can be factored and reduced.

If you get a constant divided by 0, then the limit does not exist.

We get $\lim\limits_{x \to 1} \dfrac{(x+1)\cancel{(x-1)}}{\cancel{(x-1)}} = \lim\limits_{x \to 1} x+1 = 1+1 = 2.$

Example 7

$$\lim_{s \to a} \frac{(s^2-a^2)}{(s-a)} = \lim_{s \to a} \frac{\cancel{(s-a)}(s+a)}{\cancel{(s-a)}} = \lim_{s \to a}(s+a) = 2a$$

We now have two ways to determine a limit. We can use the limit theorems for sums, differences, products, and quotients, or we can factor and then use the limit theorems. For completeness, here is the mathematical definition of a limit:

The limit of $f(x)$ as x approaches c is the number L if given any radius e > 0 about L there exists a radius g > 0 about c such that for all x:

$$0 < |x - c| < g \text{ implies that } |f(x) - L| < e$$

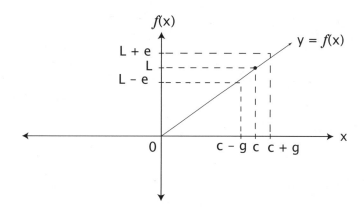

The intended height of $f(x)$ as x approaches c is L. As an example, consider $y = 4/5x$ or $f(x) = 4/5x$ if c = 5 and g = .2 then c − g = 4.8, c + g = 5.2.

$$f(c - g) = 4/5(4.8) = 3.84 \; (L - e = 4 - .16)$$

$$f(c + g) = 4/5(5.2) = 4.16 \; (L + e = 4 + .16)$$

$$0 < |x - 5| < .2 \text{ implies } |f(x) - 4| < .16$$

As the value of 'g' gets smaller, then 'e' gets smaller.

LESSON 8

Limits and Continuity

If the numerical value of a variable, v, becomes and remains great without bounds, we say that v becomes infinite. The notation for infinity looks like the number eight lying on its side (∞). If the variable 'v' becomes positively infinite, then we use $v = +\infty$; if the variable 'v' becomes negatively infinite, then we use $v = -\infty$. Don't forget that infinity is not a number, but a notation to denote that the number can always get larger. It would be wrong to say that a limit *equals* positive or negative infinity. Instead, we would say that the value of the limit approaches infinity.

It should be noted that some texts refer to a limit value of $+\infty$, $-\infty$, or ∞ as a nonexistent limit.

Example 1

$f(x) = \dfrac{1}{x^2}$ Find the $\lim\limits_{x \to 0} \dfrac{1}{x^2}$.

We can see from the drawing below that the left-hand and right-hand limits agree, so the limit = ∞.

$\lim\limits_{x \to 0^+} \dfrac{1}{x^2} = \infty$

$\lim\limits_{x \to 0^-} \dfrac{1}{x^2} = \infty$

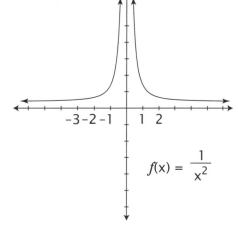

$f(x) = \dfrac{1}{x^2}$

When x approaches 0^+, it means that we look at values of x which get closer and closer to zero from the positive side. In other words, we need to determine the value of the function when x = 2, x = 1, x = .1, x = .01, etc. In this case, the value of the function gets infinitely large. Likewise, x approaches 0^- means that we look at values of x which get closer and closer to zero from the negative side. In other words, we need to determine the value of the function when x = –2, x = –1, x = –.1, x = –.01, etc. Again the value of the function gets infinitely large. We can say that the value of the limit is the same from the right side and the left side and that it approaches infinity.

Example 2

$$f(x) = \frac{1}{x} \qquad \text{Find the } \lim_{x \to 0} \frac{1}{x}.$$

The left-hand limit is $-\infty$; the right-hand limit is $+\infty$. Therefore, the limits do not agree and the original limit does not exist.

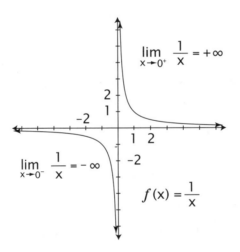

Example 3

$$\lim_{x \to 2^-} \frac{x}{(x-2)} = -\infty$$

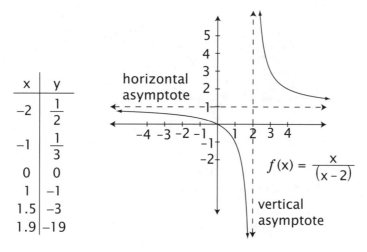

x	y
-2	$\frac{1}{2}$
-1	$\frac{1}{3}$
0	0
1	-1
1.5	-3
1.9	-19

Example 4

$$\lim_{x \to 2^+} \frac{x}{(x-2)} = +\infty$$

x	y
10	1.25
4	2
3	3
2.5	5
2.1	21

Summary

$$\lim_{x \to 2} \frac{x}{(x-2)} = \text{dne}$$

Limits can approach infinity. These can be evaluated by graphing or by a simple polynomial manipulation. Some limits occur frequently and are characterized by one of the six following forms where 'c' is a nonzero constant:

1. $\lim_{x \to \infty} cx = \infty$

2. $\lim_{x \to \infty} c + x = \infty$

3. $\lim_{x \to \infty} \frac{x}{c} = \infty$

4. $\lim_{x \to \infty} c^x = \infty \qquad c > 1$

5. $\lim_{x \to \infty} \frac{c}{x} = 0 \qquad c > 1$

6. $\lim_{x \to \infty} c^{-x} = 0 \qquad c > 1$

Example 5

$\lim_{r \to \infty} 3r = \infty$ This is similar to rule #1.

Example 6

$\lim_{r \to \infty} \frac{-2}{r} = 0$ This is similar to rule #5.

Example 7

$\lim_{r \to \infty} \frac{r}{\sqrt{2}} = \infty$ This is similar to rule #3.

Example 8

$\lim_{r \to \infty} 3^r = \infty$ This is similar to rule #4.

LIMITS OF RATIONAL EXPRESSIONS WITH POLYNOMIALS

When limits involve rational expressions that include polynomials, it is often helpful to divide by the highest power of x found in the entire expression. Then when the limit is taken, all terms with a denominator of x raised to any power will approach zero and will simplify the limit.

Example 9

$$\lim_{t \to \infty} \frac{4t+3}{t^2-5} \quad \text{Dividing by } t^2. \quad \lim_{t \to \infty} \frac{\frac{4}{t}+\frac{3}{t^2}}{1-\frac{5}{t^2}} = \frac{0}{1} = 0$$

Example 10

$$\lim_{x \to \infty} \frac{x^2-3x}{2x^2+4} \quad \text{Dividing by } x^2. \quad \lim_{x \to \infty} \frac{1-\frac{3}{x}}{2+\frac{4}{x^2}} = \frac{1}{2}$$

Example 11

$$\lim_{x \to \infty} \frac{6x^3-5x^2+3}{2x^3+4x-7} \quad \text{Dividing by } x^3. \quad \lim_{x \to \infty} \frac{6-\frac{5}{x}+\frac{3}{x^3}}{2+\frac{4}{x^2}-\frac{7}{x^3}} = \frac{6}{2} = 3$$

MORE TECHNIQUES TO EVALUATE LIMITS

We've seen that limits can be evaluated by graphing, factoring, and dividing by the highest power of x in rational functions. Some other techniques involve the use of trigonometric identities or the combining of fractions or the multiplication of an appropriate *conjugate*.

Example 12

$$\lim_{x \to 0} \frac{\sin(x)}{\tan(x)}$$

Remember $\tan(x) = \frac{\sin(x)}{\cos(x)}$.

$$\lim_{x \to 0} \frac{\sin(x)}{\frac{\sin(x)}{\cos(x)}} = \lim_{x \to 0} \cos(x) = \cos(0) = 1$$

Example 13

$$\lim_{x \to 0} \frac{1-\cos(x)}{2\sin^2(x)} = \lim_{x \to 0} \frac{1-\cos(x)}{2(1-\cos^2(x))}$$

$$= \lim_{x \to 0} \frac{\cancel{(1-\cos(x))}}{2\cancel{(1-\cos(x))}(1+\cos(x))}$$

$$= \lim_{x \to 0} \frac{1}{2(1+\cos(x))}$$

$$= \lim_{x \to 0} \frac{1}{2(1+1)} = \frac{1}{4}$$

Example 14

$$\lim_{x \to 1} \frac{x-1}{\sqrt{x}-1} = \lim_{x \to 1} \frac{(x-1)(\sqrt{x}+1)}{(\sqrt{x}-1)(\sqrt{x}+1)}$$

$$= \lim_{x \to 1} \frac{(x-1)(\sqrt{x}+1)}{(x-1)}$$

$$= \lim_{x \to 1} \sqrt{x}+1 = 2$$

Example 14 illustrates using a conjugate to aid in the evaluation of a limit. Be sure to leave the numerator in factored form when employing this method.

AP note: A limit that is used frequently is $\lim_{x \to 0} \frac{\sin(x)}{x} = 1$.

Memorize this in preparation for the AP exam.
There are some homework problems using this fact.

> *We have seen six ways to evaluate limits:*
> 1. Graphing.
> 2. Substitution: if $f(x)$ is continuous at 'a', substitute 'a' for x.
> 3. Factoring.
> 4. Dividing a rational polynomial by the highest power of x.
> 5. Conjugate application to the denominator.
> 6. Application of trigonometric identities or definitions.

LIMITS, CONTINUITY, AND ASYMPTOTES

We've seen that vertical asymptotes occur when the denominator is zero for simplified fractions. Another way to say this is to note that if $f(x)$ has a *vertical asymptote* at $x = a$, then $f(x)$ approaches ∞ (either positive or negative or both) when x approachs a.

Horizontal asymptotes can occur as well. If $\lim\limits_{x \to +\infty} f(x) = n$ or $\lim\limits_{x \to -\infty} f(x) = n$, where n is a real number, then $y = n$ is the horizontal asymptote. An illustrative example follows.

Example 15

$\lim\limits_{x \to +\infty} \dfrac{x}{(x-2)} \; \lim\limits_{x \to +\infty} \dfrac{1}{\left(1 - \dfrac{2}{x}\right)} = 1$

$\lim\limits_{x \to -\infty} \dfrac{x}{(x-2)} \; \lim\limits_{x \to -\infty} \dfrac{1}{\left(1 - \dfrac{2}{x}\right)} = 1$

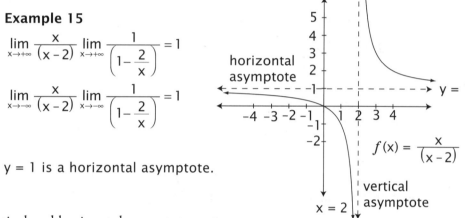

$y = 1$ is a horizontal asymptote.

Vertical and horizontal asymptote summary:

1. The functions which are most likely to have asymptotes are rational functions.

2. *Vertical asymptotes* – occur when the denominator of the rational function in simplified form is 0.

3. *Horizontal asymptotes* – occur when the $\lim\limits_{x \to \infty}$ and/or the $\lim\limits_{x \to -\infty}$ of the function exists.

Another way to consider horizontal asymptotes is to observe the degree of the numerator and the denominator of the rational function. Horizontal asymptotes occur when:

1. The degree of the numerator is less than the degree of the denominator. The asymptote is $y = 0$.

2. The degree of the numerator is equal to the degree of the denominator. The asymptote is $y = \frac{a}{b}$, where 'a' is the coefficient of the term with the highest degree in the numerator and 'b' is the coefficient of the term with the highest degree in the denominator.

Example 16

Find the vertical and horizontal asymptotes for:

$$f(x) = \frac{(x^2 - 8x + 15)}{(x^2 - 3x - 10)}$$

Factoring, we get $\frac{(\cancel{x-5})(x-3)}{(\cancel{x-5})(x+2)}$.

This will mean that we have a hole in the graph at x = 5.
There will be a vertical asymptote at x = -2.

$$\lim_{x \to \infty} \frac{x^2 - 8x + 15}{x^2 - 3x - 10} = 1 \qquad \lim_{x \to -\infty} \frac{x^2 - 8x + 15}{x^2 - 3x - 10} = 1$$

Therefore, there is one horizontal asymptote at y = 1.

The other approach is to observe that a = 1 and b = 1 and since the degrees are the same, we have a horizontal asymptote at $y = \frac{1}{1} = 1$.

Plotting points:

x	y
4	$\frac{1}{6}$
2	$-\frac{1}{4}$
1	$-\frac{1}{3}$
0	$-1\frac{1}{2}$
−1	−4
−3	6
−4	$3\frac{1}{2}$
−6	$2\frac{1}{4}$

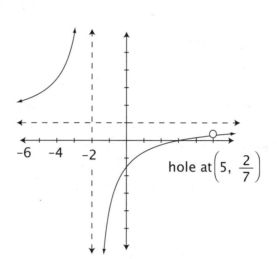

hole at $\left(5, \frac{2}{7}\right)$

If the limit as x approaches a is the same as the value of the function when x = a, then the function is continuous at x = a. Now we can put together our ideas of limits and continuity.

A function will be continuous at x = a if: $\lim_{x \to a} f(x) = f(a)$. That is, if the limit as x approaches 'a' is the same as the value of the function when x = a, then the function is continuous at x = a.

Example 17

$$f(x) = \frac{1}{3}x \quad \text{if } x = 6, \quad f(6) = \frac{1}{3}(6) = 2$$

$$\lim_{x \to 6} \frac{1}{3}x = 2$$

If this is not true, then the function is discontinuous at x = a. We will see in the next chapter why the existence of a limit is so valuable.

Recapping, there are three ways that a limit of a function may not exist at a:

1. Both the left-hand and right-hand limits exist, but they are different. (As in examples 2 and 3.)

2. The value of the limit is + or – infinity. (Some texts allow infinity as the final answer, but the AP exam does not. You must state in that case the limit does not exist.) We will allow infinity (plus and/or minus) as the final answer in this text.

3. A left-hand or a right-hand limit does not exist, or the function oscillates around the point a. See example below.

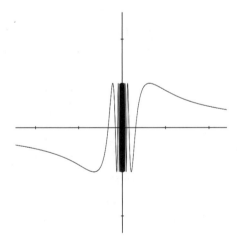

LESSON 9

Definition of a Derivative

RATE OF CHANGE

Miles per hour, miles per gallon, and price per ton all represent everyday rates of change. When we have a function $y = f(x)$ and the function changes uniformly, that is, the change in y corresponds to a given change in x, we have a simple rate of change $\frac{d}{dx}(y) = m$. You'll recognize this as the slope of a line or $y = mx + b$. Lines are uniform functions. Most functions do not change uniformly. When a function does not change uniformly, $\frac{\Delta y}{\Delta x}$ represents the average rate of change in the function.

Velocity is a prime example of a non-uniform function. If we travel 80 miles in two hours, then the average rate of change is $\frac{80 \text{ miles}}{2 \text{ hours}} = 40$ m/hr. This does not mean that we traveled at 40 m/hr for every minute of that trip. We could have traveled at 70 m/hr for a few miles, taken a quick break, and then traveled on at 60 m/hr. We could have encountered traffic and stoplights along the way as well.

INSTANTANEOUS RATE OF CHANGE

If we let Δx approach 0, then $\frac{\Delta y}{\Delta x}$ becomes the instantaneous rate of change in y with respect to x.

It is written as $\lim_{\Delta \to 0} \frac{\Delta y}{\Delta x}$.

Keep in mind: The slope of a tangent line at a given point, the instantaneous rate of change at a given time, and a derivative are all equal to each other.

AVERAGE RATE OF CHANGE

Temperature readings were recorded in northern Idaho for each hour of a day in March. We will measure time (H for hours) by the number of hours past midnight. The domain of x is [0, 24]. The temperatures (measured in Celsius) were observed and recorded in the following chart.

Example 1

H	D(°C)	H	D(°C)	H	D(°C)
0	6.0	9	7.5	18	16.2
1	5.1	10	8.6	19	14.6
2	4.3	11	10.0	20	12.0
3	3.7	12	12.2	21	10.2
4	4.0	13	14.2	22	9.3
5	4.2	14	15.1	23	8.1
6	4.8	15	16.0	24	6.6
7	5.6	16	16.8		
8	6.2	17	17.4		

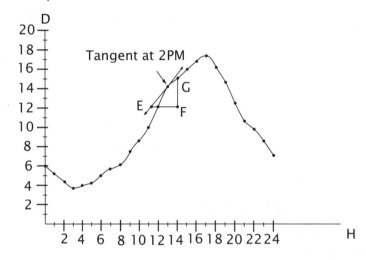

Find the average rate of change in Idaho's temperature with respect to time:

 a. From midnight to 4 a.m.
 b. From 11 a.m. to 3 p.m.
 c. Estimate the instantaneous rate of change at 2 p.m.

A. The change in temperature, ΔD, from midnight to 4 a.m. was:

$$\Delta D = D_F - D_I$$

D_F is the final temperature; D_I is the initial temperature.

$$4.0 - 6.0 = -2$$

The average rate of temperature with respect to time was:

$$\frac{\Delta D}{\Delta H} = -\frac{2}{4} \text{ hours} = -.5$$

On average, the temperature dropped half a degree an hour from midnight to 4 a.m.

B. $\Delta D = 16.0 - 10.0 = 6 \qquad \frac{\Delta D}{\Delta H} = \frac{6}{4} = 1.5$

On average, the temperature rose 1.5 degrees every hour from 11 a.m. to 3 p.m.

C. A tangent line is drawn on the map and a triangle is shown to estimate that tangent line. The slope of the hypotenuse will be close to the slope of the tangent line.

$$M = \frac{\text{rise}}{\text{run}} = \frac{\overline{FG}}{\overline{EF}} = \frac{(16.0 - 12.2)}{3} = 1.27$$

The instantaneous rate of the temperature at 2 p.m. with respect to time is approximately 1.27°C per hour.

DEFINITION OF A DERIVATIVE OF ONE VARIABLE

The fundamental definition of a derivative is:

$$f'(x) = \frac{dy}{dx} = \lim_{h \to 0} \frac{f(x+h) - f(x)}{h}$$

When the limit of this ratio exists, the function is said to be differentiable, or to possess a derivative.

Example 2 — Proof of the derivative formula

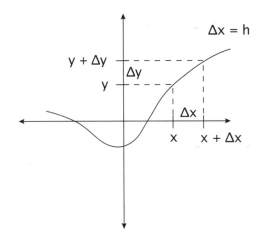

Given the function $y = f(x)$, consider x to have a fixed value.

Let x take on an increment Δx; then the function y takes on an increment Δy and the new value of the function is $y + \Delta y = f(x + \Delta x)$.

Solve for y: $\Delta y = f(x + \Delta x) - f(x)$.

Divide both sides by Δx: $\dfrac{\Delta y}{\Delta x} = \dfrac{f(x + \Delta x) - f(x)}{\Delta x}$.

Now take the limit of both sides as $\Delta x \to 0$.

$$\lim_{\Delta x \to 0} \frac{\Delta y}{\Delta x} = \lim_{\Delta x \to 0} \frac{f(x + \Delta x) - f(x)}{\Delta x}$$

This defines the derivative of y with respect to x.

Most modern texts will exchange h for Δx yielding:

$$y' = f'(x) = \frac{dy}{dx} = \lim_{h \to 0} \frac{f(x+h) - f(x)}{h}$$

The other ways of denoting a derivative are $\frac{dy}{dx}$, $f'(x)$, or y'. Expressions representing a derivative are slope of a tangent line at a point and instantaneous rate of change.

HOW TO FIND A DERIVATIVE USING THE DEFINITION

Example 3
Find $\frac{dy}{dx}$ when $y = 3x^2 + 5$.

Step 1 In the function, replace x with x + h.

$$f(x+h) = 3(x+h)^2 + 5$$
$$= (x^2 + 2xh + h^2) + 5$$
$$= 3x^2 + 6xh + 3h^2 + 5$$

Step 2 Subtract the given value of the function from this new value.

$$f(x+h) - f(x) = 3x^2 + 6xh + 3h^2 + 5 - (3x^2 + 5)$$
$$= 6xh + 3h^2$$

Step 3 Divide the remainder by h.

$$\frac{f(x+h) - f(x)}{h} = \frac{6xh + 3h^2}{h} = 6x + 3h$$

Step 4 Find the limit as $h \to 0$.

$$\lim_{h \to 0} \frac{f(x+h) - f(x)}{h} = \lim_{h \to 0} (6x + 3h) = 6x$$

Example 4

$y = 2x - x^2$

Step 1 $\quad f(x + h) = 2(x + h) - (x + h)^2$
$\qquad\qquad\quad = 2x + 2h - x^2 - 2xh - h^2$

Step 2 $\quad f(x+h) - f(x) = 2x + 2h - x^2 - 2xh - h^2 - (2x - x^2)$
$\qquad\qquad\qquad\qquad = 2h - 2xh - h^2$

Step 3 $\quad \dfrac{f(x+h) - f(x)}{h} = \dfrac{2h - 2xh - h^2}{h}$
$\qquad\qquad\qquad\qquad = 2 - 2x - h$

Step 4 $\quad \lim\limits_{h \to 0} \dfrac{f(x+h) - f(x)}{h} = \lim\limits_{h \to 0}(2 - 2x - h^2) = 2 - 2x$

Example 5

$y = \dfrac{-4}{x}$

Step 1 $\quad f(x + h) = \dfrac{-4}{(x+h)}$

Step 2 $\quad f(x + h) - f(x) = \dfrac{-4}{(x+h)} - \dfrac{(-4)}{x}$

$\qquad$ Finding a common denominator: $\dfrac{-4x + 4(x+h)}{x(x+h)} = \dfrac{4h}{x(x+h)}$

Step 3 $\quad \dfrac{f(x+h) - f(x)}{h} = \dfrac{4h}{x(x+h)} \div h = \dfrac{4}{x(x+h)}$

Step 4 $\quad \lim\limits_{h \to 0} \dfrac{f(x+h) - f(x)}{h} = \lim\limits_{h \to 0} \dfrac{4}{x(x+h)} = \dfrac{4}{x^2}$ or $4x^{-2}$

Example 6

$$y = 4x + 1$$

$$\lim_{h \to 0} \frac{f(x+h) - f(x)}{h} = \lim_{h \to 0} \frac{[4(x+h) + 1] - [4x + 1]}{h}$$

$$= \lim_{h \to 0} \frac{4x + 4h + 1 - 4x - 1}{h}$$

$$= \lim_{h \to 0} \frac{4h}{h} = \lim_{h \to 0} 4 = 4$$

Example 7

Find the distance between the points (0, −1) and (2, −3).

The distance between two points is derived from the Pythagorean theorem. The distance d between two points (x_1, y_1) and (x_2, y_2) is given by:

$$d = \sqrt{(x_2 - x_1)^2 + (y_2 - y_1)^2}$$

$$d = \sqrt{(0-2)^2 + (-1-(-3))^2}$$

$$= \sqrt{4+4} = \sqrt{8} = 2\sqrt{2}$$

The *greatest integer function* is written as $f(x) = [x]$ and is defined as follows: For each x-value, $f(x)$ is the greatest integer less than or equal to x. The graph of the greatest integer function looks like this.

Figure 1

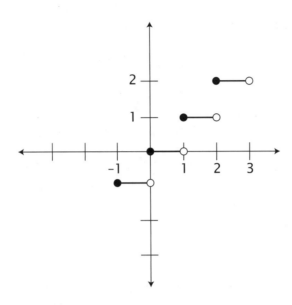

This function has numerous discontinuities and many applications in life. The United States postal service uses a *step function* to determine postal rates.

Example 8

Graph $f(x) = [x] + 1$ for x in $[-1, 1]$.

x	y
-1	0
$-\frac{1}{2}$	0
0	1
$\frac{1}{2}$	1
1	2

LESSON 10

Derivative Rules

First we will list seven rules for differentiation, and then give examples for each rule. The letters u and v are functions of x and C is a constant.

Figure 1

Derivative Rules

1. $\frac{d}{dx}(C) = 0$

2. $\frac{d}{dx}(x) = 1$

3. $\frac{d}{dx}(u + v) = \frac{d}{dx}(u) + \frac{d}{dx}(v)$

4. $\frac{d}{dx}(Cv) = C\frac{d}{dx}(v)$

5. $\frac{d}{dx}(uv) = u\frac{d}{dx}(v) + v\frac{d}{dx}(u)$ **Product rule**

6. $\frac{d}{dx}v^n = nv^{n-1}\frac{d}{dx}(v)$ **Power rule**

7. $\frac{d}{dx}\left(\frac{u}{v}\right) = \frac{v\frac{d}{dx}(u) - u\frac{d}{dx}(v)}{v^2}$ **Quotient rule**

> **RULE 1** $\quad \frac{d}{dx}(C) = 0$
>
> The derivative of a constant is zero.

Example 1

$\frac{d}{dx}(4) = 0$

y = 4 is a horizontal line.

The slope of the line tangent to y = 4 is 0.

This makes sense graphically. If y equals the constant C, then we have a horizontal line, and the slope of the tangent to that line will also be horizontal and will have the resultant slope of zero.

> **RULE 2** $\quad \frac{d}{dx}(x) = 1$
>
> The derivative of a variable with respect to itself is 1.

Example 2

$\frac{d}{dx}(x) = 1$

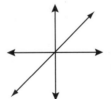

The slope of y = x is 1.

The slope of the line tangent to y = x is 1.

The graph of y = x has a slope of 1.

RULE 3 $\quad \frac{d}{dx}(u+v) = \frac{d}{dx}(u) + \frac{d}{dx}(v)$

The derivative of a sum is the sum of the derivatives.

Example 3

$$\frac{d}{dx}(x+3) = \frac{d}{dx}(x) + \frac{d}{dx}(3) = 1 + 0 = 1$$

RULE 4 $\quad \frac{d}{dx}(Cv) = C\frac{d}{dx}(v)$

The derivative of a constant times a function equals the constant times the derivative of the function.

Example 4

$$\frac{d}{dx}(4x) = 4 \cdot \frac{d}{dx}(x) = 4 \cdot 1 = 4$$

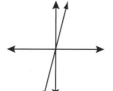

Slope is 4

The slope of the tangent to y = 4x is also 4.

RULE 5 $\quad \frac{d}{dx}(uv) = u\frac{d}{dx}(v) + v\frac{d}{dx}(u)$

The derivative of the product of two functions is the first term times the derivative of the second term plus the second term times the derivative of the first term.

Example 5

$$\frac{d}{dx}(4x)(x+3) = 4x \cdot \frac{d}{dx}(x+3) + (x+3) \cdot \frac{d}{dx}(4x)$$
$$= 4x(1) + (x+3)4 \cdot \frac{d}{dx}(x)$$
$$= 4x + (x+3)(4)$$
$$= 4x + 4x + 12$$
$$= 8x + 12$$

RULE 6 $\quad \frac{d}{dx}v^n = nv^{n-1}\frac{d}{dx}(v)$

The derivative of a function raised to a constant is the product of the constant, the function raised to the original power minus one, and the derivative of the function.

Example 6A

$$\frac{d}{dx}(x^5) = 5 \cdot x^4 \cdot \frac{d}{dx}(x) = 5x^4$$

Example 6B

$y = x^2$

$$\frac{d}{dx}(x^2) = 2x^1(1) = 2x$$

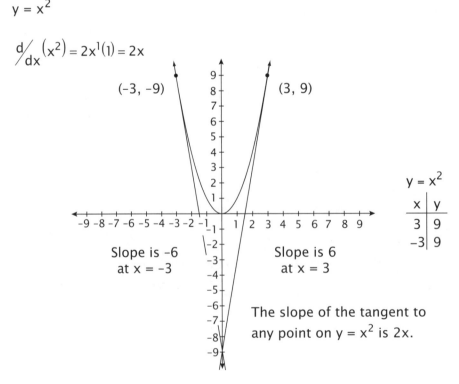

(-3, -9)   (3, 9)

Slope is -6 at x = -3

Slope is 6 at x = 3

$y = x^2$

x	y
3	9
-3	9

The slope of the tangent to any point on $y = x^2$ is $2x$.

Example 6C
y = x^{-3}

$$\frac{d}{dx}(x^{-3})$$
$$-3x^{-4}(1) = -3x^{-4} \text{ or } \frac{-3}{x^4}$$

Example 6D
y = (2x−1)^5

$$\frac{d}{dx}(2x-1)^5 = 5(2x-1)^4 \cdot \frac{d}{dx}(2x-1)$$
$$= 5(2x-1)^4 \cdot \left[\frac{d}{dx}(2x) - \frac{d}{dx}(1)\right]$$
$$= 5(2x-1)^4 \cdot 2$$
$$= 10(2x-1)^4$$

Example 6E
y = −x^2

$$\frac{d}{dx}-x^2 = (-1)\frac{d}{dx}(x^2)$$
$$= (-1)(2x^1)(1) = -2x$$

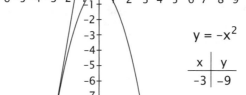

(−3, −9)

y = −x^2

x	y
−3	−9

Slope, or derivative, is −2x.
At x = −3, slope is + 6.

> **RULE 7** $\quad \dfrac{d}{dx}\left(\dfrac{u}{v}\right) = \dfrac{v\dfrac{d}{dx}(u) - u\dfrac{d}{dx}(v)}{v^2}$
>
> To find the derivative of the quotient of two functions (u/v), multiply the denominator (v) by the derivative of the numerator (du/dx). Subtract the product of the numerator (u) and the derivative of the denominator (dv/dx). Divide the entire result by the denominator squared (v^2).

Example 7

$$\dfrac{d}{dx}\dfrac{(2x+1)}{(4-3x)} = \dfrac{(4-3x)\dfrac{d}{dx}(2x+1) - (2x+1)\dfrac{d}{dx}(4-3x)}{(4-3x)^2}$$

$$= \dfrac{(4-3x)(2) - (2x+1)(-3)}{(4-3x)^2}$$

$$= \dfrac{8 - 6x + 6x + 3}{(4-3x)^2} = \dfrac{11}{(4-3x)^2}$$

More Examples:

Differentiate.

A. y = 5 - 2x

$y' = \dfrac{d}{dx}(5) - \dfrac{d}{dx}(2x)$ **Rule III** 5 + (-2x) Same as sum formula

$= 0 - \dfrac{d}{dx}(2x)$ **Rule I** Derivative of a Constant

$= 0 - 2\dfrac{d}{dx}(x)$ **Rule IV** Derivative of a constant times a function

$= 0 - 2$ **Rule II** Derivative of x

$= -2$

B. $f(x) = 2x^3 - 3x^2 + 7$

$$f'(x) = \frac{d}{dx}(2x^3) - \frac{d}{dx}(3x^2) + \frac{d}{dx}(7)$$
$$= 2 \cdot \frac{d}{dx}(x^3) - 3\frac{d}{dx}(x^2) + 0$$
$$= 2 \cdot (3x^2) \cdot \frac{d}{dx}(x) - 3(2x) \cdot \frac{d}{dx}(x) \quad \textbf{Power Rule VI}$$
$$= 6x^2 - 6x$$

C. $y = \dfrac{x^2}{3}$

$$y' = \frac{1}{3}\frac{d}{dx}(x^2)$$
$$= \frac{1}{3} \cdot 2x \cdot \frac{d}{dx}(x)$$
$$= \frac{2}{3}x$$

D. $s = \sqrt{t}$ First rewrite the square root as an exponent.

$$s = t^{\frac{1}{2}}$$
$$s' = \frac{1}{2}t^{-\frac{1}{2}}\frac{d}{dt}(t)$$
$$s' = \frac{1}{2\sqrt{t}} = \frac{1}{2\sqrt{t}} \cdot \frac{\sqrt{t}}{\sqrt{t}} = \frac{\sqrt{t}}{2t}$$

E. $S = \sqrt{1-t}$

Notice that for the first time we have to consider the derivative of the inside of the square root. It is not negligible.

$$S = (1-t)^{\frac{1}{2}}$$

$$S' = \frac{1}{2}(1-t)^{-\frac{1}{2}} \cdot \frac{d}{dt}(1-t)$$

$$= \frac{1}{2}(1-t)^{-\frac{1}{2}} \left[\frac{d}{dt}(1) - \frac{d}{dt}(t) \right]$$

$$= \frac{1}{2}(1-t)^{-\frac{1}{2}} [0-1]$$

$$= \frac{1}{2}(1-t)^{-\frac{1}{2}} (-1)$$

$$= \frac{-1}{2\sqrt{1-t}}$$

$$= \frac{-1}{2\sqrt{1-t}} \cdot \frac{\sqrt{1-t}}{\sqrt{1-t}} = \frac{-\sqrt{1-t}}{2(1-t)}$$

F. $y = (2-x)\sqrt{x}$

$$y = 2\sqrt{x} - x\sqrt{x}$$

$$y = 2x^{\frac{1}{2}} - x^{\frac{3}{2}}$$

$$y' = 2\frac{d}{dx}\left(x^{\frac{1}{2}}\right) - \frac{d}{dx}\left(x^{\frac{3}{2}}\right)$$

$$y' = 2\left(\frac{1}{2}x^{-\frac{1}{2}}\right)(1) - \left(\frac{3}{2}x^{\frac{1}{2}}\right)(1)$$

$$y' = x^{-\frac{1}{2}} - \frac{3}{2}x^{\frac{1}{2}} \quad \text{Factor out the common term with the smallest power.}$$

$$y' = x^{-\frac{1}{2}}\left(1 - \frac{3}{2}x\right) = \frac{1 - \frac{3}{2}x}{\sqrt{x}} = \frac{\sqrt{x}\left(1 - \frac{3}{2}x\right)}{x}$$

G. $y = (2-x)\left(x^{\frac{1}{2}}\right)$ Product rule is used here.

$y' = (2-x)\left(\frac{1}{2}x^{-\frac{1}{2}}\right)(1) + \left(x^{\frac{1}{2}}\right)(-1)$

$y' = \left(1 - \frac{1}{2}x\right)x^{-\frac{1}{2}} - x^{\frac{1}{2}}$

$y' = x^{-\frac{1}{2}}\left(1 - \frac{1}{2}x - x\right)$

$y' = x^{-\frac{1}{2}}\left(1 - \frac{3}{2}x\right) = \dfrac{1 - \frac{3}{2}x}{\sqrt{x}} = \dfrac{\sqrt{x}\left(1 - \frac{3}{2}x\right)}{x}$

H. $y = \dfrac{4-2x}{3-x}$ Quotient rule is used here.

$y' = \dfrac{(3-x)\cdot \frac{d}{dx}(4-2x) - (4-2x)\cdot \frac{d}{dx}(3-x)}{(3-x)^2}$

$y' = \dfrac{(3-x)\cdot\left[\frac{d}{dx}(4) - \frac{d}{dx}(2x)\right] - (4-2x)\cdot\left[\frac{d}{dx}(3) - \frac{d}{dx}(x)\right]}{(3-x)^2}$

$y' = \dfrac{(3-x)\cdot\left[0 - 2\frac{d}{dx}(x)\right] - (4-2x)\cdot[0-1]}{(3-x)^2}$

$y' = \dfrac{-6 + 2x + 4 - 2x}{(3-x)^2} = \dfrac{-2}{(3-x)^2}$

$= \dfrac{4-2x}{3-x} = (4-2x)(3-x)^{-1}$ Product Rule is used here.

$y' = (4-2x)\frac{d}{dx}(3-x)^{-1} + (3-x)^{-1}\frac{d}{dx}(4-2x)$

$y' = (4-2x)\left[-1(3-x)^{-2}\right](-1) + (3-x)^{-1}(-2)$

$y' = +(4-2x)(3-x)^{-2} - 2(3-x)^{-1}$ Factor out the common term with the smallest power.

$y' = (3-x)^{-2}\left[4 - 2x - 2(3-x)^1\right]$

$y' = \dfrac{-2}{(3-x)^2}$

CALCULUS   DERIVATIVE RULES - LESSON 10

In order to make our final fraction more compact, we factor out similar terms raised to the lowest exponential power present. In the example above, the term (3 – x) appears twice. Once it has a power of –1 and the other time it has an exponent of –2. Factor it out with the –2 value for the exponent. You should check your work using the distributive property.

This lesson requires practice. You need to work enough problems to become proficient with each of the formulas. You will know you have mastered this lesson when you are able to use all of these rules without consulting the table of formulas.

You can perform all your quotient problems with the product rule which alleviates the need to memorize another formula. For example:

$$y = \frac{\sqrt{x+2}}{x-3} \text{ is the same as } y = \left(\sqrt{x+2}\right)(x-3)^{-1}$$

Either the product rule or the quotient rule will yield the correct answer. Your answer may be in a different form than the one given in the answer key. As long as they are equivalent, you are fine.

LESSON 11

Chain Rule

The Chain rule is probably the most frequently used rule for calculating derivatives. It is used for computing the derivative of composite functions (functions of functions). The general rule is:

> **Chain Rule**
> $$\frac{dy}{dx} = \frac{dy}{du} \cdot \frac{du}{dx}$$ where y is a function of u and u is a function of x.

Example 1
y = 6u + 2 and u = 2x

Substituting: u = 2x into y = 6u + 2
$$y = 6(2x) + 2$$
$$y = 12x + 2$$
$$\frac{dy}{dx} = 12$$

Using the chain rule: $\frac{dy}{dx} = \frac{dy}{du} \cdot \frac{du}{dx}$

Where y is a function of u and u is a function of x:
$$\frac{dy}{dx} = \frac{d}{du}(6u+2) \cdot \frac{d}{dx}(2x)$$
$$\frac{dy}{dx} = 6 \cdot 2 = 12$$

Example 2

$y = 3u^2 + 5$ and $u = 4x$

$$\frac{dy}{dx} = \frac{dy}{du} \cdot \frac{du}{dx}$$

$$\frac{dy}{dx} = \frac{d}{du}(3u^2 + 5) \cdot \frac{d}{dx}(4x)$$

$$\frac{dy}{dx} = (6u)(1) \cdot (4) = 24u$$

I have u in the answer but I am looking for the derivative of y with respect to x, so I replace u with 4x.

$$\frac{dy}{dx} = 24(4x) = 96x$$

Substituting first and then finding the derivative we arrive at:

$$y = 3(4x)^2 + 5$$
$$y = 3 \cdot 16x^2 + 5 = 48x^2 + 5$$
$$\frac{dy}{dx} = 96x$$

Example 3

$y = \sqrt{u}$ and $u = 3x - 2$

Chain Rule

$$\frac{dy}{dx} = \frac{dy}{du} \cdot \frac{du}{dx}$$

$$\frac{dy}{dx} = \frac{d}{du}\left(u^{\frac{1}{2}}\right) \cdot \frac{d}{dx}(3x-2)$$

$$\frac{dy}{dx} = \frac{1}{2}u^{-\frac{1}{2}}(1) \cdot (3)$$

$$\frac{dy}{dx} = \frac{3}{2}u^{-\frac{1}{2}} = \frac{3}{2\sqrt{u}}$$

Replacing 3x − 2 for u:

$$dy/dx = \frac{3}{2\sqrt{3x-2}}$$

Substituting, then finding the derivative:

$$y = \sqrt{3x-2} = (3x-2)^{\frac{1}{2}}$$

$$dy/dx = \frac{1}{2}(3x-2)^{-\frac{1}{2}}(3)$$

$$dy/dx = \frac{3}{2\sqrt{3x-2}}$$

> The chain rule may be extended infinitely.
>
> $$dy/dx = dy/du \cdot du/dt \cdot dt/dx$$

Example 4

$y = 4u^2$, $u = (3 − 2t)$, and $t = 2x$

To find $dy/dx = dy/du \cdot du/dt \cdot dt/dx$

$$dy/dx = (8u)(-2)(2) = -32u$$

Replacing u with (3 − 2t):

$$dy/dx = -32(3-2t) = -96 + 64t$$

Replacing t with (2x):

$$dy/dx = -96 + 64(2x) = -96 + 128x$$

Substituting first:
$$y = 4[3-2(2x)]^2$$
$$y = 4[9-12(2x)+4(2x)^2]$$
$$y = 4[9-24x+16x^2]$$
$$y = 36-96x+64x^2$$
$$dy/dx = -96+128x$$

HIGH ORDER DERIVATIVES

We will soon observe applications of the first derivative. The second derivative will give us more information about the shape of the graph. The second derivative is simply the derivative of the first derivative. f' is the first derivative and is pronounced "f prime." f'' is the second derivative and is read "f double prime."

Example 5

If $f(x) = 3x^3 - 7x^2 + 3$, what is $f''(x)$?

$$f(x) = 3x^3 - 7x^2 + 3$$

$$f'(x) = 9x^2 - 14x$$

$$f''(x) = 18x - 14$$

Example 6

If $f(x) = 8\sqrt{x}$, what is $f''(4)$?

$$f(x) = 8\sqrt{x} \text{ or } 8x^{\frac{1}{2}}$$

$$f'(x) = 4x^{-\frac{1}{2}} = \frac{4}{\sqrt{x}}$$

$$f''(x) = -2x^{-\frac{3}{2}} = \frac{-2}{\left(\sqrt{x}\right)^3}$$

$$f''(4) = \frac{-2}{\left(\sqrt{4}\right)^3} = -\frac{2}{8} = -\frac{1}{4} \quad \text{Replacing x with 4.}$$

Sometimes it is necessary to take a higher order derivative. The third derivative is obtained by taking the derivative three times. The fourth derivative is obtained by taking the derivative four times, and so on.

Example 7

Find $f'''(x)$ if $f(x) = 2x^5 - 3x^3 + x^2 - 7$.

$$f'(x) = 10x^4 - 9x^2 + 2x$$

$$f''(x) = 40x^3 - 18x + 2$$

$$f'''(x) = 120x^2 - 18$$

Example 8

What is $f''(-1)$ in example 7?

If $f''(x) = 40x^3 - 18X + 2$, then: $f''(x) = 40(-1)^3 - 18(-1) + 2$
$f''(x) = -40 + 18 + 2 = -20$

APPEARANCE OF DIFFERENTIABLE FUNCTIONS

What does a differentiable function look like? We already know that functions which have vertical tangents are not differentiable at the point where the vertical tangent occurs.

Example 9

$$f(x) = \frac{1}{x} = x^{-1}$$

$$f'(x) = -x^{-2} = -\frac{1}{x^2}$$

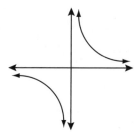

In example 9, the original function is undefined when x = 0. There is a vertical asymptote at x = 0. Since the derivative definition includes a limit, we can see that the limit will not exist at x = 0 either. All other values of x are differentiable.

There are two other similar looking shapes which will produce a nondifferentiable point. Consider:

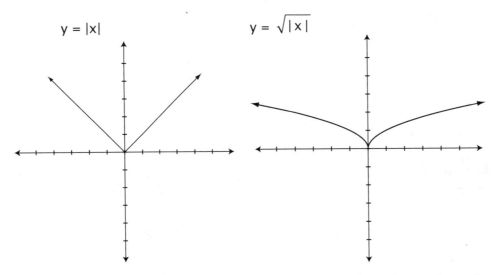

The graph on the left produces a "corner" at x = 0. The right graph produces a "point" or *cusp* at x = 0. Neither graph is differentiable at x = 0. At x = 0 neither graph has a curve, so there are an infinite number of values for the derivative. Visually, an infinite number of tangent lines can be drawn, each with a different slope at x = 0. This means that the derivative does not exist there.

We can deduce that not all continuous graphs are differentiable. It is true, however, that all differentiable functions are continuous. It is important to note that removable discontinuities are not differentiable at their discontinuity.

> Where functions are NOT differentiable:
>
> 1. Vertical tangents
> 2. Corners
> 3. Cusps
> 4. Discontinuities

1.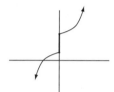

   Vertical Tangent

2.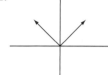

   Corner

3.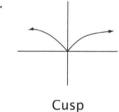

   Cusp

4.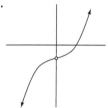

   Discontinuity

All four graphs are nondifferentiable at x = 0.

Example 10

Are the following functions differentiable for all values of x? If not, list the x-value(s) where the derivative of the function does not exist.

a. $f(x) = \dfrac{2}{(x-3)}$ (when x = 3)

b. $f(x) = x^2 - 7x - 3$ (a parabola; all values work)

c. $f(x) = \dfrac{3}{x^2 + 5x + 6}$ (when x = -2, -3)

LESSON 12

Derivatives of Trig Functions

Let's begin with a discussion of the derivative of $f(x) = \sin(x)$. The graph is drawn below for $\sin(x)$.

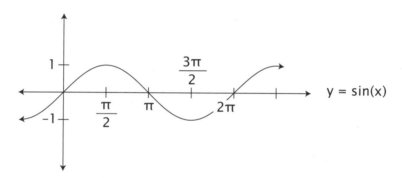

Consider the slope of the tangent line along this curve. At the lowest and highest points of $y = \sin(x)$, the slope of the tangent line is zero because the tangent is a horizontal line. At $x = 0$ and $x = 2\pi$, the slope of the tangent line is one. At $x = \pi$, the slope of the tangent line is -1. Plotting these points and filling in the intermediate values we get the following curve. Therefore the derivative of $y = \sin(x)$ is $\cos(x)$.

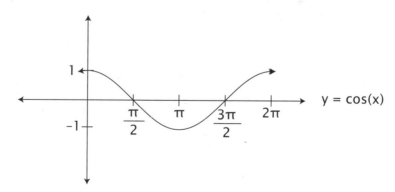

Here is a list of the derivatives of the six trigonometric functions where "u" is a function of x.

$$\frac{d}{dx} \sin(u) = \cos(u) \frac{du}{dx}$$

$$\frac{d}{dx} \cos(u) = -\sin(u) \frac{du}{dx}$$

$$\frac{d}{dx} \tan(u) = \sec^2(u) \frac{du}{dx}$$

$$\frac{d}{dx} \cot(u) = -\csc^2(u) \frac{du}{dx}$$

$$\frac{d}{dx} \sec(u) = \sec(u) \tan(u) \frac{du}{dx}$$

$$\frac{d}{dx} \csc(u) = -\csc(u) \cot(u) \frac{du}{dx}$$

Example 1
$y = \sin(4x)$

Let $u = 4x$; $\frac{du}{dx} = 4$.

$$y = \sin(u)$$
$$y' = \cos(u) \cdot \frac{du}{dx}$$
$$= (\cos(4x))(4)$$
$$= 4\cos(4x)$$

Example 2
$y = 2\cos(7x^2)$

Let $u = 7x^2$, $\dfrac{du}{dx} = 14x$.

$$y = 2\cos(u)$$
$$y' = 2 \cdot \dfrac{d}{dx}(\cos(u))$$
$$= 2(-\sin(u))\dfrac{du}{dx}$$
$$= -2\sin(7x^2) \cdot (14x)$$
$$= -28x\sin(7x^2)$$

Example 3
$y = \tan\sqrt{1-2x}$

Let $u = (1-2x)^{\frac{1}{2}}$; $\dfrac{du}{dx} = \dfrac{1}{2}(1-2x)^{-\frac{1}{2}}(-2) = \dfrac{-1}{\sqrt{1-2x}}$.

$$y = \tan(u)$$
$$y' = \sec^2(u) \cdot \dfrac{du}{dx}$$
$$= \sec^2\sqrt{1-2x} \cdot \left(\dfrac{-1}{\sqrt{1-2x}}\right) = \dfrac{-\sec^2(\sqrt{1-2x})}{\sqrt{1-2x}}$$

You can also use the Chain Rule for examples 1–3. See examples 4–6.

Example 4 (Example 1 revisited)
$y = \sin(4x)$

Using the Chain Rule:

$$y = \sin(u) \qquad u = 4x$$
$$\dfrac{dy}{du} = \cos(u) \qquad \dfrac{du}{dx} = 4$$

$$\dfrac{dy}{dx} = \dfrac{dy}{du} \cdot \dfrac{du}{dx}$$
$$= \cos(u) \cdot (4)$$
$$= 4\cos(4x)$$

Example 5 (Example 2 revisited)
$y = 2\cos(7x^2)$

Using the Chain Rule:

$$y = 2u \qquad u = \cos(v) \qquad v = 7x^2$$
$$\frac{dy}{du} = 2 \qquad \frac{du}{dv} = -\sin(v) \qquad \frac{dv}{dx} = 14x$$

$$\begin{aligned}\frac{dy}{dx} &= \frac{dy}{du} \cdot \frac{du}{dv} \cdot \frac{dv}{dx} \\ &= (2)(-\sin(v))(14x) \\ &= -28x\sin(v) \\ &= -28x\sin(7x^2) \qquad \text{Replace } v \text{ with } 7x^2.\end{aligned}$$

Example 6 (Example 3 revisited)
$y = \tan\sqrt{1-2x}$

Using the Chain Rule:

$$y = \tan(u) \qquad u = \sqrt{v} \qquad v = 1 - 2x$$
$$\frac{dy}{du} = \sec^2(u) \qquad \frac{du}{dv} = \frac{1}{2}v^{-\frac{1}{2}} \qquad \frac{dv}{dx} = -2$$

$$\begin{aligned}\frac{dy}{dx} &= \frac{dy}{du} \cdot \frac{du}{dv} \cdot \frac{dv}{dx} \\ &= \sec^2(u)\left(\frac{1}{2}v^{-\frac{1}{2}}\right)(-2) \\ &= \sec^2(\sqrt{v})\left(-\frac{1}{\sqrt{v}}\right) \qquad \text{Replace } u \text{ with } \sqrt{v}. \\ &= \frac{-\sec^2(\sqrt{1-2x})}{\sqrt{1-2x}} \qquad \text{Replace } v \text{ with } 1-2x.\end{aligned}$$

Example 7

$y = x^2 \sec(2x)$ Use the product rule.

$$y' = x^2 \cdot \tfrac{d}{dx}(\sec(2x)) + \sec(2x) \cdot \tfrac{d}{dx}(x^2)$$
$$= x^2 \sec(2x)\tan(2x) \cdot \tfrac{d}{dx}(2x) + (\sec(2x))(2x)$$
$$= 2x^2 \sec^2(2x)\tan(2x) + 2x(\sec(2x))$$
$$= 2x(\sec(2x))[x\,\tan(2x) + 1]$$

Example 8

$y = 3\cot\left(\dfrac{x}{3}\right)$

$$y' = 3\,\tfrac{d}{dx}\left(\cot\left(\tfrac{x}{3}\right)\right)$$
$$= 3\left(-\csc^2\left(\tfrac{x}{3}\right)\right)\tfrac{d}{dx}\left(\tfrac{x}{3}\right)$$
$$= -3\left(\csc^2\left(\tfrac{x}{3}\right)\right)\left(\tfrac{1}{3}\right)$$
$$= -\csc^2\left(\tfrac{x}{3}\right)$$

Proof for derivative of the cos(u):

We know from trigonometry that the cosine function and the sine function are the same with a phase shift of $\tfrac{\pi}{2}$. That is $\cos(u) = \sin\left(\tfrac{\pi}{2} - u\right)$ and $\sin(u) = \cos\left(\tfrac{\pi}{2} - u\right)$ where u is a function of x.

$$\tfrac{d}{dx}\sin\left(\tfrac{\pi}{2} - u\right) = \cos\left(\tfrac{\pi}{2} - u\right)(-1) = -\cos\left(\tfrac{\pi}{2} - u\right)$$

Using the trigonometric difference formula we get:

$$-\left[\cos\left(\tfrac{\pi}{2}\right)\cos(u) + \sin\left(\tfrac{\pi}{2}\right)\sin(u)\right] = -[0 \cdot \cos(u) + (1)\sin(u)]$$
$$= -\sin(u)$$

The derivative for tan(u) can be derived by using the quotient rule for $\dfrac{\sin(u)}{\cos(u)}$.

Note: It is important to have a plan for evaluating the derivative. Example 9 gives examples of problems whose first rule applied is the sum or difference formula or the product rule, etc.

Example 9 Recognizing types of problems

I. Sum and difference formula problems.

$$y = \sin(2x) - (\tan(x))^2$$
$$y = 4\sec(3x) + \tan(x) - (\sec(x))^5$$

II. Product rule problems.

$$y = (\sin(2x))(\cos(4x^2))$$
$$y = \cos(x-1)(\tan^2(x))$$

III. Power rule problems.

$$y = \tan^4(\sin(x^2))$$
$$y = 4[\sin(x) + \cos(2x)]^3$$

IV. Quotient rule problems.

$$y = \frac{\sin(2x)}{\cos(3x)}$$

V. Trigonometry problems.

$$y = \tan(\sin(3x))$$
$$y = \cos(\sec^2(4x-1))$$

LESSON 13

Derivative of e^x and $\ln(x)$

The easiest derivative to compute is the derivative of e^x. The derivative of e^x is e^x.

If u is a function of x, then: $\frac{d}{dx}(e^x) = e^x$.

If u is a function of x, then: $\frac{d}{dx}(e^u) = e^u \cdot \frac{du}{dx}$.

Example 1
Find the derivative of $y = e^{3x}$ using $u = 3x$ and $\frac{du}{dx} = 3$.

$$y' = e^{3x} \cdot 3 = 3e^{3x}$$

Essentially, you copy the function and multiply this by the derivative of the exponent.

Example 2
Find the derivative of $y = 2e^{(1-x)} - e^{4x}$.

$$y' = 2e^{(1-x)} \cdot \frac{d}{dx}(1-x) - e^{4x} \cdot \frac{d}{dx}(4x)$$
$$y' = -2e^{(1-x)} - 4e^{4x} \quad \text{(difference of two functions)}$$

Example 3
Find the derivative of $y = x^2 e^x$.

$$y' = x^2 \frac{d}{dx}(e^x) + e^x \frac{d}{dx}(x^2)$$
$$y' = x^2 e^x + e^x 2x = e^x(x^2 + 2x)$$

It is helpful to factor out all possible terms when arriving at the final answer. In future lessons we will be determining what x-values make these derivatives equal to zero. Having a final answer as a set of factors will simplify that task.

Example 4
Find $\frac{du}{dx}$ if $y = e^{\sin(2x)}$. Let $u = \sin(2x)$ and $\frac{du}{dx} = 2\cos(2x)$.

Here we have an exponential function with a trigonometric exponent.

$$y' = e^{\sin(2x)}(2\cos(2x)) = 2e^{\sin(2x)}\cos(2x)$$

Example 5
Find $\frac{du}{dx}$ if $y = \frac{4x}{e^x}$.

This problem may be worked out with either the product formula or quotient formula. We chose the quotient rule.

$$y' = \frac{e^x(4) - 4x(e^x)}{(e^x)^2} = \frac{4e^x(1-x)}{e^{2x}} = \frac{4(1-x)}{e^x}$$

Now we will look at the derivative of the inverse function, $\ln(x)$.

$$\frac{d}{dx}(\ln(x)) = \frac{1}{x}$$

If u is a function of x, then: $\frac{d}{dx}(\ln(u)) = \frac{1}{u} \cdot \frac{du}{dx}$.

Using what we learned about the derivative of e^x, here is a proof for the new formula for the derivative of $\ln(x) = \frac{1}{x}$.

If y = ln(x), $e^y = x$ and $x = e^y$:

$$\frac{d}{dy}(x) = \frac{d}{dy}(e^y)$$ Take the derivative with respect to y.

$$\frac{d}{dy} = e^y$$

$$\frac{dy}{dx} = \frac{1}{e^y}$$ Take the reciprocal of both sides.

$$\frac{d}{dx}\ln(x) = \frac{1}{x}$$ Replace e^y with x and y with ln(x).

Example 6
Find $\frac{du}{dx}$ if y = ln(3x). Let u = 3x and $\frac{du}{dx} = 3$.

$$\frac{dy}{dx} = \frac{1}{3x} \cdot (3) = \frac{1}{x}$$

Example 7
Find the derivative of $f(x) = \ln(x^2 + 1)$. Let $u = x^2 + 1$ and $\frac{du}{dx} = 2x$.

$$f'(x) = \left(\frac{2x}{x^2+1}\right)$$

Example 8
Find the derivative of f(x) = ln(sin(x)). Let u = sin(x) and $\frac{du}{dx} = \cos(x)$.

$$f(x) = \ln(\sin(x))$$
$$f'(x) = \frac{1}{u}\frac{du}{dx} = \frac{1}{\sin(x)} \cdot \frac{\cos(x)}{1} = \frac{\cos(x)}{\sin(x)} = \cot(x)$$

Example 9

Find y' if $y = \ln(x^3) + \ln(1-x)$.

$$y' = \frac{1}{x^3} \cdot \frac{d}{dx}(x^3) + \left(\frac{1}{1-x}\right)\left[\frac{d}{dx}(1-x)\right]$$

$$y' = \frac{3x^2}{x^3} + \left(\frac{-1}{1-x}\right) = \frac{3}{x} - \frac{1}{1-x}$$

Example 10

Find the derivative of $y = \frac{\ln(2x)}{x^2}$.

$$f'(x) = \frac{x^2 \cdot \frac{1}{2x} \cdot 2 - [\ln(2x)](2x)}{x^4}$$

$$f'(x) = \frac{x - 2x[\ln(2x)]}{x^4}$$

$$f'(x) = \frac{1 - 2[\ln(2x)]}{x^3}$$

LESSON 14

Implicit Differentiation

All of our discussion in calculus has been surrounding *explicit* functions of the form y = *f*(x). Functions can also be *implicit*. Such an equation may not be solvable for y. Consider x − y − cos(xy) = xy. These are often not true functions because there may be multiple y-values for a single x-value, but they do have graphs and derivatives.

Implicit differentiation is a tool used to calculate a derivative of an implicit function. The method is simple. Every term is differentiated on both sides of the equation. Most implicit differentiation is taken with respect to x, but it is permissible to differentiate implicitly with respect to other variables.

The derivative of y with respect to x is $\frac{dy}{dx}$ or y'. For convenience we will use y'. The derivative of z with respect to x is z' and the derivative of w with respect to x would be w' etc. Here's a simple example.

Example 1
Find the derivative with respect to x two ways: explicitly and implicitly.

$$2x + 4y = 8$$

Explicitly: $2x + 4y = 8$
$4y = -2x + 8$
$y = -1/2x + 2$
$y' = -1/2$

Implicitly: $2x + 4y = 8$

Step 1 Take the derivative of both sides of the equation.

$$2 + 4y' = 0$$

Remember $\frac{d}{dx}(x) = 1$ and $\frac{d}{dx}(y) = y'$.

Step 2 Solve for y'. $4y' = -2$
$$y' = -\frac{1}{2}$$

As we expected, the answers are the same.

Example 2
Differentiate $5y^2 = 3x + y$ implicitly with respect to x.

Step 1 $5(2y)y' = 3(1) + y'$

Step 2   In order to factor out y', you will need to move all the terms containing y' to one side of the equation.

$$10y\,y' - y' = 3$$
$$y'(10y - 1) = 3$$

Step 3 $$y' = \frac{3}{10y - 1}$$

Example 3
Differentiate $4y + \sqrt{y} = 2x - 3$ with respect to x.

Step 1   Differentiate every term with respect to x.

$$4y' + \frac{1}{2}y^{-\frac{1}{2}}y' = 2$$

Step 2   Factor out y'.

$$y'\left(4 + \frac{1}{2}y^{-\frac{1}{2}}\right) = 2$$

Step 3   Solve for y'.

$$y' = \frac{2}{4 + \frac{1}{2\sqrt{Y}}}$$

Example 4
Implicitly differentiate with respect to x.

$$x^2 + y^2 = 2ay$$

Step 1 $2x + 2y\,y' = 2ay'$

Step 2 $2x = 2ay' - 2y\,y'$
$2x = y'(2a - 2y)$

Step 3 $y' = \dfrac{2x}{2a - 2y} = \dfrac{x}{a - y}$

Example 5
Implicitly differentiate with respect to x.

$$(\sin(y))(e^x) = 2x$$
$$(\sin(y))(e^x) + e^x(\cos(y))y' = 2$$
$$e^x \cos(y) \cdot y' = 2 - \sin(y) \cdot e^x$$
$$y' = \frac{2 - (\sin(y))(e^x)}{e^x \cos(y)}$$
$$= \frac{2 - e^x \sin(y)}{e^x \cos(y)}$$

Example 6

Consider this circle centered at the origin with a radius of 1.
The equation is $x^2 + y^2 = 1$.

The derivative gives the slope of the tangent line to the curve. Therefore the tangent at (0, 1) and (0, -1) should be horizontal and zero. The tangent at (1, 0) and (-1, 0) should be vertical and undefined.

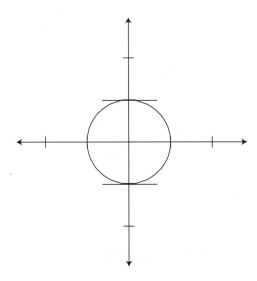

Implicitly differentiating with respect to x we get $2x + 2y\, y' = 0$:

$$2x = -2y\, y'$$

$$y' = \frac{-2x}{2y} = -\frac{x}{y}$$

When we substitute the point (0, 1) into this derivative we get $-\frac{0}{1} = 0$ as we would expect.

When we substitute (1, 0) into the derivative we get $\frac{1}{0}$ = undefined.

Example 7

Determine which points on $x^2 + y^2 = 1$ cause the slope of the tangent line to be equal to one.

$$1 = \frac{-x}{y} \qquad y = -x$$

Substituting into the original formula we have:

$$x^2 + (-x)^2 = 1$$

$$2x^2 = 1$$

$$x^2 = 1/2$$

$$x = \pm\sqrt{\frac{1}{2}} = \pm\frac{1}{\sqrt{2}} = \pm\frac{\sqrt{2}}{2}$$

Since x and y need to have opposite signs, the points where the slope of the tangent line equals zero are $\left(\frac{\sqrt{2}}{2}, -\frac{\sqrt{2}}{2}\right)$ and $\left(-\frac{\sqrt{2}}{2}, \frac{\sqrt{2}}{2}\right)$. This makes sense graphically.

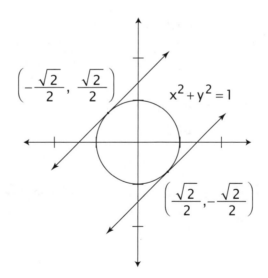

Example 8
Find the lines that are tangent and normal to ln(y) - x = -2 at (2, 1).

Note: A normal line is perpendicular to the tangent line.

$$\frac{1}{y}y' - 1 = 0$$
$$\frac{y'}{y} = 1$$
$$y' = y$$

Slope of tangent line at (2, 1) is 1.

Plugging the point and slope into the point-slope formula of the equation of a line we get 1 = (1)(2) + b,
$$b = -1.$$

Therefore the equation of the tangent line to ln(y) - x = -2 at (2, 1) is y = (1)x - 1 or y = x - 1.

The normal line would be perpendicular to the line y = x - 1. That means the slope of the normal line is -1. Plugging the point and the slope of -1 into the point-slope equation we get:

$$1 = (-1)(2) + b$$
$$1 = -2 + b$$
$$b = 3$$

The equation of the normal line through (2, 1) is y = -x + 3.

LESSON 15

Graphing with the 1st Derivative

We have depended upon the plotting of points in order to construct a graph of a given curve or function. In chapter 2 we learned about the basic shapes of quadratics, which aided us as well. The first and second derivatives reveal more key information about the appearance of a curve.

Consider $f(x) = x^2$. We know that it is a "smile" parabola with a minimum value of the function occurring at $(0, 0)$. The first derivative is $f'(x) = 2x$. The first derivative is 0 when $x = 0$, negative when x is negative, and positive when x is positive. The first derivative gives us a formula for the slope of the tangent line to the curve for every x-value.

$f(x) = x^2$

$f'(x) = 2x$

When $x < 0$,

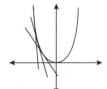

x	$f'(x)$
-3	-6
-2	-4
-1	-2
0	0

When $x > 0$, $f'(x) > 0$

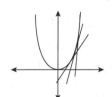

x	$f'(x)$
0	0
+1	+2
+2	+4
+3	+6

Notice that the $f(x)$ values are decreasing when x is negative and are increasing when x is positive. This is exactly the information that is given by the first derivative. In addition, when the first derivative is equal to zero the potential for a minimum or maximum value exists. In this example a minimum value exists when $f'(x) = 0$.

Let's consider $f(x) = -x^2$. This parabola is a "frown" and has a maximum at (0, 0). The first derivative is −2x. We can tell by the first derivative that the function will be decreasing for positive values of x and increasing for negative values of x. Again there is an extreme point, this time a maximum value, when the first derivative is equal to zero.

$f(x) = -x^2$

$f'(x) = -2x$

when $x < 0, f'(x) > 0$

when $x > 0, f'(x) < 0$

x	$f'(x)$
−3	+6
−2	+4
−1	+2
0	0

x	$f'(x)$
0	0
+1	−2
+2	−4
+3	−6

First Derivative Test

If x is a maximum of f, then f' is positive to the left of x and negative to the right of x.

If x is a minimum of f, then f' is negative to the left of x and positive to the right of x.

Finding the minimum and maximum values of a given function is often the goal. Another broader term for maximum and minimum is *extrema*. We've seen in the examples above that when the first derivative is equal to zero, a maximum or minimum MAY exist. However, there are other places to consider. The following test helps us to graph complex equations by identifying significant points or critical points on the graph.

> To find the extreme points on a function, find:
> 1. The points where $f' = 0$.
> 2. The points where f' does not exist.
> 3. The endpoints (covered more in a later chapter).

These are called *critical points* and they are the only places where the function will have a maximum or a minimum.

There could be more than one maximum or minimum value for a given function. We distinguish between these by calling them *local* or *absolute (global)* maximums or minimums. The absolute or global maximum or minimum values are the highest or lowest values on the curve. See the graph.

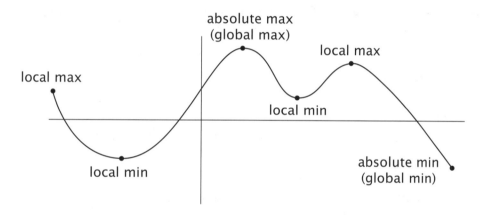

Example 1

Find the intervals where $f(x) = 2x^3 - 9x^2 - 108x + 3$ is increasing and decreasing. Also give the coordinates of any maximum or minimum values, if they exist.

Step 1   Find the critical points.

$$f'(x) = 6x^2 - 18x - 108$$

$$0 = 6(x^2 - 3x - 18) \qquad x - 6 = 0 \qquad x + 3 = 0$$
$$0 = 6(x - 6)(x + 3) \qquad x = 6 \qquad x = -3$$

The critical points are $x = 6$ and $x = -3$.

Step 2   Test all the intervals to the left and right of the critical values.

When x = -10, $f'(-10)$ is positive.
$$6(-10)^2 - 18(-10) - 108 = +672$$

When x = 0, $f'(0)$ is negative.
$$6(0)^2 - 18(0) - 108 = -108$$

When x = 10, $f'(10)$ is positive.
$$6(10)^2 - 18(10) - 108 = +312$$

When evaluating points in the first derivative, the sign of the answer is all that is needed. Put the value into the factored form of the equation and record the sign for each factor. For example, when x = –10 and (–10 – 6)(–10 + 3), we get a negative times a negative times or (–)(–) = (+) which is a positive. The positive 6 in front of the parenthesis does not impact the answer.

Alternate Step 2

Test just the signs using the factors of the first derivative.

When x = -10, $f'(-10)$ is positive.
$$f'(-10) = 6(-10 - 6)(-10 + 3) = (-)(-) = (+)$$

When x = 0, $f'(0)$ is negative.
$$f'(0) = 6(0 - 6)(0 + 3) = (-)(+) = (-)$$

When x = 10, $f'(10)$ is positive.
$$f'(10) = 6(10 - 6)(10 + 3) = (+)(+) (-)$$

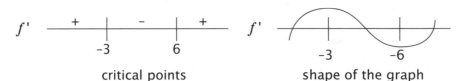

intervals

critical points   shape of the graph

Step 3   Make conclusions. Place critical points back into original equation. There is a local minimum at (-3, 192).

$$f(-3) = 2(-3)^3 - 9(-3)^2 - 108(-3) + 3 = (-3, 192)$$

There is a local maximum at (-3, 192).

$$f(6) = 2(6)^3 - 9(6)^2 - 108(6) + 3 = (6, -537)$$

There is a local minimum at (6, -537).

$f(x)$ is increasing on (-∞, -3) and (6, ∞).
$f(x)$ is decreasing on (-3, 6)

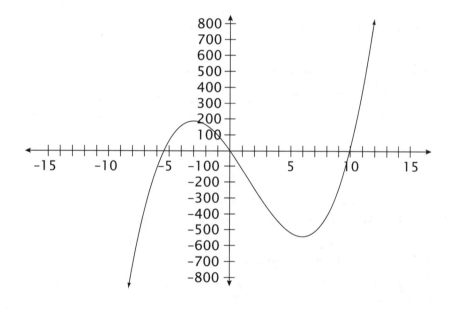

Example 2

Find the intervals where $f(x) = 3x^4 - 4x^3 - 12x^2 + 5$ is increasing and decreasing. Give the coordinates of any maximum or minimum values, if they exist.

Step 1 Find the critical points.
$$f'(x) = 12x^3 - 12x^2 - 24x$$

$$f'(x) = 12x(x^2 - x - 2)$$
$$0 = 12x(x - 2)(x + 1)$$

$$12x = 0 \quad x - 2 = 0 \quad x + 1 = 0$$
$$x = 0 \quad\quad x = 2 \quad\quad x = -1$$

The critical points are x = 0, x = 2, and x = 1

Step 2 Test all the intervals to the left and right of the critical values.

When x = -2, $f'(-2)$ is negative.
$$f'(-2) = 12(-2)3 - 12(-2)2 - 24(-2) = -96$$

When x = -.5, $f'(-.5)$ is positive.
$$f'(-.5) = 12(-.5)^3 - 12(-.5)^2 - 24(-.5) = +7.5$$

When x = 1, $f'(1)$ is negative.
$$f'(1) = 12(1)^3 - 12(1)^2 - 24(1) = -24$$

When x = 3, $f'(3)$ is positive.
$$f'(3) = 12(3)^3 - 12(3)^2 - 24(3) = +144$$

Alternate Step 2

Or test just the signs using the factors of the first derivative.

When $x = -2$, $f'(-2)$ is negative.
$$f'(-2) = 12(-2)(-2-2)(-2+1) = (-)(-)(-)= (-)$$

When $x = -.5$, $f'(-.5)$ is positive.
$$f'(-.5) = 12(-.5)(-.5-2)(-.5+1) = (-)(-)(+) = (+)$$

When $x = 1$, $f'(1)$ is negative.
$$f'(1) = 12(1)(1-2)(1+1) = (+)(-)(+) = (-)$$

When $x = 3$, $f'(3)$ is positive.
$$f'(3) = 12(3)(3-2)(3+1) = (+)(+)(+) = (+)$$

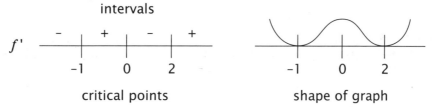

Step 3 Make conclusions.

There is local minimum at $(-1, 0)$. $(2, -27)$ is an absolute minimum.

$$f(2) = 3(2)^4 - 4(2)^3 - 12(2)^2 + 5 = -27$$
$$f(-1) = 3(-1)^4 - 4(-1)^3 - 12(-1)^2 + 5 = 0$$

There is a local maximum at $(0, 5)$.

$$f(0) = 3(0)^4 - 4(0)^3 - 12(0)^2 + 5 = 5$$

$f(x)$ is increasing on $(-1, 0)$ and $(2, \infty)$.
$f(x)$ is decreasing on $(-\infty, -1)$ and $(0, 2)$.

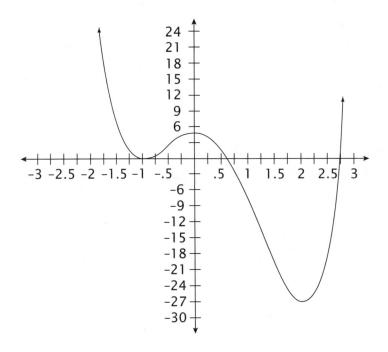

Example 3

Find the critical points and intervals of increasing and decreasing.

$$f(x) = \frac{2}{x^2+1}$$

$$f'(x) = \frac{(x^2+1)(0) - 2(2x)}{(x^2+1)^2} = \frac{-4x}{(x^2+1)^2}$$

Step 1   Find the critical points.

$$0 = \frac{-4x}{(x^2+1)^2}$$

There is no x-value which makes the denominator equal to 0.

The critical point is x = 0.

Step 2   Test all the intervals to the left and right of the critical value(s), which in this problem is x = 0.

When x = -1, $f'(-1)$ is positive.

$$f'(-1) = f'(-1) = \frac{-4(-1)}{(1+1)^2} = \frac{+}{+}$$

When x = +1, f '(1) is negative.

$$f'(1) = f'(1) = \frac{-4(1)}{(1+1)^2} = \frac{-}{+}$$

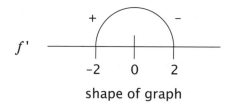

shape of graph

Step 3   Make conclusions.

There is an absolute maximum at (0, 2). $f(0) = \frac{2}{(0^2+1)} = \frac{2}{1}$

$f(x)$ is increasing on $(-\infty, 0)$.
$f(x)$ is decreasing on $(0, \infty)$.

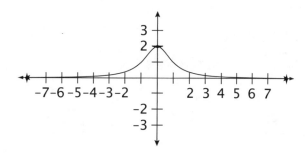

Example 4

Find the intervals where $f(x) = x^4 - 8x^2$ is increasing and decreasing. Draw the graph of $f(x) = x^4 - 8x^2$ and label all extrema.

Step 1 Find the critical points.
$$f'(x) = 4x^3 - 16x$$

$$f'(x) = 4x(x^2 - 4)$$
$$0 = 4x(x - 2)(x + 2)$$

$$4x = 0 \quad x - 2 = 0 \quad x + 2 = 0$$
$$x = 0 \quad x = 2 \quad x = -2$$

The critical points are $x = 0$, $x = 2$, and $x = -2$.

Step 2 Test the signs using the factors of the first derivative.

When $x = -3$, $f'(-3)$ is negative.
$$f'(-3) = 4(-3)(-3 - 2)(-3 + 2) = (-)(-)(-) = (-)$$

When $x = 1$, $f'(-1)$ is positive.
$$f'(-1) = 4(-1)(-1 - 2)(-1 + 2) = (-)(-)(+) = (+)$$

When $x = 1$, $f'(1)$ is negative.
$$f'(1) = 4(1)(1 - 2)(1 + 2) = (+)(-)(+) = (-)$$

When $x = 3$, $f'(3)$ is positive.
$$f'(3) = 4(3)(3 - 2)(3 + 2) = (+)(+)(+) = (+)$$

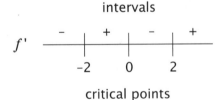
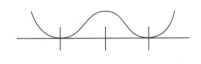

Step 3   Make conclusions.

There is a local minimum at (-2, -16) and (2, -16).

Indeed, these are both absolute minimums because their function values are the same, we have more than one absolute minimum.

$f(-2) = (-2)^4 - 8(-2)^2 = 16 - 32 = -16$
$f(2) = (2)^4 - 8(2)^2 = 16 - 32 = -16$

There is a local maximum at (0,0).

$f(0) = (0)^4 - 8(0)^2 = 0 - 0 = 0$

$f(x)$ is increasing on (-2, 0) and (2, ∞).
$f(x)$ is decreasing on (-∞, -2) and (0, 2).

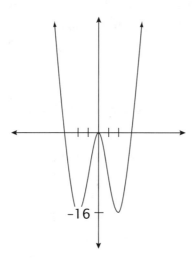

$f(x) = (x)^4 - 8(x)^2$

Example 5
Find a so that $ax^2 - 4x - 2$ has a maximum value when $x = 2$.

$f(x) = ax^2 - 4x - 2$

$f'(x) = 2ax - 4$

A maximum is when $f'(x) = 0$, so $0 = 2ax - 4$.

When $x = 2$,
$$0 = 2a(2) - 4$$
$$0 = 4a - 4$$
$$4 = 4a$$
$$1 = a$$

Check: When $a = 1$,
$$f'(x) = 2(1)x - 4$$
$$0 = 2x - 4$$
$$4 = 2x$$
$$2 = x$$

Series and Summations

A *series* is a sequence that is added together. It may be symbolized with either a capital S or a Greek sigma. We opt for the *sigma*. On the top and bottom of the sigma are two numbers that indicate the scope of the series.

Example 6

$$\sum_{k=1}^{5} \{2k+1\}$$

This represents the scope as *k*, moving from 1 to 5 to generate five terms. The numbers 1 through 5 are placed in $2k + 1$ to find the value of each term in the sequence. The letter *k* is called an index. It is the variable in the sequence.

$$\sum_{k=1}^{5} \{2k+1\} = \{2(1) + 1 + 2(2) + 1 + 2(3) + 1 + 2(4) + 1 + 2(5) + 1\}$$
$$= \{3 + 5 + 7 + 9 + 11\}$$
$$= 35$$

Example 7
Compute.

$$\sum_{i=1}^{4} 2i = 2(1) + 2(2) + 2(3) + 2(4)$$
$$= 2 + 4 + 6 + 8 = 20$$

Example 8
Compute.

$$\sum_{i=0}^{5} (i-1) = (0 - 1) + (1 - 1) + (2 - 1) + (3 - 1) + (4 - 1) + (5 - 1)$$
$$= -1 + 0 + 1 + 2 + 3 + 4 = 9$$

LESSON 16

Graphing with the 2nd Derivative
Concavity and Inflection

The first derivative tells us where the graph is increasing or decreasing. The second derivative tells us how the graph is bending. We use the term *concave up* to refer to a graph where the graph has a shape of a "U." Consequently, *concave down* refers to a graph where the shape is reversed, or "∩." Consider $f(x) = x^2$ again. It is obvious that the graph has the shape of a capital U for all of its values and would be considered concave up. This means that the first derivative is always increasing. As you observe the slope of the tangent lines on the graph, notice how they increase as they get farther from the origin and decrease as they approach zero. As you move from left to right the values of the first derivative are always increasing. The chart of values documents this trend.

Figure 1

$f(x) = x^2$

$f'(x) = 2x$

$f''(x) = +2$

Concave UP everywhere.

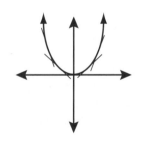

x	$f'(x)$
-3	-6
-2	-4
-1	-2
0	0
+1	+2
+2	+4
+3	+6

The values of the first derivative are always increasing.

This may be seen again in the graph of $f(x) = -x^2$. This time the shape is clearly concave downward. This indicates the first derivative is decreasing.

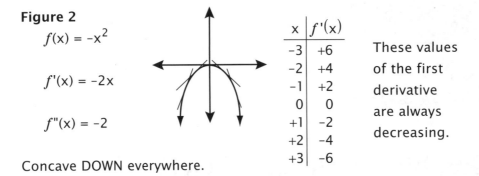

Figure 2

$f(x) = -x^2$

$f'(x) = -2x$

$f''(x) = -2$

Concave DOWN everywhere.

x	$f'(x)$
-3	+6
-2	+4
-1	+2
0	0
+1	-2
+2	-4
+3	-6

These values of the first derivative are always decreasing.

When the curve changes from concave upward to concave downward (or downward to upward) in the same graph, the point of change is called an *inflection point*. Consider $f(x) = x^3$. For values of x less than zero, the first derivative is steadily decreasing. At zero the first derivative is zero. For x values greater than zero, the first derivative is steadily increasing. Graphs can also have combinations of concavity within the same graph, as in the next example.

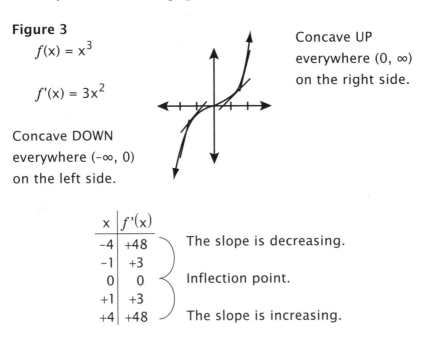

Figure 3

$f(x) = x^3$

$f'(x) = 3x^2$

Concave DOWN everywhere $(-\infty, 0)$ on the left side.

Concave UP everywhere $(0, \infty)$ on the right side.

x	$f'(x)$
-4	+48
-1	+3
0	0
+1	+3
+4	+48

The slope is decreasing.

Inflection point.

The slope is increasing.

Let's examine these three examples mathematically. If $f(x) = x^2$, then $f'(x) = 2x$ and $f''(x) = 2$. Notice that no matter what x value we choose, the second derivative is always positive. In figure 1, we saw that the graph is concave up everywhere.

If $f(x) = -x^2$, then $f'(x) = -2x$ and $f''(x) = -2$. Notice that no matter what x value we choose, the second derivative is always negative. And this graph (figure 2) is concave down for all values of x.

Finally, if $f(x) = x^3$, then $f'(x) = 3x^2$ and $f''(x) = 6x$. Now we have options for the second derivative, and they depend on the value of x. When x is negative, the second derivative is negative, and correspondingly when x is positive, the second derivative is positive. Therefore the graph is concave down for x < 0 and concave up for x > 0. At zero, the second derivative is also zero. That is the *inflection point* where the concavity changed from negative to positive. Zero is an inflection point for $f(x) = x^3$.

> **Definition of Concavity**
> The graph of $f(x)$ is concave down on any interval where $f''(x) < 0$.
> The graph of $f(x)$ is concave up on any interval where $f''(x) > 0$.

A classic example of a graph with many inflection points is $f(x) = \sin(x)$. It may be seen from the graph that there are an infinite number of inflection points at every multiple of pi.

Figure 4

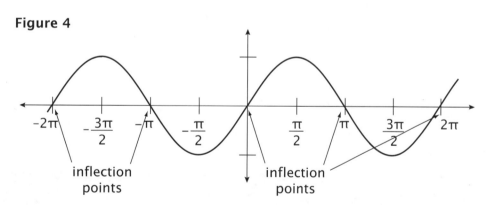

$f(x) = \sin(x)$ on $[-2\pi, 2\pi]$

Example 1

Sketch the curve of $f(x) = x^3 - 3x^2 + 1$.

$f'(x) = 3x^2 - 6x = 3x(x - 2)$, $x = 0$, $x = 2$ are critical points.
$f''(x) = 6x - 6 = 6(x - 1)$, $x = 1$ is a possible inflection point.

x	-1	0	1	2	3
$f(x)$	-3	1	-1	-3	1
$f'(x)$	+	0	-	0	+

        max min

inf. pt.

$f''(x)$	-	-	0	+	+

concave down   concave up
$(-\infty, 1)$ $(1, +\infty)$

Notice that when x = 0, the 2nd derivative was negative and we had a maximum. Also when x = 2, the second derivative was positive and we had a minimum.

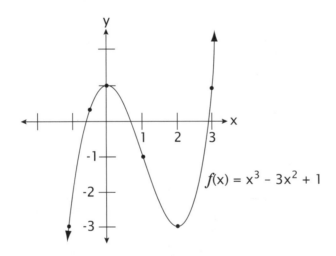

$f(x) = x^3 - 3x^2 + 1$

2nd Derivative Test

If f is a twice differentiable function and there is a c such that $f'(c) = 0$ then:
1. If $f''(x) > 0$, then $f(c)$ is a minimum.
2. If $f''(x) < 0$, then $f(c)$ is a maximum.
3. If $f''(x) = 0$, there is no conclusion. Apply the first derivative test.

Example 2

Sketch the curve of $f(x) = x^4 - 4x^3 + 1$

$f'(x) = 4x^3 - 12x^2 = 4x^2(x - 3)$; $x = 0, 3$ are critical points.

$f''(x) = 12x^2 - 24x = 12x(x - 2)$; $x = 0, 2$ are possible inflection pts.

x	−1	0	1	2	3	4
$f(x)$	6	1	−2	−15	−26	1
$f'(x)$	−	0	−	−	0	+

min

inf. pt.   inf. pt.

$f''(x)$	+	0	−	0	+	+

concave up   concave down   concave up
$(-\infty, 0)$ $(0, 2)$ $(2, +\infty)$

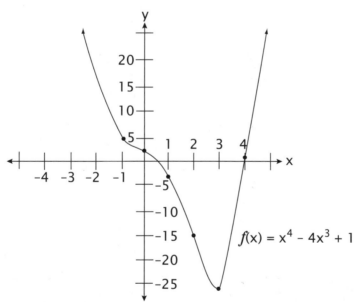

absolute or global minimum at $(3, -26)$

Example 3

Sketch a continuous curve such that

$f(1) = 3$ and $f'(x) < 0$ for $x < 1$ and $f'(x) > 0$ for $x > 1$.

This graph has one definite point, namely (1, 3).

Now all points $x < 1$ are decreasing and all points $x > 1$ are increasing. That would form a parabolic shape. Answers will vary. The "u" shape can be wide or narrow.

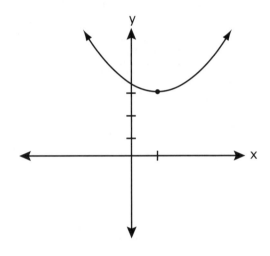

Example 4

Given the function $f(x)$, sketch $f'(x)$ and $f''(x)$.

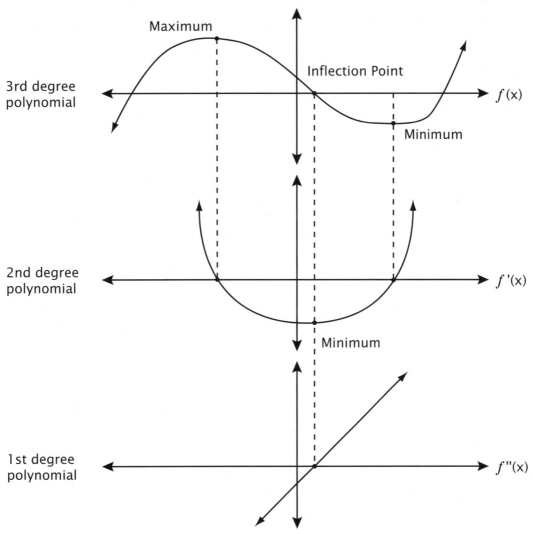

Note of Caution

When potential critical points and inflection points are found by setting the first or second derivative to zero, be advised that there MAY or MAY NOT be maximums, minimums, or inflection points. Check to see that the first derivative changes around any critical points. Also check to see if the second derivative changes around an inflection point. If there is no change, then you do not have anything except a point on the graph.

There are functions whose 2nd derivative is difficult to obtain. When this occurs, it is often wise to use the first derivative exclusively to obtain desired maximum and minimum values.

Bonus: Asymptotes

We have learned how to find vertical and horizontal asymptotes which aid our ability to sketch a curve. Reduced rational functions in which the degree of the numerator is one larger than the degree of the the denominator will have an oblique asymptote. *Oblique asymptotes* are asymptotes that are neither horizontal or vertical.

Example 5

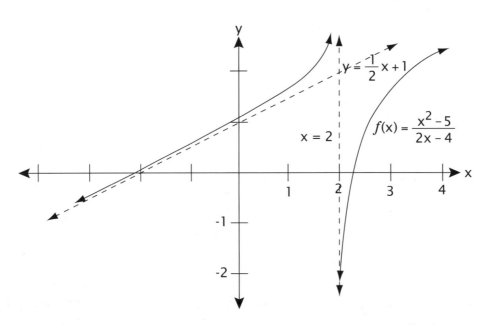

$f(x) = \dfrac{x^2 - 5}{2x - 4}$

$$\begin{array}{r} \frac{1}{2}x + 1 \\ 2x-4 \overline{\smash{\big)}\, x^2 + 0x - 5} \\ \underline{x^2 - 2x} \\ 2x - 5 \\ \underline{2x - 4} \\ -1 \end{array}$$

There is an oblique asymptote at $f(x) = \frac{1}{2}x + 1$.

We can rewrite the quotient, with the remainder as:

$f(x) = \dfrac{x}{2} + 1 - \dfrac{1}{2x - 4}$

Now it's easy to see $\dfrac{1}{2x-4}$ will approach 0 as x approaches infinity.

Also, we have a vertical asymptote at x = 2 since x = 2 would make the denominator 0. A sketch of this graph is shown.

LESSON 17

Mean Value Theorem; L'Hôpital's Rule

A theorem is a statement which can be proven true through accepted mathematical arguments such as theorems and postulates. The Mean Value Theorem has numerous applications.

> **Mean Value Theorem (MVT)**
> Suppose we have a graph $y = f(x)$ which is continuous on [a, b] and differentiable on (a, b). There must be at least one number c between a and b such that the tangent line at c is parallel to chord ab.

Figure 1

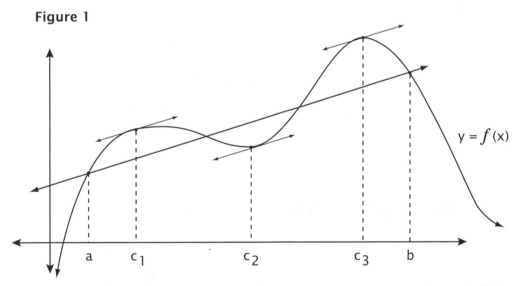

One possible graph is shown above. Draw a line by connecting $f(a)$ with $f(b)$. The Mean Value Theorem states that there must be at least one number c between a and b such that the tangent line at c is parallel to chord ab. In the graph above, there are three such places labeled as c_1, c_2 and c_3. It is interesting to note that the MVT states that at least one point c exists, but it doesn't tell us how to locate that point.

Example 1
Find any value of c which satisfies the MVT for $f(x) = x^2 + 2x + 2$ on [0, 1].

Solution
The MVT states that there is a point X between 0 and 1 where $f'(x) = m$ when m is the slope of the segment made by connecting $f(0)$ with $f(1)$.

It is important to note that the conditions of the MVT are met. In this case we have a parabola which is continuous on [0, 1] and differentiable on (0, 1).

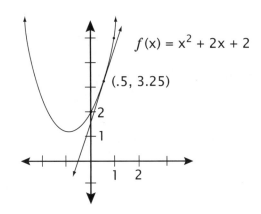

Slope of segment between $f(0)$ and $f(1)$: $f(0) = 2$ (0, 2)
$f(1) = 5$ (1, 5)

$$m = \frac{5-2}{1-0} = \frac{3}{1} = 3$$

$f'(x) = 2x + 2$

The MVT states that there must be a point c where $2x + 2 = 3$;

If $2x + 2 = 3$, then $x = \frac{1}{2}$.

That point is $\left(\frac{1}{2}, 3\frac{1}{4}\right)$ or (.5, 3.25).

Example 2

Find any value of c which satisfies the MVT for $f(x) = \sqrt[3]{x}$ or $x^{\frac{1}{3}}$ on [1, 8].

Solution

The coordinates of the endpoints are (1, 1) and (8, 2).

The slope between these points is $m = \frac{[2-1]}{[8-1]} = \frac{1}{7}$.

$$f'(x) = \frac{1}{3}x^{-\frac{2}{3}} = \frac{1}{3x^{\frac{2}{3}}}$$

The function is differentiable and continuous on [1, 8], so there is at least one point where:

$$\frac{1}{7} = \frac{1}{3x^{\frac{2}{3}}}$$

$$3x^{\frac{2}{3}} = 7$$

$$x = \left(\frac{7}{3}\right)^{\frac{3}{2}}$$

$$x \approx 3.56$$

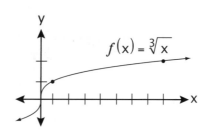

The point, which satisfies the MVT for $f(x) = \sqrt[3]{x}$ on [1, 8], is approximately [3.56, 1.53].

If we take another look at the MVT, we see that it claims that the average rate of change over the interval is equal to the instantaneous rate of change at one or more points.

$\frac{[f(b) - f(a)]}{(b-a)}$ is the average rate of change on [a, b].

$f'(c)$ is the instantaneous rate of change at point c.

Example 3
Buddy Quick arrived at a toll booth after traveling two hours on a road where the speed limit was 60 mph. He had covered 150 miles. He was cited for speeding. Use the MVT to explain why.

Solution
Buddy's average rate of change for that trip was 150 miles/2 hours = 75 miles per hour. Therefore, according to the MVT, there must be some time during that trip where he was travelling at 75 mph.

Rolle's Theorem is a special case of the MVT.

> **Rolle's Theorem**
> In Rolle's Theorem, $f(a) = f(b) = 0$. The conclusion is that there exists at least one point c in (a, b) where $f'(c) = 0$.

Rolle's Theorem is used in advanced algebra classes to prove the existence of a root of a polynomial.

Figure 2

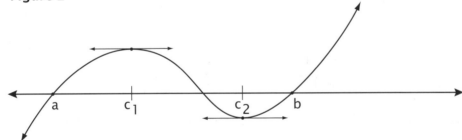

Example 4

Show that $x - \dfrac{3}{x} = 0$ has exactly one real solution on [1, 3].

Look at what the function is doing at these endpoints.

$f(1) = 1 - 3 = -2$
$f(3) = 3 - 1 = 2$

Somewhere between $x = 1$ and $x = 3$ the function must cross the x-axis at least one time.

Look at what the first derivative is doing: $f'(x) = 1 + \dfrac{3}{x^2}$

The first derivative is always positive and never zero.

Therefore, there are no max or min values and the function is always increasing. This means that between $x = 1$ and $x = 3$, the function must cross the x-axis exactly one time. That is where the real solution to this polynomial exists.

To find the real solution on [1, 3] we solve for x.

$$x\left(x - \frac{3}{x}\right) = 0$$
$$x^2 - 3 = 0$$
$$x = \pm\sqrt{3}$$

$x = \sqrt{3}$ is the only real solution on [1, 3].

L'Hôpital's Rule (also spelled L'Hospital's Rule) connects limits with derivatives.

> **L'Hôpital's Rule (LR)**
> Suppose $f(a) = g(a) = 0$ and $f'(a)$ and $g'(a)$ exist with $g'(a)$ not equal to 0, then:
> $$\lim_{x \to a} \frac{f(x)}{g(x)} = \lim_{x \to a} \frac{f'(a)}{g'(a)}$$

We are seeking to find the limit of a fraction where both the numerator and the denominator have a value of 0 when 'a' is inserted for x.

Example 5

$$\lim_{x \to 0} \frac{3x + \sin(x)}{2x}$$

All the prerequisites are satisfied.

Here $f(x) = 3x + \sin(x)$ and $f(0) = 0$; $g(x) = 2x$ and $g(0) = 0$.

Also $g'(x) = 2$, so $g'(0)$ exists and is not equal to 0 and $f'(x) = 3 + \cos(x)$.

Differentiate each part of the fraction separately and evaluate the limit.

$$\lim_{x \to 0} \frac{3 + \cos(x)}{2} = \frac{4}{2} = 2$$

L'Hôpital's rule can be extended to limits that involve $\frac{\infty}{\infty}$ as well as $\frac{0}{0}$.

This rule can also be applied as many times as necessary.

Example 6

$$\lim_{x \to \infty} \frac{\ln^2(x+1)}{2x^3} = \frac{\infty}{\infty}$$

Applying LR,

$$\lim_{x \to \infty} \frac{2[\ln(x+1)]\left(\frac{1}{x+1}\right)(1)}{6x^2} = \lim_{x \to \infty} \frac{2\ln(x+1)}{(x+1)6x^2} \text{ or } \lim_{x \to \infty} \frac{2\ln(x+1)}{6x^3+6x^2} = \frac{\infty}{\infty}$$

Applying LR again

$$\lim_{x \to \infty} \frac{2\left(\frac{1}{x+1}\right)}{18x^2+12x} = \lim_{x \to \infty} \frac{2}{(x+1)(18x^2+12x)} = \frac{2}{\infty} = 0$$

Remember, you can only apply LR when the limit form is $\frac{\text{infinity}}{\text{infinity}}$ or $\frac{0}{0}$.

Odd and Even Functions
A function $f(x)$ is an even function if $f(-x) = f(x)$ for all x in the domain.
A function $f(x)$ is an odd function if $f(-x) = -f(x)$ for all x in the domain.

In other words, all *even functions* are symmetric about the y-axis and all *odd functions* are symmetric about the origin.

Example 7
Is $f(x) = \frac{2}{x^2}$ even, odd, or neither?

$f(-x) = \frac{2}{(-x)^2} = \frac{2}{x^2}$ so $f(-x) = f(x)$ and the function is even.

Conclusion: You can see that functions which have only even powers will be even functions.

Example 8

Is $f(x) = x^3 + \dfrac{1}{x}$ even, odd, or neither?

$$f(-x) = -x^3 - \dfrac{1}{x}$$

$$-f(x) = -x^3 - \dfrac{1}{x} \text{ so this function is odd.}$$

You will see with the next example that functions with only odd powers are not always odd functions.

Example 9

Is $f(x) = 2 - \dfrac{1}{x}$ even, odd, or neither?

There are no even powers, so it is not an even function.

$$f(-x) = 2 + \dfrac{1}{x}$$

$$-f(x) = -2 + \dfrac{1}{x}$$

Since $f(-x)$ does not equal $-f(x)$, this function is neither odd nor even.

LESSON 18

Physics Applications

The position of an object or person is given by $x(t)$, $s(t)$, or $d(t)$. The *instantaneous velocity* $v(t)$ is represented as the first derivative of the distance, $x'(t)$. The *acceleration* of the object is given by the first derivative of the velocity, $v'(t)$, which is the second derivative of the distance $x''(t)$.

$$\text{ft} \quad\quad x(t) = \text{distance}$$

$$\frac{\text{ft}}{\text{sec}} \quad\quad x'(t) = v(t) = \text{velocity}$$

$$\frac{\text{ft}}{\text{sec}^2} \quad\quad x''(t) = v'(t) = a(t) = \text{acceleration}$$

Note: In this book we will use the term velocity, which can be positive or negative. The sign indicates direction. When other books refer to *speed*, they are discussing the absolute value of velocity, $|v(t)|$.

Example 1
Joe has finished his last exam in his college career. He is so elated that he opens his window (which is 10 m above the ground) and throws his pencil in the air with an initial velocity of 4 m/sec. The distance formula for this problem is:

$$x(t) = -4.9\, t^2 + v_0 t + x_0$$

where v_0 is the initial velocity and x_0 is the initial distance.

a. For how many seconds will the pencil continue to travel upward?
b. What will be the maximum height that the pencil will achieve?
c. What is the total flight time for the pencil?
d. How long will it take for the pencil to fall to the ground?
e. What is the pencil's velocity when it hits the ground?
f. What is the pencil's acceleration when it hits the ground?

a. The pencil will continue upward until the velocity becomes zero. Then it will change direction and begin to fall.

$$x(t) = -4.9t^2 + v_0 t + x_0$$
$$x(t) = -4.9t^2 + 4t + 10$$
$$v(t) = x'(t) = -9.8t + 4$$
$$0 = -9.8t + 4$$
$$t = \frac{4}{9.8} = .41 \text{ seconds}$$

The pencil will continue upward for .41 seconds.

b. The maximum height is achieved when the velocity is zero.

The time is .41 from part a.

$$x(t) = -4.9t^2 + 4t + 10$$
$$x(.41) = -4.9(.41)^2 + 4(.41) + 10$$
$$x(.41) = -.82 + 1.64 + 10 = 10.82$$

The maximum height of the pencil is 10.82 meters.

c. The total flight time for the pencil ends when the pencil hits the ground. At that time, the distance is zero.

$$x(t) = -4.9t^2 + 4t + 10$$
$$0 = -49t^2 + 40t + 100 \quad \text{Multiply by 10 to remove decimals.}$$

$$x = \frac{-40 \pm \sqrt{40^2 - 4(-49)(100)}}{2(-49)}$$

$$x = \frac{-40 \pm \sqrt{21{,}200}}{-98} = \frac{-40 + 145.6}{-98} = -1.08$$

$$= \frac{-40 - 145.6}{-98} = 1.89$$

Disregard the negative answer.
The total flight time is 1.89 seconds.

  d. From part a, we know that .41 seconds was used to travel upward, so the trip downward was 1.89s - .41s = 1.48 seconds.

  e. From part c we know that the pencil will hit the ground when t = 1.89 seconds.

  $v(1.89)$ = the velocity at impact
  $v(1.89) = -9.8(1.89) + 4 = -14.52$ m/sec

  f. $a(t) = v'(t) = -9.8$, therefore the pencil has an acceleration of 9.8 m/sec^2 in a downward motion.

Note: Gravity's acceleration is written as -9.8 m/sec^2 or -32 ft/sec^2

Example 2

Susan was traveling at a constant speed of 24 m/sec when she missed a school zone sign. A police car accelerated from rest at a constant acceleration of 2 m/sec^2 in order to catch up with her car. The distance formula for the police car is given by:

$$d = d_o + v_o t + \frac{1}{2} a t^2 \quad \text{where } d_o = \text{initial distance and } v_o = \text{initial velocity.}$$

  a. How long will it take for the police car to catch up with Susan's car, assuming that Susan maintains a constant velocity of 24 m/sec?
  b. How far did the police car travel in order to catch up with Susan?
  c. What was the velocity of the police car when it reached Susan's car?

a. $d = d_o + v_o t + \frac{1}{2}at^2$

The policeman's initial distance and velocity are both zero since he accelerated from rest. The policeman's acceleration is 2 m/sec^2.

$$d_{police} = \frac{1}{2}(2)t^2 = t^2$$

rate x time = distance for Susan

Susan's velocity is 24 m/sec.

$$d_{Susan} = 24 \frac{m}{sec}(t) = 24t$$

When the policeman catches up to Susan, the distance traveled will be the same for each.

$$t^2 = 24t$$
$$t = 24 \text{ sec}$$

It will take 24 seconds for the police car to catch up with Susan.

b. $d_{police} = t^2 = (24)^2 = 576$ meters

The police car traveled 576 meters in 24 seconds to catch up with Susan.

c. $v(t) = d'(t) = 2t$
    $v(24) = 2(24) = 48$ m/sec

The police car was traveling at 48 m/sec when he caught up with Susan.

LESSON 19

Economics Applications

Maximizing profits and minimizing costs are common business goals. Several economic terms need to be defined in order to use calculus effectively.

$C(x)$ = cost function is the cost of producing x items. The letter x denotes the number of sales, or how many items are produced. Costs can include royalty payments, production, printing, and distribution costs, etc.

$c(x)$ = average cost: $c(x) = C(x)/x$

$p(x)$ = price function, or the price of an item. This is sometimes referred to as a demand function and can be labeled as $d(x)$.

$R(x)$ = revenue function: $R(x) = x \cdot p(x)$

$P(x)$ = profit function: $P(x) = R(x) - C(x)$

Marginal cost, marginal revenue, and marginal profit refer to the derivative for the cost, revenue, and profit functions respectively.

Example 1
Given the cost function $C(x) = 500 + 20x - .1x^2 + .001x^3$:

a. Find the average cost function.
b. Find the marginal cost function.
c. Find the minimum value of the marginal cost.

a. Find the average cost function.

$$c(x) = \frac{C(x)}{x} = \frac{500 + 20x - .1x^2 + .001x^3}{x}$$

$$c(x) = \frac{500}{x} + 20 - .1x + .001x^2$$

b. The marginal cost function is $C'(x) = 20 - .2x + .003x^2$.

c. The minimum value of the marginal cost: the minimum value of $C'(x)$ will occur where $C''(x) = 0$.

$$C'(x) = 20 - .2x + .003x^2$$
$$C''(x) = -.2 + .006x$$
$$0 = -.2 + .006x$$
$$.2 = .006x$$
$$x = \frac{100}{3} = 33\frac{1}{3} \qquad \text{Multiply by 1,000 and solve for x.}$$

$C'''(x) = .006$ so $\left(33\frac{1}{3}\right)$ would be a minimum.

Example 2

FunTime Computer Company has produced a new computer program with an estimated price of $p(x) = 50 - .001x$, where x is the number of sales per year and $p(x)$ is the price in dollars. The cost of manufacturing is 10 dollars per program plus fixed costs of $100,000.

a. Where is the price = 0? Draw a sketch.
b. Find the revenue function. Draw a sketch.
c. Maximize the revenue function.
d. Determine the cost function.
e. What is the cost of the 10,001st item?
f. What is the cost to produce an additional item?
   (The definition of the marginal cost function $C'(x)$.)
g. Determine the profit function.
h. Find the maximum profit.

The price curve (which is a line) starts at a price of $50 for a quantity of zero and will fall as x increases until it reaches zero.

a. Where is the price equal to zero?

$$50 = .001x$$

The domain for $p(k)$ is $[0, 50k]$.
The price = 0, when x = 50,000.

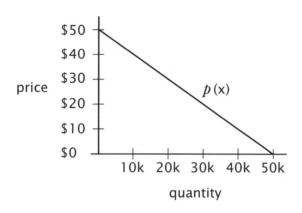

b. Find the revenue function.
$$R(x) = x \cdot p(x)$$

$$R(x) = 50x - .001x^2$$ This graph is a frown parabola.

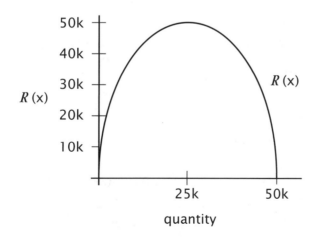

c. Maximize the revenue function.

$$R'(x) = 50 - .002x$$
$$0 = 50 - .002x$$
$$50 = .002x$$
$$x = 25,000$$

$R''(x) = -.002$ so 25,000 is a maximum.

$$R(x) = 50x - .001x^2$$
$$R(25,000) = 50(25,000) - .001(25,000)^2$$
$$= 1,250,000 - 625,000 = 625,000$$

d. Determine the cost function.

$$C(x) = 100,000 + 10x$$

100,000 is a fixed cost, and $10 per program is 10x. This is a line.

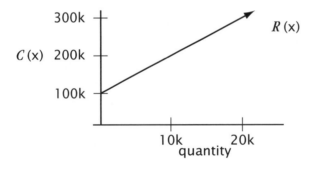

e. What is the cost of the 10,001st item?

To determine the cost of the nth item, subtract the cost of the $(n - 1)^{th}$ item from the nth item.

$$C(10,001) - C(10,000) = \text{Cost of 10,001st item}$$
$$C(10,001) = 100,000 + 10(10,001) = 200,010$$
$$C(10,000) = 100,000 + 10(10,000) = 200,000$$

Cost of $10,001^{st}$ item = 200,010 - 200,000 = \$10

f. What is the cost to produce an additional item?

That is the definition of the marginal cost function ($C'(x)$).

$$C(x) = 100{,}000 + 10x$$
$$C'(x) = 10$$

It costs $10 to produce the next item regardless of which item we are producing.

Note: Usually the cost function is more complex and we would have to substitute a value for x into the equation. $C'(n)$ is the approximate cost of the $(n + 1)^{st}$ item.

g. Determine the profit function.

$$P(x) = R(x) - C(x)$$
$$P(x) = 50x - .001x^2 - (100{,}000 + 10x)$$
$$P(x) = 40x - .001x^2 - 100{,}000$$

h. Find the maximum profit.

$$P(x) = 40x - .001x^2 - 100{,}000$$
$$P'(x) = 40 - .002x$$
$$0 = 40 - .002x$$
$$40 = .002x$$
$$x = 20{,}000$$

$$P'(x) = 40 - .002x$$
$$P''(x) = -.002x$$
$P''(20{,}000)$ is a max, since $P''(x)$ is negative.

$$P(20{,}000) = 40(20{,}000) - .001(20{,}000)^2 - 100{,}000$$
$$= 800{,}000 - 400{,}000 - 100{,}000 = 300{,}000$$

If you manufacture 20,000 computer games you will realize a profit of $300,000.

Break–Even Point

The ***break-even point*** is where the cost function equals the revenue function. Companies are interested in this point because it tells them when they begin to make a profit.

Example 3

NeverFail Printer Company has just produced their best printer ever. The cost to produce one printer is $100.00. The startup costs are $90,000.00 and each printer will sell for $250.00. Where will the break-even point occur?

$$R(x) = 250x$$
$$C(x) = 90{,}000 + 100x$$

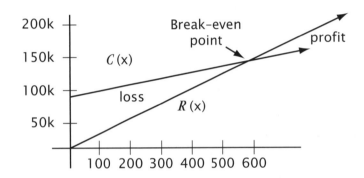

The break-even point will occur when $R(x) = C(x)$.

$$R(x) = C(x).$$

$$250x = 90{,}000 + 100x$$
$$x = 600$$

If we put 600 back into the original equations, it checks.

$$250(600) = 90{,}000 + 100(600)$$
$$150{,}000 = 150{,}000$$

LESSON 20

Optimization

One of the most important applications of the first derivative is its use in optimization problems. Manufacturers desire to maximize profits and minimize costs (unless it is the government). This lesson has several examples of how to do this. Here are some guidelines to make these applications easier to solve so you can minimize wasteful energy and maximize your time.

1. Read each problem carefully and thoroughly before you begin to solve it.

2. Draw a sketch of what the problem looks like.

3. Define the variables and label your drawing. Define your boundaries.

4. Make equations related to the drawing and determine which equation is to be maximized or minimized. You will usually have two equations. One is a *constraint equation* and the other is an *optimization equation*. The constraint equation is used to solve for one variable which will then be substituted into the optimization equation.

5. Once you have made the substitution of the results of the constraint equation into the optimization equation, take the derivative. The equation should be a function of only one variable.

6. Use the second derivative to verify that you have found a maximum or minimum value. You can also use the first derivative test to find the maximum or minimum.

7. Answer the original question.

Example 1

Find two nonnegative numbers whose sum is six. The product of one number and the square of the other number should be a maximum.

Step 3: x = the first number, y = the second number.

$x \geq 0$, $y \geq 0$ boundary of x [0, 6]; boundary of y [0, 6]

Step 4: Constraint (sum): $x + y = 6$
Optimization (product): max $P = xy^2$

Solving the constraint equation for y, we get y = 6 − x. Substitute this into the optimization equation, we have max $P = x(6 - x)^2$.

Step 5:

$$P = x(6-x)^2$$
$$P = x(36-12x+x^2)$$
$$P = 36x - 12x^2 + x^3$$
$$P = x^3 - 12x^2 + 36x$$
$$P' = 3x^2 - 24x + 36$$
$$P' = 3(x^2 - 8x + 12)$$
$$P' = 3(x-6)(x-2), \text{ so } x = 6, x = 2$$

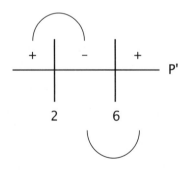

Step 6:
$$P' = 3x^2 - 24x + 36$$
$$P'' = 6x - 24$$
$$P''(6) = 6(6) - 24 = 12$$
$$P''(2) = 6(2) - 24 = -12 \quad \text{max}$$
There is a maximum at x = 2.

Step 7: When x = 2, y = 4, because x + y = 6.

Closure: max xy^2 is $2 \cdot 4^2 = 32$

Example 2

What is the area of the largest rectangular corral that can be built with two perpendicular partitions creating three sections, using 1,000 feet of fencing?

Step 2:

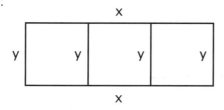

Step 3: x = width of the rectangle,
y = the height of the rectangle
boundary of x [0, 500], boundary of y [0, 250]

Step 4: Constraint (fencing): 1000 = 2x + 4y
Optimization (area): max A = xy

Solving the constraint equation for x, we get x = 500 - 2y.

Substituting this into the optimization equation, we have:
$$\max A = (500 - 2y)y$$
$$A = 500y - 2y^2$$

Step 5: A = 500y - 2y²
A' = 500 - 4y
0 = 500 - 4y
4y = 500
y = 125   Thus the critical point is at y = 125.

Step 6: A" = -4, so y = 125 as a max is confirmed.
A"(125) is a maximum.

If y = 125, then x = 500 - 2(125) = 250.

Step 7: Area = (125)(250) = 31,250 square feet

Example 3

A rectangular box with a square base and an open top is to be made. Find the volume of the largest box that can be designed from 1200 square feet of material.

Step 2:

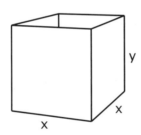

Step 3: x = the length of a side on the square base;
y = the height of the box.

Step 4: Constraint (material): 1200 = x² + 4xy
(Each side has an area xy and the base area is x².)
Optimization (volume): max V = x²y

Solving the constraint equation for y, we get 4xy = 1200 - x².

$$y = \frac{1200 - x^2}{4x}$$

Substituting this into the optimization equation, we have step 5.

Step 5: $\max V = x^2 \left(\dfrac{1200 - x^2}{4x} \right) = 300x - \dfrac{x^3}{4}$

$$V' = 300 - \dfrac{1}{4}(3x^2)$$

$$V' = 300 - \dfrac{3}{4}x^2 = 0$$

$$\dfrac{3}{4}x^2 = 300$$

$$x^2 = 400$$

$$x = \pm 20$$

The critical values are ± 20, but the negative value makes no sense.

Step 6: $V'' = -\dfrac{3}{2}x$; $V''(20)$ is negative, so $x = 20$ is a maximum.

Step 7: $V(20) = 300(20) - \dfrac{(20)^3}{4}$

$\qquad\qquad = 6000 - 2000$

$\qquad\qquad = 4000$ cubic feet

Closure $\quad y = \dfrac{1200 - (20)^2}{4(20)}$

$\qquad\qquad = \dfrac{1200 - 400}{80} = 10$

Base + 4 sides = 400 + 800

$\qquad\qquad\qquad = 1200$

Example 4

A sheet of cardboard 4 feet by 6 feet is to be made into a box by cutting squares of equal size from each corner and folding up the edges. What is the volume of the largest box that could be made?

Step 2:

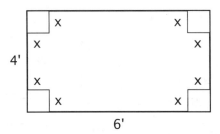

Step 3: x is the length of one side of the square, to be cut from the corners. The boundary of x is [0, 2].

Step 2 again:

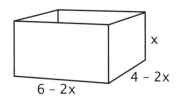

Step 4: No constraint equation.
$$\text{Maximum Volume} = (6 - 2x)(4 - 2x)(x)$$
$$= 4x^3 - 20x^2 + 24x$$

Step 5: $V' = 12x^2 - 40x + 24$
$$0 = 4(3x^2 - 10x + 6)$$
$$0 = 3x^2 - 10x + 6$$

Using the quadratic formula to find the root:

$$\frac{10 \pm \sqrt{100 - 4(3)(6)}}{2(3)} = \frac{5 \pm \sqrt{7}}{3}$$

$\frac{5 + \sqrt{7}}{3}$ is approximately 2.55

$\frac{5 - \sqrt{7}}{3}$ is approximately .78

Step 6: $V' = 12x^2 - 40x + 24$

$V'' = 24x - 40$

$V''(.78) = 24(.78) - 40 = -21.28$ Since this is negative, $x = .78$ is the max.

$V''(2.55) = 24(2.55) - 40 = 21.2$

Step 7: $6 - 2x = 6 - 2(.78) = 4.44$
$4 - 2x = 4 - 2(.78) = 2.44$
$x = .78$

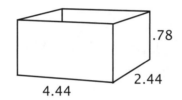

Closure $V = (4.44)(2.44)(.78)$
              $= 8.45 \text{ ft}^3$

Example 5
Find the point (x, y) on the graph of $y = \sqrt{x}$ which is closest to $(6, 0)$.

Step 2:

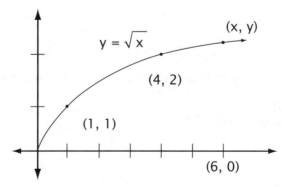

Step 3: boundary of x $[0, 6]$
        boundary of y $[0, \sqrt{6}]$

Step 4: Constraint $y = \sqrt{x}$.

$$\text{Minimum Distance} = \sqrt{(x-6)^2 + (y-0)^2}$$

$$\text{Minimum Distance} = \sqrt{(x-6)^2 + y^2}$$

Substitute constraint:

$$\text{Minimum Distance} = \sqrt{(x-6)^2 + (\sqrt{x})^2}$$

$$\text{Minimum Distance} = \sqrt{(x-6)^2 + x} = (x^2 - 11x + 36)^{\frac{1}{2}}$$

Step 5: $D' = \dfrac{1}{2}(x^2 - 11x + 36)^{-\frac{1}{2}}(2x - 11)$

$$D' = \frac{2x - 11}{2(x^2 - 11x + 36)^{\frac{1}{2}}}$$

$$0 = \frac{2x - 11}{2(x^2 - 11x + 36)^{\frac{1}{2}}}$$

$$0 = 2x - 11$$

$$\frac{11}{2} = x$$

$$x = 5.5$$

Step 6: $D'(1) = \text{neg}$
$D'(10) = \text{pos}$

We are evaluating the first derivative of points on either side of x = 5.5 in order to establish that x = 5.5 is a minimum.

Step 7: 5.5 is a minimum.
Since x = 5.5, then $y = \sqrt{5.5}$.

The point (x, y) is $(5.5, \sqrt{5.5})$.

$$\text{Minimum Distance} = \sqrt{(5.5 - 6)^2 + 5.5}$$

which is approximately 2.4.

LESSON 21

Related Rates

Each *related rate* problem states the rate of change for one variable and requires that we find the rate of change for another variable. These variables will have a known relationship. Here are the steps for solving a related rate problem.

1. Separate the general information from the particular information. General information is always true. Particular information is only true at a particular time and will be used later in the problem solving process. It is a common mistake to use the particular information too soon.

2. Draw and label a sketch using only the general information.

3. Identify the known rate and desired rate.

4. Use your knowledge of geometry or trigonometry to relate the known and desired rates.

5. Differentiate implicitly with respect to time.

6. Substitute the particular information to determine the desired rate.

Example 1

Two cars begin at the same starting point. At the same time, one car travels due north at a rate of 30 ft/sec and the second travels due east at a rate of 40 ft/sec. At what rate is the distance between them increasing after 1 second?

1. General information:

    Car 1 has a rate of 30 ft/sec.
    Car 2 has a rate of 40 ft/sec.

2.

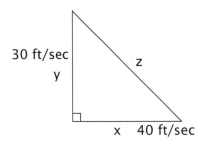

3. We were given dx/dt and dy/dt and we desire dz/dt when 1 second has elapsed.

4. They are related by the Pythagorean theorem.

    $x^2 + y^2 = z^2$, so $z = \sqrt{x^2 + y^2}$

5. Differentiating implicitly with respect to time:

    $$2x\frac{dx}{dt} + 2y\frac{dy}{dt} = 2z\frac{dz}{dt}$$

6. $$2x\frac{dx}{dt} + 2y\frac{dy}{dt} = 2\left(\sqrt{x^2+y^2}\right)\frac{dz}{dt}$$
$$x\frac{dx}{dt} + y\frac{dy}{dt} = \left(\sqrt{x^2+y^2}\right)\frac{dz}{dt}$$
$$40(40) + 30(30) = \left(\sqrt{2,500}\right)\frac{dz}{dt} \quad \text{After 1 second, } x = 40, y = 30.$$
$$\frac{dz}{dt} = 50 \, \text{ft}/\text{sec}$$

Example 2

A man is walking at 5 mi/hr toward the foot of a 60 ft flagpole. At what rate is he approaching the top of the flagpole when he is 80 ft from the foot of the flagpole?

1. General information:

    Velocity of the man is 5 mi/hr; flagpole is 60 ft tall.

2. Particular information:

    Man will be 80 ft from the pole at some time.

    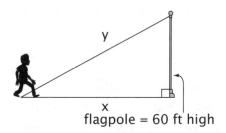
    flagpole = 60 ft high

    Note: $5 \, \frac{\text{mi}}{\text{hr}} \cdot \frac{5,280 \, \text{ft}}{\text{mile}} = 26,400 \, \frac{\text{ft}}{\text{hr}}$

    (We'll use this measurement because our problem is using feet and not miles.)

3. We were given $\frac{dx}{dt}$ and we desire $\frac{dy}{dt}$.

4. Using the Pythagorean theorem, we get: $y^2 = x^2 + 3,600$.

5. Differentiating:

$$2y\frac{dy}{dt} = 2x\frac{dx}{dt}$$

$$y\frac{dy}{dt} = x\frac{dx}{dt}$$

6. Solving for y from step 4 we get:

$$y = \sqrt{x^2 + 3{,}600}$$

Substituting into step 5 we get:

$$\sqrt{x^2 + 3{,}600} \cdot \frac{dy}{dt} = x \cdot \frac{dx}{dt}$$

When x = 80 we get:

$$\sqrt{80^2 + 3{,}600} \cdot \frac{dy}{dt} = 80 \cdot \frac{26{,}400 \text{ ft}}{\text{hr}}$$

$$\frac{dy}{dt} = 21{,}120 \text{ ft/hr}$$

$$= 4 \text{ mi/hr} \qquad 1 \text{ mi} = 5{,}280 \text{ ft}$$

Example 3

Helium gas is escaping from a spherical balloon at a rate of 1,000 in³/min. At the instant when the radius is 10 inches, at what rate is the radius decreasing? At what rate is the surface area decreasing?

1. General information:

   Gas is escaping at a rate of 1,000 in³/min.

   Particular information:

   At some point the radius will be 10 inches.

2.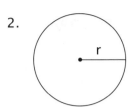

3. We know $\frac{dV}{dt}$. We want $\frac{dS}{dt}$ and $\frac{dr}{dt}$.

4. $V = \frac{4}{3}\pi r^3$ $S = 4\pi r^2$ (S = surface area)

5. Differentiating: $\frac{dV}{dt} = \frac{4}{3}\pi(3r^2)\frac{dr}{dt}$

    $= 4\pi r^2 \frac{dr}{dt}$

6. $-1{,}000\frac{in^3}{min} = 4\pi(10\ in)^2 \frac{dr}{dt}$

    $\frac{dr}{dt} = \frac{-2.5}{\pi}\frac{in}{min} \approx -.80\frac{in}{min}$

    $\frac{dS}{dt} = 8\pi r \frac{dr}{dt}$

    $\frac{dS}{dt} = 8\pi(10\ in)\left(\frac{-2.5}{\pi}\frac{in}{min}\right)$

    $= -200\frac{in^2}{min}$

The surface area is decreasing at 200 in²/min.

Example 4

Suppose we have two resistors, R_1 and R_2, connected in parallel. The total resistance, R, is:

$$\frac{1}{R} = \frac{1}{R_1} + \frac{1}{R_2}$$

If R_1 is increasing at a rate of .3 ohm/min and R_2 is decreasing at a rate of .5 ohm/min, at what rate is R changing when $R_1 = 20$ ohm and $R_2 = 50$ ohms?

To avoid confusion with multiple Rs, let $R_1 = N$, and $R_2 = W$

$$\frac{1}{R} = \frac{1}{N} + \frac{1}{W} \text{ or } R^{-1} = N^{-1} + W^{-1}$$

Differentiating implicitly with respect to time we find:

$$R^{-1} = N^{-1} + W^{-1}$$

$$-R^{-2}\frac{dR}{dT} = -N^{-2}\frac{dN}{dT} - W^{-2}\frac{dW}{dT} \text{ or } -\frac{1}{R^2}\frac{dR}{dT} = \frac{-1}{N^2}\frac{dN}{dT} - \frac{1}{W^2}\frac{dW}{dT}$$

$$\frac{dR}{dT} = R^2 \left(\frac{1}{N^2}\frac{dN}{dT} + \frac{1}{W^2}\frac{dW}{dT} \right)$$

To find R, we recall $N = 20$ and $W = 50$.

$$\frac{1}{R} = \frac{1}{N} + \frac{1}{W}$$

$$\frac{1}{R} = \frac{1}{20} + \frac{1}{50} \qquad \text{Substituting.}$$

$$\left(\frac{1}{R} = \frac{1}{20} + \frac{1}{50} \right) \frac{100R}{1} \qquad \text{Multiplying by 100R.}$$

$$100 = 5R + 2R$$

$$100 = 7R$$

$$R = \frac{100}{7}$$

Replacing all the values into the implicitly differentiating formula for dR/dT:

$$\frac{dR}{dT} = \left(\frac{100}{7} \right)^2 \left(\frac{1}{20^2}(.3) + \frac{1}{50^2}(-.5) \right)$$

$$\frac{dR}{dT} = \frac{10{,}000}{49}(.00075 - .0002)$$

$$= .11 \, \frac{\text{ohm}}{\text{min}}$$

R is increasing at a rate of .11 ohm/min.

LESSON 22

Antiderivatives

We've spent the first 21 chapters developing and using the concept of a derivative. Now we will investigate the concept of an *antiderivative*.

Let's consider the derivative rules in reverse.

Rule 1 $\quad \dfrac{d}{dx}(C) = 0$ where C is a constant.

Therefore the anti-derivative of 0 equals any constant.

Rule 2 $\quad \dfrac{d}{dx}(x) = 1$

Therefore the anti-derivative of 1 is x (with respect to x).

Example 1
If $f(x) = x + 3$, then $f'(x) = 1$.

If $f(x) = x - 2$, then $f'(x) = 1$.

If $f(x) = x + C$, then $f'(x) = 1$ where C is a constant.

Therefore the antiderivative of 1 is x + C. This is an interesting point. A derivative of a function produces a single function. The antiderivative of a function produces an infinite number of functions.

The notation we use to refer to an antiderivative is the ***indefinite integral***, $\int$. $\int f(x)dx$ means the antiderivative of $f(x)$ with respect to x. Therefore, using the example from the paragraph above.

$$\int 1\, dx = x + C \qquad \int 1\, dy = y + C$$

because $\frac{d}{dx}(x + C) = 1$ and $\frac{d}{dy}(y + C) = 1$

Rule 3 $\quad \frac{d}{dx}(f(x)+g(x)) = \frac{d}{dx}(f(x)) + \frac{d}{dx}(g(x))$

It is also true that the antiderivative of the sum is the sum of the antiderivatives.

$$\int (f(x)+g(x))dx = \int (f(x))dx + \int (g(x))dx$$

Rule 4 $\quad \frac{d}{dx}x^n = nx^{n-1}$

The power rule in reverse is: $\int x^n dx = \frac{1}{n+1}x^{n+1} + C$ for all $n \neq -1$.

Example 2

$$\int x^3 dx = \qquad\qquad \text{when } n = 3$$

$$= \frac{1}{3+1}x^{3+1} + C = \frac{x^4}{4} + C$$

This is true because $\frac{d}{dx}\left(\frac{x^4}{4} + C\right) = \frac{1}{4} \cdot 4x^3 + 0 = x^3$.

Example 3

$$\int r^{\frac{1}{3}} dr = \qquad \text{when } n = \frac{1}{3}$$

$$= \frac{1}{\frac{4}{3}} r^{\frac{4}{3}} + C = \frac{3}{4} r^{\frac{4}{3}} + C$$

This is true because $\frac{d}{dx}\left(\frac{3}{4} r^{\frac{3}{4}} + C\right) = \frac{3}{4} \cdot \frac{4}{3} r^{\frac{1}{3}} = r^{\frac{1}{3}}$.

> **Rule 5** $\quad \frac{d}{dx} C \cdot f(x) = C \cdot \frac{d}{dx} f(x)$ where C is a constant.
>
> This rule is also true for antiderivatives: $\int C \cdot f(x) = C \cdot \int f(x)$.
>
> The antiderivative of a constant times a function is the constant times the antiderivative of the function.

Example 4

$$\int 2r^{\frac{1}{3}} = 2 \int r^{\frac{1}{3}}$$

$$= 2\left(\frac{3}{4}\right) r^{\frac{4}{3}} + C$$

$$= \frac{3}{2} r^{\frac{4}{3}} + C$$

Example 5

Even though both derivatives would legitimately have their own constant, we can combine them into one final C as long as we are not asked to evaluate them further as in examples 3 and 4.

$$\int \left(3y^4 - 2\sqrt{y}\right) dy = \int \left(3y^4 - 2y^{\frac{1}{2}}\right) dy$$

$$= 3 \int y^4 dy - 2 \int y^{\frac{1}{2}} dy$$

$$= 3\left(\frac{1}{5}\right) y^5 - 2\left(\frac{2}{3}\right) y^{\frac{3}{2}} + C$$

$$= \frac{3}{5} y^5 - \frac{4}{3} y^{\frac{3}{2}} + C$$

Example 6

Find the equation of a curve which passes through (2, 3) if the slope of the tangent at any point is 3x + 2.

We know that the derivative of the function we are seeking is 3x + 2. We will take the antiderivative of that function in order to obtain the original function.

$$\int (3x + 2)dx = \int 3x\, dx + \int 2\, dx$$

$$= 3\int x^1 dx + 2\int x^0 dx$$

$$= 3 \cdot \frac{1}{2}x^2 + 2x + C$$

$y = \frac{3}{2}x^2 + 2x + C$ is the family of parabolas we are seeking.

Using the point (2, 3), we have:

$$3 = \frac{3}{2}(2)^2 + 2 \cdot 2 + C$$

$$3 = 6 + 4 + C$$

$$-7 = C$$

The specific parabola is $y = \frac{3}{2}x^2 + 2x - 7$.

Example 7

Find the curve for which y" = 2 and which passes through (3, 1) and (-1, 2).

$$y' = \int 2dx.$$

$$= 2\int dx$$

$$= 2x + C_1$$

$$y = \int (2x + C_1)dx$$
$$= 2\int x\,dx + C_1 \int dx$$
$$= 2\,\frac{x^2}{2} + C_1 x + C_2$$
$$= x^2 + C_1 x + C_2$$

We know that this curve $y = x^2 + C_1 x + C_2$ passes through (3, 1) and (-1, 2) so substituting we get:

$$1 = 9 + 3C_1 + C_2 \quad\text{and}\quad 2 = 1 - C_1 + C_2$$
$$3C_1 + C_2 = -8 \quad\text{and}\quad -C_1 + C_2 = 1$$

Using elimination:

$$3C_1 + C_2 = -8$$
$$C_1 - C_2 = -1$$
$$\overline{4C_1 = -9}$$
$$C_1 = -\frac{9}{4}$$

Since $C_1 - C_2 = -1$ then:

$$\left(-\frac{9}{4}\right) - C_2 = -1$$
$$-C_2 = \frac{5}{4}$$
$$C_2 = -\frac{5}{4}$$

Therefore our curve is $y = x^2 - \frac{9}{4}x - \frac{5}{4}$ or $4y = 4x^2 - 9x - 5$.

Example 8

Find the curve for which y" = 6 and which has a critical point at (1, 4).

$$y'' = 6$$
$$y' = \int 6\,dx$$
$$= 6x + C_1$$

When y' = 0 and x = 1, we have:

$$0 = 6(1) + C_1$$
$$C_1 = -6$$

So, y' = 6x − 6

$$y = \int (6x - 6)\,dx$$
$$= 3x^2 - 6x + C_2$$

When x = 1 and y = 4, we have:

$$4 = 3(1) - 6(1) + C_2$$
$$C_2 = 7$$

The desired polynomial is $3x^2 - 6x + 7$.

LESSON 23

Integration Formulas

These integration formulas follow as an extension of our rules for differentiation.

Fundamental Integration Formulas

1. $\int du = u + C$

2. $\int (du + dr + \ldots + dv) = \int du + \int dr + \int \ldots + \int dv$

3. $\int c\, du = c \int du \quad$ where "c" is a constant

4. $\int u^n du = \dfrac{u^{n+1}}{n+1} + C \quad (n \neq -1)$

5. $\int \dfrac{du}{u} = \ln u + C$

6. $\int e^u du = e^u + C$

7. $\int \cos(u)\, du = \sin(u) + C$

8. $\int \sin(u)\, du = -\cos(u) + C$

9. $\int \sec^2(u)\, du = \tan(u) + C$

10. $\int \csc^2(u)\, du = -\cot(u) + C$

11. $\int \sec(u)\tan(u)\, du = \sec(u) + C$

12. $\int \csc(u)\cot(u)\, du = -\csc(u) + C$

We can now employ the substitution method of integration.

Example 1

$$\int 2\cos(2x)\, dx$$

Let $u = 2x$; $\dfrac{du}{dx} = 2$; $du = 2dx$.

Substituting we get:

$$\int \cos(u)\, du = \sin(u) + C$$
$$= \sin(2x) + C$$

Let's check our work: $\dfrac{d}{dx}(\sin(2x) + C) = 2\cos(2x)$

Example 2

$$\int 3\sec(3x)\tan(3x)\, dx$$

Let $u = 3x$; $\dfrac{du}{dx} = 3$; $du = 3dx$.

Substituting we get:

$$\int \sec(u)\cdot\tan(u)\, du = \sec(u) + C$$
$$= \sec(3x) + C$$

Check: $\dfrac{d}{dx}(\sec(3x) + C) = \sec(3x)\tan(3x)\cdot 3$

Example 3 shows how to integrate when the needed constant is not present.

Example 3

$$\int \sec^2(2x)\, dx$$

Let $u = 2x$; $\dfrac{du}{dx} = 2$; $du = 2dx$.

This problem is missing the 2 needed for a complete du value. It can be inserted in the following way. The needed value is placed in the integral and the reciprocal is placed outside. This is the same as multiplication by 1 and does not change the value of the integral.

$$\frac{1}{2}\int \sec^2(2x) \cdot dx \cdot 2 = \frac{1}{2}\int \sec^2(u)du$$

$$= \frac{1}{2}\tan(u) + C$$

$$= \frac{1}{2}\tan(2x) + C$$

Check: $\frac{d}{dx}\left[\frac{1}{2}\tan(2x)+C\right] = \frac{1}{2}\sec^2(2x) \cdot 2 = \sec^2(2x)$

Example 4

$$\int 2x(1 + x^2)^8 dx$$

You can use this same idea with polynomials. You could use Pascal's triangle to write out all the terms, but let's try substitution.

Let $u = 1 + x^2$; $\frac{du}{dx} = 2x$; $du = 2xdx$

Rearranging we get:

$$= \int (1 + x^2)^8 2xdx$$

$$= \int u^8 du$$

$$= \frac{u^9}{9} + C$$

$$= \frac{1}{9}(1 + x^2)^9 + C$$

Check:

$$\frac{d}{dx}\left[\frac{1}{9}(1+x^2)^9 + C\right] = \frac{1}{9} \cdot 9(1+x^2)^8 \cdot 2x$$

$$= 2x(1+x^2)^8$$

Example 5

$$\int x(1-x^2)^7 dx$$

Let $u = 1 - x^2$; $\frac{du}{dx} = -2x$; $du = -2xdx$.

This time we do not have the needed constant.
Multiplying by one, we get:

$$-\frac{1}{2}\int -2x(1-x^2)^7 dx = -\frac{1}{2}\int (1-x^2)^7(-2xdx)$$

$$= -\frac{1}{2}\int u^7 du$$

$$= -\frac{1}{2} \cdot \frac{u^8}{8} + C$$

$$= -\frac{1}{16}(1-x^2)^8 + C$$

Check:

$$\frac{d}{dx}\left[-\frac{1}{16}(1-x^2)^8 + C\right] = -\frac{1}{16} \cdot 8(1-x^2)^7(-2x)$$

$$= x(1-x^2)^7$$

The tricky part will be determining what u will be. If you pick something that is incorrect for u, try again. These problems will become simpler with practice.

Example 6

$$\int e^{5x} dx$$

Let $u = 5x$; $\frac{du}{dx} = 5$; $du = 5dx$.

Multiply by 1.

$$\frac{1}{5}\int 5e^u du = \frac{1}{5}e^u + C = \frac{1}{5}e^{5x} + c$$

Check:

$$\frac{d}{dx}\left(\frac{1}{5}e^{5x} + C\right) = \frac{1}{5}e^{5x} \cdot 5 = e^{5x}$$

Here we have an example where we have the wrong constant. Simply move the wrong constant out in front of the integral and supply the correct constant by multiplying by 1.

Example 7

$$\int \frac{\cos(2\theta)}{\sin^2(2\theta)} \cdot 5 d\theta$$

Let $u = (\sin(2\theta))$; $\frac{du}{d\theta} = \cos(2\theta) \cdot 2$; $du = \cos(2\theta) \cdot d\theta$.

$$5\int \frac{\cos(2\theta)}{\sin(2\theta)^2} \cdot d\theta = \frac{1}{2} \cdot 5 \int \frac{\cos(2\theta)}{\sin(2\theta)^2} \cdot 2 d\theta$$

$$= \frac{5}{2} \int \frac{du}{u^2}$$

$$= \frac{5}{2} \int u^{-2} du$$

$$= \frac{5}{2} \frac{u^{-1}}{-1} + C$$

$$= -\frac{5}{2} u^{-1} + C$$

$$= \frac{-5}{2(\sin(2\theta))} + C$$

Check:

$$\frac{d}{dx}\left(\frac{-5}{2\sin(2\theta)} + C\right) = \frac{5}{2}(\sin(2\theta))^{-2} \cos(2\theta)(2)$$

$$= \frac{5 \cos(2\theta)}{\sin^2(2\theta)}$$

Example 8

$$\int \frac{(x^2-1)^2}{x^2} dx$$

None of the formulas we have learned cover this form of the integral. Thankfully the exponent is small. We can square the numerator and divide by the denominator.

$$\int \frac{(x^2-1)^2}{x^2} dx = \int \left(\frac{x^4 - 2x^2 + 1}{x^2} \right) dx$$

$$= \int x^2 dx - \int 2 dx + \int x^{-2} dx$$

$$= \frac{x^3}{3} - 2x + \frac{x^{-1}}{-1} + C$$

$$= \frac{1}{3}x^3 - 2x - \frac{1}{x} + C$$

Check:

$$\frac{d}{dx}\left(\frac{1}{3}x^3 - 2x - \frac{1}{x} \right) = \frac{1}{3}(3x^2) - 2 - (-x^{-2})$$

$$= x^2 - 2 + \frac{1}{x^2}$$

$$= \frac{x^4 - 2x^2 + 1}{x^2} = \frac{(x^2-1)^2}{x^2}$$

LESSON 24

Area Under a Curve

Let's look at the area between $f(x) = x^2$ and the x-axis on $[0, 2]$.

Figure 1

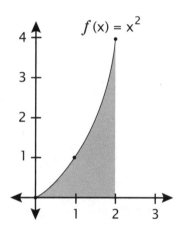

We can estimate the area by dividing the interval $[a, b]$ into n subintervals.

Then $\Delta x = \dfrac{b-a}{n}$ where Δx = width of interval.

In this case $b = 2$ and $a = 0$, so $\Delta x = \dfrac{2-0}{n} = \dfrac{2}{n}$.

The height of each rectangle will be $f(x)$.

Dividing into n = 4 intervals and using the right hand endpoint to estimate the height of the rectangle would look like:

Figure 2

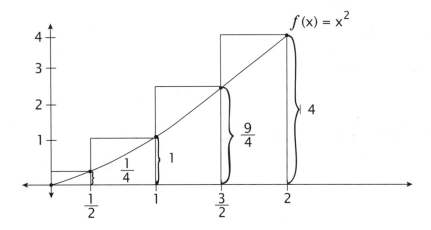

$$\text{Area} = \frac{1}{2}\left(\frac{1}{4}\right) + \frac{1}{2}(1) + \frac{1}{2}\left(\frac{9}{4}\right) + \frac{1}{2}(4)$$
$$= \frac{1}{8} + \frac{1}{2} + \frac{9}{8} + 2$$
$$= \frac{5}{4} + 2\frac{1}{2} = 3\frac{3}{4}$$

Let's divide that interval into four pieces and use the left hand endpoint to estimate the height of the rectangle.

Figure 3

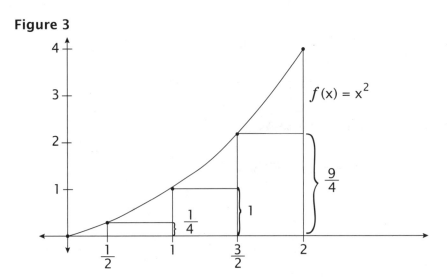

$$\text{Area} = \frac{1}{2}(0) + \frac{1}{2}\left(\frac{1}{4}\right) + \frac{1}{2}(1) + \frac{1}{2}\left(\frac{9}{4}\right)$$
$$= \frac{1}{8} + \frac{1}{2} + \frac{9}{8}$$
$$= \frac{5}{4} + \frac{1}{2} = \frac{7}{4} = 1\frac{3}{4}$$

The first estimate was too high. The second estimate was too low. To get a more accurate estimate we would need to divide the region into more sections, but the exact answer is given by the *definite integral*.

Definite Integrals

Definite integrals can be created from indefinite integrals by adding what are known as *limits* at the top and bottom of the integral sign. To solve this type of problem, we take the integral, and evaluate the resulting function using each limit as a value of the independent variable of the function. Once this is done, the value from the lower limit is subtracted from the value for the upper limit.

Here is an integral with an upper limit of 2 and a lower limit of 0. The 2 and 0 represent the highest and lowest values of x from our example problem.

$$\int_0^2 x^2 dx$$

In the following step, we have integrated the function, and have indicated the upper and lower limits on the vertical bar following the result of integration.

$$\int_0^2 x^2 dx = \left.\frac{x^3}{3}\right|_0^2$$

Substitute 2 for X:
$$\frac{(2)^3}{3} = \frac{8}{3}$$

Substitute 0 for X:
$$\frac{(0)^3}{3} = 0$$

Subtract:
$$\frac{8}{3} - 0 = \frac{8}{3} \text{ or } 2\frac{2}{3}$$

If we could continue to slice the area under the curve into smaller and smaller pieces and add them up, then we would have the exact area. A clever mathematician, Georg Riemann, was able to show that if f is continuous on [a, b], then:

$$\int_a^b f(x)dx = \lim_{n\to\infty} \sum_{i=1}^n f(x_i)\, \Delta x_i$$

These sums are referred to as ***Riemann sums***.

Integral Properties

1. $\int_a^a f(x)dx = 0$
   If the limits are the same, then there is no area under the curve.

2. $\int_a^b cf(x)dx = c\int_a^b f(x)dx$
   The constant can be factored out.

3. $\int_a^b \left[f(x) \pm g(x)\right] dx = \int_a^b f(x)dx \pm \int_a^b g(x)dx$
   The sum of definite integrals is the definite integral of the sum.

4. $\int_a^b f(x)dx = -\int_b^a f(x)$
   The limits on the definite integral can be interchanged, but a negative sign is inserted.

1. $\int_a^b f(x)dx = \int_a^m f(x)dx + \int_m^b f(x)dx$

We can integrate a function over adjacent intervals [a, m] and [m, b]. This is extremely useful.

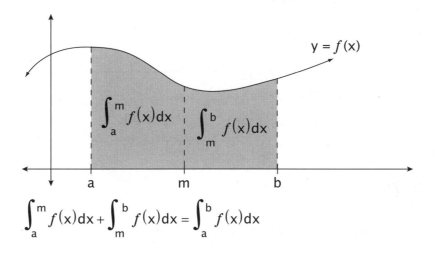

$\int_a^m f(x)dx + \int_m^b f(x)dx = \int_a^b f(x)dx$

Example 1

$\int_3^0 x^2 dx$

Solution

$\int_3^0 x^2 dx = -\int_0^3 x^2 dx = -\frac{x^3}{3}\bigg|_0^3 = -\frac{1}{3}(27-0) = -9$

Example 2

$\int_0^3 (4x^2 - 3)dx$

Solution

$\int_0^3 (4x^2 - 3)dx = 4\frac{x^3}{3}\bigg|_0^3 - 3(x)\bigg|_0^3$

$= \left[\frac{4}{3}(27) - \frac{4}{3}(0)\right] - [3(3) - 3(0)]$

$= 36 - 9 = 27$

Example 3

$$\int_{45}^{45} \frac{\sin(x) - e^x}{\cos(x)} dx$$

Solution

The answer is 0 from property #1.

Example 4

If $\int_{-2}^{1} f(x)dx = 3$ and $\int_{4}^{8} f(x)dx = 7$

and $\int_{4}^{1} f(x)dx = -2$, then what is $\int_{-2}^{8} f(x)dx$?

Solution

$$\int_{1}^{4} f(x)dx = -\int_{4}^{1} f(x)dx = -(-2) = 2$$

$$\int_{-2}^{8} f(x)dx = \int_{-2}^{1} f(x)dx + \int_{1}^{4} f(x)dx + \int_{4}^{8} f(x)dx =$$

$$3 \quad + \quad 2 \quad + \quad 7 \quad = 12$$

The Fundamental Theorem of Calculus relates derivatives to antiderivatives. Here is one part of that theorem.

Fundamental Theorem of Calculus

If f is continuous at every point on [a, b] and F is an antiderivative of f on [a, b] then:

$$\int_{a}^{b} f(x)dx = F(b) - F(a)$$

Figure 4

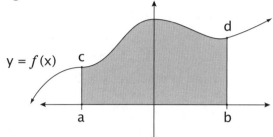

The $\int_a^b f(x)dx$ gives the area bounded by the curve y = f(x) and the x-axis from x = a and x = b. This area is shaded.

If the graph of a function lies both above and below the x-axis, then the integral will give positive values to the areas above the x-axis and negative values to the areas below the x-axis.

Figure 5

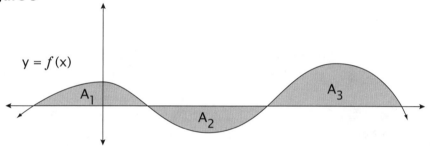

Area between y = f(x) and the x-axis would be $A_1 + |A_2| + A_3$

You will want to take the absolute value of the area below the curve in order to achieve an answer for the total area between the curve and the x-axis.

Example 5

Find the area between the curve y = 2x³ − 2x and the x-axis.
Draw a sketch.

To determine where a graph crosses the x-axis, set y = 0.

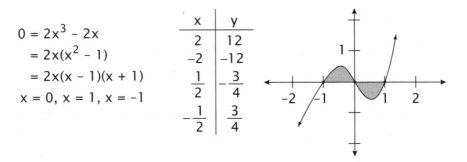

$$0 = 2x^3 - 2x$$
$$= 2x(x^2 - 1)$$
$$= 2x(x - 1)(x + 1)$$
$$x = 0, \ x = 1, \ x = -1$$

x	y
2	12
−2	−12
$\frac{1}{2}$	$-\frac{3}{4}$
$-\frac{1}{2}$	$\frac{3}{4}$

$$\int_{-1}^{0} (2x^3 - 2x)\,dx = \left(2\frac{x^4}{4} - 2\frac{x^2}{2}\right)\Big|_{-1}^{0}$$
$$= \left(\frac{1}{2}x^4 - x^2\right)\Big|_{-1}^{0}$$
$$= 0 - \left(\frac{1}{2} - 1\right) = \frac{1}{2}$$

Since this graph is symmetrical, the area between the curve y = 2x³ − 2x and the x-axis would be $(2)\left(\frac{1}{2}\right) = 1$.

LESSON 25

Definite Integrals

This chapter gives more practice in evaluating definite integrals using substitution and other formulas discussed in previous chapters.

Example 1

Let $u = \sin(3x)$; $\dfrac{du}{dx} = \cos(3x) \cdot 3$; $du = 3\cos(3x)\,dx$

$$\int_{-\frac{\pi}{2}}^{\frac{\pi}{2}} 5\sin^4(3x)\cos(3x)\,dx \cdot 3$$

$$= 5\left(\frac{1}{3}\right)\int_{-\frac{\pi}{2}}^{\frac{\pi}{2}} (\sin(3x))^4 \cos(3x)\,dx \cdot 3$$

$$= \frac{5}{3}\int_{-\frac{\pi}{2}}^{\frac{\pi}{2}} u^4\,du$$

$$= \frac{5}{3}\left(\frac{u^5}{5}\right)\Bigg|_{-\frac{\pi}{2}}^{\frac{\pi}{2}}$$

$$= \frac{1}{3}\left[\sin\left(\frac{3\pi}{2}\right)\right]^5 - \frac{1}{3}\left[\sin\left(-\frac{3\pi}{2}\right)\right]^5$$

$$= -\frac{1}{3} - \frac{1}{3} = -\frac{2}{3}$$

Example 2

Let $u = 1+\sqrt{y} = 1+y^{\frac{1}{2}}$; $\frac{du}{dy} = \frac{1}{2}y^{-\frac{1}{2}}$; $du = \frac{dy}{2\sqrt{y}}$

$$\int_1^4 \frac{dy}{3\sqrt{y}\left(1+\sqrt{y}\right)^3} = \frac{1}{3}\int_1^4 \frac{dy}{\sqrt{y}\left(1+\sqrt{y}\right)^3}$$

$$= (2)\left(\frac{1}{3}\right)\int_1^4 \frac{dy}{\sqrt{y}\left(1+\sqrt{y}\right)^3}\left(\frac{1}{2}\right)$$

$$= \frac{2}{3}\int_1^4 u^{-3}du$$

$$= \frac{2}{3}\left(\frac{u^{-2}}{+2}\right)\Big|_1^4$$

$$= -\frac{1}{3}\left(\frac{1}{\left(1+\sqrt{y}\right)^2}\right)\Big|_1^4$$

$$= -\frac{1}{3}\left(\frac{1}{\left(1+\sqrt{4}\right)^2} - \frac{1}{\left(1+\sqrt{1}\right)^2}\right)$$

$$= -\frac{1}{3}\left(\frac{1}{9} - \frac{1}{4}\right)$$

$$= -\frac{1}{3}\left(-\frac{5}{36}\right) = \frac{5}{108}$$

Example 3

Let $u = 3x$; $\frac{du}{dx} = 3$; $du = 3 \cdot dx$

$$\int_0^1 e^{3x}\,dx = \left(\frac{1}{3}\right)\int_0^1 e^{3x}\,dx\,(3)$$

$$= \frac{1}{3}\int_0^1 e^u\,du$$

$$= \frac{1}{3}e^u\Big|_0^1$$

$$= \frac{1}{3}e^{3x}\Big|_0^1$$

$$= \frac{1}{3}\left(e^3 - e^0\right)$$

$$= \frac{1}{3}\left(e^3 - 1\right)$$

Example 4

$$\int_1^e \frac{dx}{x} = \ln(x) \Big|_1^e$$
$$= \ln(e) - \ln(1)$$
$$= 1 - 0 = 1$$

Example 5

Find the area between one arc of y = 2cos(x) and the x axis. Sketch the graph.

y = 2cos(x)

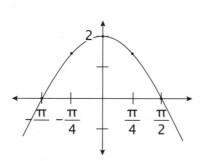

x	y
0	2
$\frac{\pi}{4}$	≈ 1.41
$-\frac{\pi}{4}$	≈ -1.41
$\frac{\pi}{2}$	0
$-\frac{\pi}{2}$	0

$$\int_{-\frac{\pi}{2}}^{\frac{\pi}{2}} 2\cos(x)\,dx = 2\int_{-\frac{\pi}{2}}^{\frac{\pi}{2}} \cos(x)\,dx$$
$$= 2\sin(x)\Big|_{-\frac{\pi}{2}}^{\frac{\pi}{2}}$$
$$= 2[1-(-1)] = 2(2) = 4$$

Example 6

$$\int_{-\frac{\pi}{8}}^{0} \tan(2\theta)\, d\theta = \int_{-\frac{\pi}{8}}^{0} \frac{\sin(2\theta)}{\cos(2\theta)}\, d\theta$$

Renaming $\tan(2\theta)$ as $\frac{\sin(2\theta)}{\cos(2\theta)}$ allows us to use substitution:

$$= \int_{-\frac{\pi}{8}}^{0} \frac{\sin(2\theta)}{\cos(2\theta)}\, d\theta$$

Let $u = \cos(2\theta)$; $\frac{du}{d\theta} = -2\sin(2\theta)$; $du = -2\sin(2\theta)\, d\theta$.

$$= -\frac{1}{2} \int_{-\frac{\pi}{8}}^{0} \frac{du}{u}$$

$$= -\frac{1}{2} \ln(\cos(2\theta)) \Big|_{-\frac{\pi}{8}}^{0}$$

$$= -\frac{1}{2} \left[\ln(\cos(0)) - \ln\left(\cos\left(-\frac{\pi}{4}\right)\right) \right]$$

$$= -\frac{1}{2} \left[0 - \ln\left(\frac{\sqrt{2}}{2}\right) \right]$$

$$= \frac{1}{2} \ln\left(\frac{\sqrt{2}}{2}\right)$$

Substitution will not work with this type of integral. Divide the numerator by the denominator and form two integrals. Then use substitution on the first integral. If you had chosen u = cos(x) then you would have u/du, not du/u, and integration would be impossible.

Example 7

$$\int_{\pi/4}^{\pi/2} \frac{\cos(x)+\sin(x)}{\sin(x)} \, dx = \int_{\pi/4}^{\pi/2} \frac{\cos(x)}{\sin(x)} \, dx + \int_{\pi/4}^{\pi/2} 1 \, dx$$

Let u = sin(x); $\frac{du}{dx}$ = cos(x); du = cos(x) dx.

$$= \int_{\pi/4}^{\pi/2} \frac{du}{u} + x \Big|_{\pi/4}^{\pi/2}$$

$$= \ln(u) \Big|_{\pi/4}^{\pi/2} + \left(\frac{\pi}{2} - \frac{\pi}{4}\right)$$

$$= \ln(\sin(x)) \Big|_{\pi/4}^{\pi/2} + \frac{\pi}{4}$$

$$= \ln\left(\sin\left(\frac{\pi}{2}\right)\right) - \ln\left(\sin\left(\frac{\pi}{4}\right)\right) + \frac{\pi}{4}$$

$$= 0 - \ln\left(\frac{\sqrt{2}}{2}\right) + \frac{\pi}{4}$$

$$= \frac{\pi}{4} - \ln\left(\frac{\sqrt{2}}{2}\right)$$

LESSON 26

Area Between Two Curves

We learned in earlier chapters that we can find the area under a curve and above the x-axis by using an integral. When the graph of the curve was above the x-axis, we had a positive result and when the graph of the curve was below the x-axis we had a negative result. In example 1 we have two curves (actually lines). If we desire to find the area bounded by the two curves and the x-axis, we will need to use two integrals.

In our first example we have the curves $y = x$ and $y = -3x + 12$. First of all, we need to find out where these graphs intersect. Using substitution we get:

$$x = -3x + 12$$
$$4x = 12$$
$$x = 3 \qquad (3, 3) \text{ is the intersection.}$$

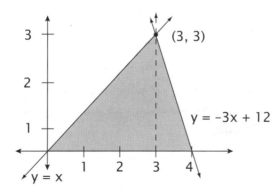

Example 1

Find the area bounded by y = x, y = -3x + 12 and y = 0.

We will divide this into two regions.

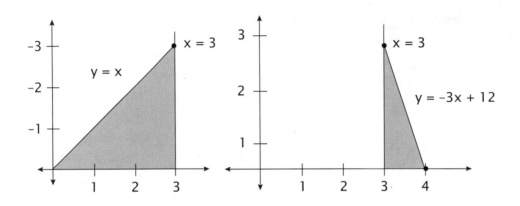

$$\int_0^3 x\,dx + \int_3^4 (-3x+12)\,dx = \frac{x^2}{2}\Big|_0^3 + \left(-3\frac{x^2}{2}+12x\right)\Big|_3^4$$

$$= \frac{9}{2} + (-24+48) - \left(-\frac{27}{2}+36\right)$$

$$= \frac{9}{2} + \frac{27}{2} + 24 - 36$$

$$= 18 - 12 = 6$$

Let's check our answer using simple geometry.

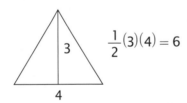

$\frac{1}{2}(3)(4) = 6$

You may wonder why we would employ calculus for such a simple task. Well, usually the curves we have are not simply lines, and simple geometry will not help us. Take a look at another example.

Example 2

Find the area bounded by $y = \sqrt{x}$, $3y = -x + 4$ and $y = 0$.

Where does $y = \sqrt{x}$ intersect with $3y = -x + 4$?

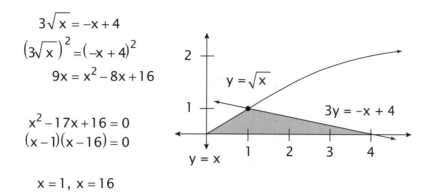

$$3\sqrt{x} = -x + 4$$
$$(3\sqrt{x})^2 = (-x+4)^2$$
$$9x = x^2 - 8x + 16$$

$$x^2 - 17x + 16 = 0$$
$$(x-1)(x-16) = 0$$

$$x = 1, \ x = 16$$

The intersection point we are seeking is $x = 1$. Now we will divide this region into two areas.

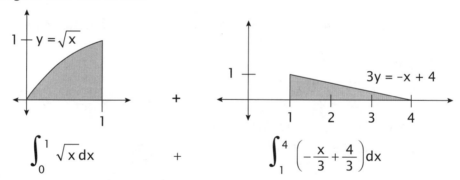

$$\int_0^1 \sqrt{x}\,dx \quad + \quad \int_1^4 \left(-\frac{x}{3} + \frac{4}{3}\right) dx$$

Note: You must solve for y before integrating.

$$= \frac{x^{\frac{3}{2}}}{\frac{3}{2}}\bigg|_0^1 + \left(-\frac{1}{3}\frac{x^2}{2} + \frac{4}{3}x\right)\bigg|_1^4$$

$$= \frac{2}{3}x^{\frac{3}{2}}\bigg|_0^1 + \left(-\frac{1}{6}x^2 + \frac{4}{3}x\right)\bigg|_1^4$$

$$= \frac{2}{3} + \left[\left(-\frac{16}{6} + \frac{16}{3}\right) - \left(-\frac{1}{6} + \frac{4}{3}\right)\right]$$

$$= \frac{14}{3} - \frac{15}{6} = \frac{28 - 15}{6}$$

$$= \frac{13}{6} = 2\frac{1}{6} \approx 2.17$$

Let's look at this problem using geometry.

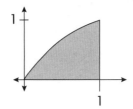

The area is approximately 1/4 of a circle with radius = 1.
The area of a circle of radius one is π, so this area would be $\frac{\pi}{4} \approx .785$.

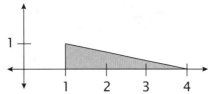

The area of this triangle is $\frac{1}{2}(3)(1) = \frac{3}{2}$

$1.5 + .785 \approx 2.29$,
which is not far from 2.17 and so our answer is reasonable.

Example 3

Find the area between $y = x^2 - 4$ and $y = 4 - x^2$

Now we are to find the area between two curves (neither are lines).

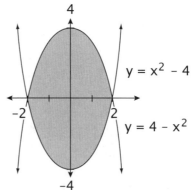

Where do $y = x^2 - 4$ and $y = 4 - x^2$ intersect?

$$x^2 - 4 = 4 - x^2$$
$$2x^2 = 8$$
$$x = \pm 2$$

The way to find the area between two curves is to subtract the bottom curve from the top curve.

$$\int_{-2}^{2} \left[(4-x^2)-(x^2-4)\right]dx$$

This should make sense.

$\int_{-2}^{2}(4-x^2)dx$ is the area above the x-axis and under $y = 4 - x^2$.

The area between the x-axis and $y = x^2 - 4$ would be negative, so subtracting it would yield a positive result.

$$\int_{-2}^{2}(8-2x^2)dx = \left(8x - 2\frac{x^3}{3}\right)\Big|_{-2}^{2}$$
$$= \left(16 - \frac{16}{3}\right) - \left(-16 + \frac{16}{3}\right)$$
$$= 32 - \frac{32}{3}$$
$$= \frac{96-32}{3} = \frac{64}{3} = 21\frac{1}{3}$$

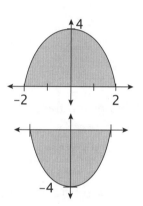

> If two functions f and g are continuous on [a, b] and $f(x) \geq g(x)$ on all points on [a, b] then the area between curves f and g on [a, b] is:
>
> $$\text{Area} = \int_{a}^{b}\left[f(x) - g(x)\right]dx$$

Example 4

Find the area between $y = x$, $y = 1$, $y = 0$, and $y = \frac{x^2}{4}$ as shown below.

An accurate drawing is necessary. To make it, we need to determine intersection points.

$y = x$ and $y = 1$ intersect at (1, 1)

$y = x$ and $y = \frac{x^2}{4}$ intersect at (0, 0) and (4, 4)

$y = 1$ and $y = \frac{x^2}{4}$ intersect at (2, 1)

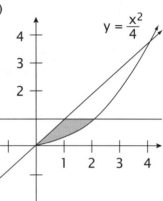

This graph has two "top" curves.

From $x = 0$ to $x = 1$, the top curve is $y = x$.

The bottom curve is $y = \frac{x^2}{4}$.

From $x = 1$ to $x = 2$ the top curve is $y = 1$. The bottom curve is $y = \frac{x^2}{4}$.

$$\int_0^1 \left(x - \frac{x^2}{4}\right)dx + \int_1^2 \left(1 - \frac{x^2}{4}\right)dx = \left(\frac{x^2}{2} - \frac{1}{4}\frac{x^3}{3}\right)\Big|_0^1 + \left(x - \frac{1}{4}\frac{x^3}{3}\right)\Big|_1^2$$

$$= \frac{1}{2} - \frac{1}{12} + \left[\left(2 - \frac{2}{3}\right) - \left(1 - \frac{1}{12}\right)\right]$$

$$= \frac{6 - 1 - 8 + 1}{12} + 1$$

$$= -\frac{1}{6} + 1 = \frac{5}{6}$$

Example 5

Find the area bounded by y = |x|, x = –2 and x = 1.

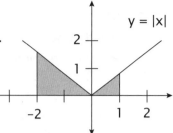

The absolute value curve has two definitions.
    y = x when x ≥ 0,
and y = –x when x < 0.
We will need to use two integrals for this problem.

$$\int_{-2}^{0} (-x)dx + \int_{0}^{1} x\,dx = -\frac{x^2}{2}\Big|_{-2}^{0} + \frac{x^2}{2}\Big|_{0}^{1}$$
$$= (0 - (-2)) + \frac{1}{2} = 2\frac{1}{2}$$

Checking our work we get:

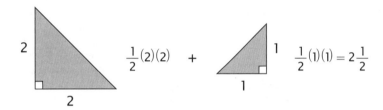

Example 6

Find the area bounded by the region $x = y^2$ and $y = x - 2$.

A graph must be drawn.

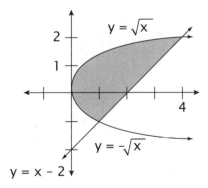

One equation is solved for y and the other is solved for x.

We change them both so that they are solved for y. Now let's find the intersection points.

$$(\sqrt{x})^2 = (x-2)^2$$
$$x = x^2 - 4x + 4$$
$$0 = x^2 - 5x + 4$$
$$0 = (x-4)(x-1)$$
$$x = 1, 4$$

$(1,-1)$ and $(4, 2)$ are the intersection points.

We must break this down into two graphs because the bottom curve changes at x = 1.

$$\int_0^1 \left[\sqrt{x} - (-\sqrt{x})\right]dx + \int_1^4 \left[\sqrt{x} - (x-2)\right]dx$$

$$= \int_0^1 2\sqrt{x}\,dx + \left(\frac{x^{\frac{3}{2}}}{\frac{3}{2}} - \frac{x^2}{2} + 2x\right)\Bigg|_1^4$$

$$= 2\frac{x^{\frac{3}{2}}}{\frac{3}{2}}\Bigg|_0^1 + \left(\frac{2x^{\frac{3}{2}}}{3} - \frac{1}{2}x^2 + 2x\right)\Bigg|_1^4$$

$$= \frac{4}{3} + \left[\left(\frac{16}{3} - 8 + 8\right) - \left(\frac{2}{3} - \frac{1}{2} + 2\right)\right]$$

$$= \frac{18}{3} - 2 + \frac{1}{2} = 4\frac{1}{2}$$

Example 7

We could also solve problem #6 by looking at the y boundaries instead of the x boundaries.

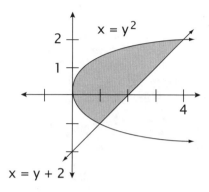

We can look at the function from right to left instead of from top to bottom.

 y = 2 forms the highest y value of the region desired.
 y = -1 forms the lowest y value.

We can integrate $\int_{-1}^{2}$ (right function) − (left function) dy. In our case we have to first solve for x in both equations. They become x = y + 2 and x = y^2. Our new integral is:

$$\int_{-1}^{2} (y+2) - (y^2) \, dy$$

This is simpler than the first method we used, and we should get the same result.

$$\int_{-1}^{2} (y + 2 - y^2) \, dy = \left(\frac{y^2}{2} + 2y - \frac{y^3}{3} \right) \Big|_{-1}^{2}$$

$$= \left(2 + 4 - \frac{8}{3}\right) - \left(\frac{1}{2} - 2 + \frac{1}{3}\right)$$

$$= 8 - \frac{8}{3} - \frac{1}{3} - \frac{1}{2} = 4\frac{1}{2}$$

In summary, here are two ways to evaluate the difference between two curves.

Case 1

We have two curves: $y = f(x)$ and $y = g(x)$ on $[a, b]$.

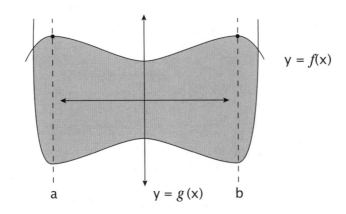

$$\int_a^b [\text{top function} - \text{bottom function}] \, dx$$

$$\int_a^b [f(x) - g(x)] \, dx$$

Case 2

We have two curves: $x = f(y)$ and $x = g(y)$ on $[c, d]$.

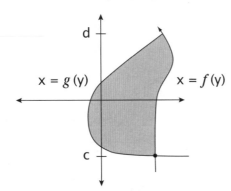

$$A = \int_c^d [\text{right function} - \text{left function}] \, dy$$

$$A = \int_c^d [f(y) - g(y)] \, dy$$

LESSON 27

Inverse Trigonometric Functions

Inverse trigonometric functions are used to find angle measurements when side measurements are known. The function y = sin(x) assumes a range of values from −1 to 1 which is repeated on every interval of length 2π. We can restrict our domain in order for the inverse function to exist. Now the function is called $\sin^{-1}(x)$. It is also referred to as arcsin(x) and invsin(x).

Example 1
a. Given y = sin(x) for all x in $\left[-\frac{\pi}{2}, \frac{\pi}{2}\right]$.

Find $\sin^{-1}(1)$, $\sin^{-1}(0)$, $\sin^{-1}(-1)$, $\sin^{-1}\left(\frac{\sqrt{2}}{2}\right)$, $\sin^{-1}\left(-\frac{\sqrt{2}}{2}\right)$.

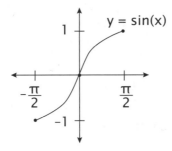

When y = sin(x) is restricted, we have an inverse function, namely y = $\sin^{-1}(x)$ with values on $\left[-\frac{\pi}{2}, \frac{\pi}{2}\right]$.

$\sin^{-1}(1) = \frac{\pi}{2}$ because $\sin\left(\frac{\pi}{2}\right) = 1$

$\sin^{-1}(0) = 0$ because $\sin(0) = 0$

$$\sin^{-1}(-1) = -\frac{\pi}{2} \qquad \text{because } \sin\left(-\frac{\pi}{2}\right) = -1$$

$$\sin^{-1}\left(\frac{\sqrt{2}}{2}\right) = \frac{\pi}{4} \qquad \text{because } \sin\left(\frac{\pi}{4}\right) = \frac{\sqrt{2}}{2}$$

$$\sin^{-1}\left(-\frac{\sqrt{2}}{2}\right) = -\frac{\pi}{4} \qquad \text{because } \sin\left(-\frac{\pi}{4}\right) = -\frac{\sqrt{2}}{2}$$

b. Include a sketch of $y = \sin^{-1}(x)$ for the given region.

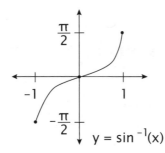

Arcsin(x) can also be written as $\sin^{-1}(x)$. Be careful not to confuse $\sin^{-1}(x)$ with csc(x). $\text{Sin}(x) = \frac{1}{\csc(x)}$. $\text{Sin}^{-1}(x)$ is the inverse, not the reciprocal, of sin(x).

Example 2

Given $y = \cos(x)$ for all x in $[0, \pi]$.

Find $\cos^{-1}(0)$, $\cos^{-1}(1)$, $\cos^{-1}(-1)$, $\cos^{-1}\left(\frac{\sqrt{3}}{2}\right)$, $\cos^{-1}\left(-\frac{\sqrt{2}}{2}\right)$.

$$\cos^{-1}(0) = \frac{\pi}{2} \qquad \text{because } \cos\left(\frac{\pi}{2}\right) = 0$$

$$\cos^{-1}(1) = 0 \qquad \text{because } \cos(0) = 1$$

$$\cos^{-1}(-1) = \pi \qquad \text{because } \cos(\pi) = -1$$

$$\cos^{-1}\left(\frac{\sqrt{3}}{2}\right) = \frac{\pi}{6} \qquad \text{because } \cos\left(\frac{\pi}{6}\right) = \frac{\sqrt{3}}{2}$$

$$\cos^{-1}\left(-\frac{\sqrt{2}}{2}\right) = \frac{3\pi}{4} \qquad \text{because } \cos\left(\frac{\pi}{2}\right) = -\frac{\sqrt{2}}{2}$$

Inverse functions also exist for tangent, cotangent, secant, and cosecant when restricted properly.

Find the derivative of $y = \sin^{-1}(u)$; u is a function of x on $\left[-\frac{\pi}{2}, \frac{\pi}{2}\right]$.

$y = \sin^{-1}(u)$

$\sin(y) = \sin[\sin^{-1}(u)] = u$ Take sine of both sides.

$u = \sin(y)$ Switch sides.

$\frac{du}{dy} = \cos(y)$ Take the derivative with respect to y.

$\frac{dy}{du} = \frac{1}{\cos(y)}$ Invert the fraction.

$\frac{dy}{dx} = \frac{dy}{du} \cdot \frac{du}{dx}$ Apply the Chain Rule.

$\frac{dy}{dx} = \frac{1}{\cos(y)} \cdot \frac{du}{dx}$ Substitute from line 4.

$\frac{dy}{dx} = \frac{1}{\sqrt{1-\sin^2(y)}} \cdot \frac{du}{dx}$ $\sin^2(y) + \cos^2(y) = 1$; $\cos(y) = \sqrt{1-\sin^2(y)}$

$\frac{dy}{dx} = \frac{1}{\sqrt{1-u^2}} \cdot \frac{du}{dx}$ Replace $u^2 = \sin^2(y)$ from line 1.

Therefore $\frac{d}{dx} \sin^{-1}(u) = \frac{1}{\sqrt{1-u^2}} \frac{du}{dx}$ $-1 < u < 1$

Here are the derivatives of the inverse trigonometric functions:

1. $\dfrac{d}{dx}\left(\sin^{-1}(u)\right) = \dfrac{1}{\sqrt{1-u^2}} \dfrac{du}{dx}$ $-1 < u < 1$

2. $\dfrac{d}{dx}\left(\cos^{-1}(u)\right) = \dfrac{-1}{\sqrt{1-u^2}} \dfrac{du}{dx}$ $-1 < u < 1$

3. $\dfrac{d}{dx}\left(\tan^{-1}(u)\right) = \dfrac{1}{1+u^2} \dfrac{du}{dx}$

4. $\dfrac{d}{dx}\left(\cot^{-1}(u)\right) = \dfrac{-1}{1+u^2} \dfrac{du}{dx}$

5. $\dfrac{d}{dx}\left(\sec^{-1}(u)\right) = \dfrac{1}{|u|\sqrt{u^2-1}} \dfrac{du}{dx}$ $|u| > 1$

6. $\dfrac{d}{dx}\left(\csc^{-1}(u)\right) = \dfrac{-1}{|u|\sqrt{u^2-1}} \dfrac{du}{dx}$ $|u| > 1$

Example 3

Find $\dfrac{d}{dx}\left(\sin^{-1}(x)^3\right)$

Let $u = x^3$; $\dfrac{du}{dx} = 3x^2$

$$\dfrac{d}{dx}\left(\sin^{-1}(x)^3\right) = \dfrac{1}{\sqrt{1-(x^3)^2}} \cdot (3x^2) = \dfrac{3x^2}{\sqrt{1-x^6}}$$

Example 4

Find $\dfrac{d}{dx}\left(\cot^{-1}(\sqrt{x})\right)$.

Let $u = \sqrt{x}$ or $x^{\frac{1}{2}}$; $\dfrac{du}{dx} = \dfrac{1}{2}x^{-\frac{1}{2}}$

$$\dfrac{d}{dx}\left(\cot^{-1}(\sqrt{x})\right) = \dfrac{-1}{1+x} \cdot \dfrac{1}{2\sqrt{x}}$$

$$= \dfrac{-1}{2\sqrt{x}\,(1+x)}$$

Example 5

Find $\dfrac{d}{dx}\left(\sec^{-1}(-2x)\right)$.

Let $u = -2x$; $\dfrac{du}{dx} = -2$

$$\dfrac{d}{dx}\left(\sec^{-1}(-2x)\right) = \dfrac{1}{|-2x|\sqrt{(-2x)^2 - 1}} \cdot (-2)$$

$$= \dfrac{-2}{2|x|\sqrt{4x^2 - 1}}$$

$$= \dfrac{-1}{|x|\sqrt{4x^2 - 1}} \qquad |x| > \dfrac{1}{2}$$

There are three trigonometric integrals which are useful.

1. $\displaystyle\int \dfrac{du}{\sqrt{1-u^2}} = \sin^{-1}(u) + C \qquad u^2 < 1$

2. $\displaystyle\int \dfrac{du}{1+u^2} = \tan^{-1}(u) + C$

3. $\displaystyle\int \dfrac{du}{u\sqrt{u^2-1}} = \sec^{-1}|u| + C \qquad u^2 > 1$

The other three trigonometric integrals are usually ignored because of their redundancy.

4. $\displaystyle\int \dfrac{du}{\sqrt{1-u^2}} = -\cos^{-1}(u) + C \qquad u^2 < 1$

5. $\displaystyle\int \dfrac{du}{1+u^2} = -\cot^{-1}(u) + C$

6. $\displaystyle\int \dfrac{du}{u\sqrt{u^2-1}} = -\csc^{-1}(u) + C \qquad u^2 > 1$

Example 6

Integrate $\int \dfrac{x\,dx}{\sqrt{1-x^4}}$.

The integral resembles the formula $\int \dfrac{du}{\sqrt{1-u^2}} = \sin^{-1}(u) + C$.

Let's try a substitution. Let $u = x^2$

$$\dfrac{du}{dx} = 2x$$
$$du = 2x\,dx$$

We have everything except the 2. That can be added.

$$\dfrac{1}{2}\int \dfrac{2x\,dx}{\sqrt{1-(x^2)^2}} = \dfrac{1}{2}\int \dfrac{du}{\sqrt{1-u^2}}$$

$$= \dfrac{1}{2}\sin^{-1}(u) + C = \dfrac{1}{2}\sin^{-1}(x^2) + C$$

Example 7

Integrate $\int \dfrac{dx}{\sqrt{16-x^2}}$.

Once again this integral appears similar to:

$\int \dfrac{du}{\sqrt{1-u^2}} = \sin^{-1}(u) + C$ but the 1 is a 16.

Let's perform some algebra on that square root.

$$\sqrt{16-x^2} = \sqrt{16\left(1-\dfrac{x^2}{16}\right)} = 4\sqrt{1-\left(\dfrac{x}{4}\right)^2}$$

Now, that looks like a formula!

Let $u = \dfrac{x}{4}$; $\dfrac{du}{dx} = \dfrac{1}{4}$; $du = \dfrac{1}{4}dx$

$$\int \dfrac{dx}{4\sqrt{1-\left(\dfrac{x}{4}\right)^2}} = \int \dfrac{du}{\sqrt{1-u^2}} = \sin^{-1}(u) + C = \sin^{-1}\left(\dfrac{x}{4}\right) + C$$

Example 8

Evaluate: $\int_0^1 \dfrac{dx}{1+x^2}$

$$\int_0^1 \dfrac{dx}{1+x^2} = \tan^{-1}(x)\Big|_0^1 = \tan^{-1}(1) - \tan^{-1}(0) = \dfrac{\pi}{4} - 0 = \dfrac{\pi}{4}$$

LESSON 28

Integration Using an Integral Table

There are books with long lists of commonly used integrals and their solutions. The formulas given assume that the constants given will not produce division by zero or roots of negative numbers. Refer to figure 1 at the end this lesson for the integral tables needed.

Example 1

$$\int \frac{x}{3x+4}\,dx$$

We can not solve this integral with a simple substitution.

If $u = 3x + 4$, then $\frac{du}{dx} = 3$, and we still have an extra x term.

The appropriate integral from the figure 1 table is #25:

$$\int \frac{u}{(au+b)}\,du = \frac{u}{a} - \frac{b}{a^2}\ln(au+b) + C$$

If $a = 3$ and $b = 4$, we get: $\frac{x}{3} - \frac{4}{9}\ln(3x+4) + C.$

Example 2

$$\int \sqrt{9-x^2}\,dx$$

The appropriate integral from figure 1 is #26:

$$\int (a^2 - u^2)^{\frac{1}{2}}\,du = \frac{u}{2}\sqrt{a^2 - u^2} + \frac{a^2}{2}\sin^{-1}\left(\frac{u}{a}\right) + C$$

With a = 3, we get:

$$\frac{x}{2}\sqrt{9-x^2} + \frac{9}{2}\sin^{-1}\left(\frac{x}{3}\right) + C$$

We can evaluate definite integrals as well as indefinite integrals using integral tables.

Example 3

$$\int_{-1}^{0} \frac{x}{3x+4}\,dx$$

Using the same formula as example 1, we get:

$$= \frac{x}{3} - \frac{4}{9}\ln(3x+4)\Big|_{-1}^{0}$$
$$= -\frac{4}{9}\ln(4) - \left(-\frac{1}{3} - \frac{4}{9}\ln(1)\right)$$
$$= -\frac{4}{9}\ln(4) + \frac{1}{3}$$

Sometimes the integral we need can not be identified easily. In that case, search for a formula which is similar and use a substitution for the variable u.

Example 4

$$\int \frac{dx}{x\sqrt{4x^2+9}}$$

The appropriate integral from figure 1 is #27:

$$\int \frac{du}{u(u^2+a^2)^{\frac{1}{2}}} = -\frac{1}{a}\ln\left(\frac{a+\sqrt{u^2+a^2}}{u}\right) + C$$

If you let u = 2x and a = 3, you can modify the original formula to be:

$$2\int \frac{dx}{2x\sqrt{(2x)^2+(3)^2}} = 2\int \frac{dx}{u\sqrt{u^2+a^2}}$$

Notice a 2 outside the integral and a 1/2 inside the integral were needed.

We still need du. If u = 2x, then du = 2dx. We've already got it!

$$= \int \frac{2dx}{u\sqrt{u^2+a^2}}$$
$$= \int \frac{du}{u\sqrt{u^2+a^2}}$$

Now apply the formula.

$$= -\frac{1}{3}\ln\left(\frac{3+\sqrt{(2x)^2+9}}{2x}\right) + C$$
$$= -\frac{1}{3}\ln\left(\frac{3+\sqrt{4x^2+9}}{2x}\right) + C$$

Lastly, some of the integrals in the integral table have yet more integrals in their answer. These are called reduction formulas. They end up reducing the power of the binomial or the trig function so that the function can be integrated. These formulas may have to be applied more than once.

Example 5

$$\int \sin^3(x)\, dx$$

The appropriate integral from figure 1 is #28:

$$\int \sin^n(u)\, du = -\frac{\sin^{n-1}(u)\cos(u)}{n} + \frac{n-1}{n}\int \sin^{n-2}(u)\, du + C$$

If n = 3, we get:

$$-\frac{\sin^2(x)\cos(x)}{3} + \frac{2}{3}\int \sin(x)\, dx + C$$

Now we can integrate again and we get:

$$-\frac{\sin^2(x)\cos(x)}{3} - \frac{2}{3}\cos(x) + C$$

Figure 1

Table of Selected Integrals

1. $\displaystyle\int \frac{u\, du}{(a+bu)^2} = \frac{1}{b^2}\left[\frac{a}{a+bu} + \ln(a+bu)\right] + C$

2. $\displaystyle\int \frac{u\, du}{(a+bu)^3} = \frac{1}{b^2}\left[-\frac{1}{a+bu} + \frac{a}{2(a+bu)^2}\right] + C$

3. $\displaystyle\int \frac{du}{u(a+bu)} = -\frac{1}{a}\ln\left(\frac{a+bu}{u}\right) + C$

4. $\displaystyle\int \frac{du}{u^2(a+bu)} = -\frac{1}{au} + \frac{b}{a^2}\ln\left(\frac{a+bu}{u}\right) + C$

5. $\displaystyle\int \frac{du}{a^2+b^2u^2} = \frac{1}{ab}\tan^{-1}\left(\frac{bu}{a}\right) + C$

6. $\displaystyle\int \frac{du}{a^2-b^2u^2} = -\frac{1}{2ab}\ln\left(\frac{a+bu}{a-bu}\right) + C$

7. $\int u(a^2 \pm b^2 u^2)^n du = \dfrac{(a^2 \pm b^2 u^2)^{n+1}}{\pm 2b^2(n+1)} + C$

8. $\int u\sqrt{a+bu}\, du = \dfrac{-2(2a-3bu)(a+bu)^{\frac{3}{2}}}{15b^2} + C$

9. $\int (u^2 \pm a^2)^{\frac{1}{2}} du = \dfrac{u}{2}\sqrt{u^2 \pm a^2} \pm \dfrac{a^2}{2}\ln\left(u + \sqrt{u^2 \pm a^2}\right) + C$

10. $\int \dfrac{du}{(u^2 \pm a^2)^{\frac{1}{2}}} = \ln\left(u + \sqrt{u^2 \pm a^2}\right) + C$

11. $\int \dfrac{du}{u(u^2 - a^2)^{\frac{1}{2}}} = \dfrac{1}{a}\sec^{-1}\left(\dfrac{u}{a}\right) + C$

12. $\int \dfrac{du}{u^2(u^2 \pm a^2)^{\frac{1}{2}}} = \dfrac{-\sqrt{u^2 \pm a^2}}{\pm a^2 u} + C$

13. $\int \dfrac{u^2 du}{(a^2 - u^2)^{\frac{1}{2}}} = -\dfrac{u}{2}\sqrt{a^2 - u^2} + \dfrac{a^2}{2}\sin^{-1}\left(\dfrac{u}{a}\right) + C$

14. $\int \dfrac{du}{u^2(a^2 - u^2)^{\frac{1}{2}}} = -\dfrac{\sqrt{a^2 - u^2}}{a^2 u} + C$

15. $\int \sqrt{\dfrac{1+u}{1-u}}\, du = -\sqrt{1-u^2} + \sin^{-1}(u) + C$

16. $\int b^{au} du = \dfrac{b^{au}}{a \ln b} + C$

17. $\int u^n e^{au} du = \dfrac{u^n e^{au}}{a} - \dfrac{n}{a}\int u^{n-1} e^{au} du$

18. $\int \ln(u)\, du = u\ln(u) - u + C$

19. $\int u^n \ln(u)\, du = u^{n+1}\left[\dfrac{\ln(u)}{n+1} - \dfrac{1}{(n+1)^2}\right] + C$

20. $\int \dfrac{du}{u\ln(u)} = \ln[\ln(u)] + C$

21. $\int \sin^2(u)\, du = \dfrac{1}{2}u - \dfrac{1}{4}\sin(2u) + C$

22. $\int \sin(mu)\sin(nu)\, du = -\dfrac{\sin(m+n)u}{2(m+n)} + \dfrac{\sin(m-n)u}{2(m-n)} + C$

23. $\int u \sin(u)\, du = \sin(u) - u\cos(u) + C$

24. $\int \tan^n(u)\, du = \dfrac{\tan^{n-1}(u)}{n-1} - \int \tan^{n-2}(u)\, du$

25. $\int u(au+b)^{-1}\, du = \dfrac{u}{a} - \dfrac{b}{a^2}\ln(au+b) + C$

26. $\int \left(a^2 - u^2\right)^{\frac{1}{2}} du = \dfrac{u}{2}\sqrt{a^2 - u^2} + \dfrac{a^2}{2}\sin^{-1}\left(\dfrac{u}{a}\right) + C$

27. $\int \dfrac{du}{u\left(u^2 + a^2\right)^{\frac{1}{2}}} = -\dfrac{1}{a}\ln\left(\dfrac{a + \sqrt{u^2 + a^2}}{u}\right) + C$

28. $\int \sin^n(u)\, du = -\dfrac{\sin^{n-1}(u)\cos(u)}{n} + \dfrac{n-1}{n}\int \sin^{n-2}(u)\, du$

29. $\int \cos^n(u)\, du = \dfrac{\cos^{n-1}(u)\sin(u)}{n} + \dfrac{n-1}{n}\int \cos^{n-2}(u)\, du$

30. $\int e^{au}\ln(u)\, du = \dfrac{e^{au}\ln(u)}{a} - \dfrac{1}{a}\int \dfrac{e^{au}}{u}\, du$

LESSON 29

Differential Equations

We will be studying first order separable differential equations. A *first order separable differential equation* has a first derivative as the highest derivative. (For example: y y' = x) Second order differential equations have a second derivative as the highest derivative. (For example y''+ 2y' = 2x). Separable differential equations of that first order can be written in the form.

$$A(y)\frac{dy}{dx} = B(x)$$

To solve a first order separable differential equation, rewrite the equation:

$$A(y)\,dy = B(x)dx$$

Then integrate both sides:

$$\int A(y)\,dy = \int B(x)\,dx$$

You will gain a general solution. That is, it will end with "+ C". If you desire a specific solution, you will need an initial condition.

Example 1

Solve $\dfrac{dy}{dx} = 2y^2 x$.

Step 1: Separate the variables: $y^{-2}dy = 2x\,dx$.

Step 2: Integrate both sides.

$$\int y^{-2}dy = \int 2x\,dx$$
$$-\dfrac{1}{y} = x^2 + C$$

Step 3: Solve for y.

$$y = -\dfrac{1}{x^2 + C}$$

Example 2

Solve $\dfrac{dy}{dx} = 2y^2 x$ when $y(3) = -\dfrac{1}{4}$

We solved this in example 1: $y = -\dfrac{1}{x^2 + C}$

Now substitute the information given by the initial condition.

$$-\dfrac{1}{4} = -\dfrac{1}{3^2 + C}$$
$$-4 = -(3^2 + C)$$
$$-5 = C$$

Therefore our particular solution is: $y = -\dfrac{1}{x^2 - 5}$

(plugging our value for C into our general solution).

Example 3

Solve $\dfrac{2}{\cos(3x)}\left(\dfrac{dy}{dx}\right) = 6\sin(3x)$ when $y(0) = 3$.

$$dy = 3\sin(3x)\cos(3x)dx$$

$$\int dy = -\int 3\sin(3x)\cos(3x)dx \cdot (-1)$$

Let $u = \cos(3x)$; $du = -3\sin(3x)dx$.

$$\int dy = -\int u\,du$$

$$y = -\dfrac{u^2}{2} + C$$

$$y = -\dfrac{1}{2}(\cos(3x))^2 + C$$

$$3 = -\dfrac{1}{2}(\cos(0))^2 + C$$

$$3 = -\dfrac{1}{2} + C$$

$$C = 3\dfrac{1}{2}$$

$$y = -\dfrac{1}{2}(\cos(3x))^2 + 3\dfrac{1}{2}$$

Example 4
Solve $3x\left(e^{x^2-\ln(2)}\right) = \dfrac{dy}{dx}$.

$$dy = 3x\left(e^{x^2-\ln(2)}\right) \cdot dx$$

$$dy = \dfrac{(3x)\left(e^{x^2}\right)dx}{e^{\ln(2)}}$$

$$dy = \dfrac{(3x)\left(e^{x^2}\right)dx}{2} \qquad \int e^u du = e^u$$

Let $u = x^2$, $\dfrac{du}{dx} = 2x$; $du = 2xdx$.

$$dy = \left(\dfrac{3}{2}\right)\left(\dfrac{1}{2}\right)2xe^{x^2}dx$$

$$\int dy = \int \dfrac{3}{4}e^u du$$

$$y = \dfrac{3}{4}e^u du \qquad y(0) = -1$$

$$-2 = \dfrac{3}{4}e^{0^2} + C$$

$$-2 = \dfrac{3}{4} + C$$

$$-2\dfrac{3}{4} = C$$

LESSON 30

Integral Application: Differential Equations

Growth and Decay Problems
A differential equation is simply an equation with one or more derivatives. Differential equation problems occur in physics, engineering, biology, and business, to name a few. The most common form of the differential equation is where the function sought is directly proportional to its derivative.

The common differential equation is:

$$\frac{dy}{dt} = Ky \qquad \text{Where K is a constant.}$$

This is the equation used for growth or decay applications. If $y(t)$ represents the number of bacteria at time t, then the instantaneous rate of change is $y'(t)$. K is referred to as the growth or decay constant. If $K > 0$, the population is increasing. If $K < 0$, the population is decreasing.

Solving the differential equation we have:

$$\frac{dy}{dt} = Ky$$

$$\frac{dy}{y} = Kdt$$

$$\frac{1}{y}dy = Kdt \qquad \text{Separate the variables.}$$

$$\int \frac{1}{y}dy = \int Kdt \qquad \text{Integrate both sides.}$$

$$\ln(y) = Kt + C$$

$$\ln_e(y) = Kt + C$$

$$y = e^{Kt + C} \text{ or } e^{Kt} \cdot e^C \qquad \text{Solve for y.}$$

when $t = 0$, $y = e^{0 + C} = e^C$

at $t = 0$ we rename e^C as C_0

Since this is the intial or original number of bacteria in the population. This is the standard growth and decay model. This letter will change names to aid our understanding. It would likely be called B_0 to denote the initial bacteria count.

> Growth and Decay Model: $y = C_0 e^{Kt}$

Example 1

Let's apply this idea to the population of a city. We know the population of a city was 30,000 in 1960 and 32,000 in 1990. Assuming the growth rate of the population is directly proportional to the size of the population, what was the projected population in 2000?

Step 1: Set up the differential equation.

$P' = KP$ or $\dfrac{dp}{dt} = KP$ where P is a function of time, t.

Step 2: Integrate to find P.

$P = P_0 e^{Kt}$ where P_0 is the initial population

This is derived from the model for growth and decay problems above.

Step 3: Find P_0.

We know $P(0) = 30,000$ and $P(30) = 32,000$.

Substituting into the model $P = P_0 e^{Kt}$:
$30,000 = P_0 e^0$ so $P_0 = 30,000$

Step 4: Find K.

$$32,000 = 30,000 e^{30K}$$
$$\frac{32,000}{30,000} = e^{30K}$$
$$1.07 = e^{30K}$$
$$\ln(1.07) = 30K$$
$$K = \frac{\ln(1.07)}{30} \approx .002255$$

Step 5: Revise the model substituting P_0 and K.

Now the model becomes: $P = 30,000\, e^{.002255t}$

Use four significant digits in answers for P.

$P(t) = 30,000\, e^{.002255t}$

Step 6: Answer the question.

From 1960 to 2000 is 40 years.

$P(40) = 30,000 e^{.002255(40)}$
$ = 32,831$

The projected population in 2000 was 32,831 people.

Example 2

An antibiotic is administered to a colony of 80 bacteria in a petri dish. The density of bacteria as a function of time is given by $B(t) = 80e^{-.5t}$ where t is time in hours.

a. How long will it take for half of the bacteria to die?
   (This is called "half life.")

$$B(t) = 80e^{-.5t}$$ When t = 0, 80 were present.
$$40 = 80e^{-.5t}$$ Number of bacteria present
$$\frac{40}{80} = \frac{1}{2} = e^{-.5t}$$ equals one-half of 80.
$$\ln \frac{1}{2} = -.5t$$
$$-.6931 = -.5t$$
$$\frac{-.6931}{-.5} = t$$
$$t \approx 1.39 \text{ hours}$$

It will take approximately 1.39 hours for half of the bacteria to die.

b. How long will it take for 90% of the bacteria to die?

$$8 = 80e^{-.5t}$$ When 90% have died, 10% remain.
$$\frac{1}{10} = e^{-.5t}$$ 10% of 80 is 8.

$$t = 4.6 \text{ hours}$$

Therefore one-half of the bacteria will be destroyed in 1.39 hours and 90% will be destroyed in 4.6 hours.

Newton's Law of Cooling:

When a hot object is placed in cool surroundings, the rate at which the object cools is proportional to the difference between the object and its surroundings. If T is the temperature at time t and T_0 is the initial temperature of the object and T_s is the temperature of the surroundings, then $T' = K(T - T_S)$; $T(0) = T_0$.

Solving for T we get:

$$\frac{dT}{dt} = K(T - T_S)$$

$$\int \frac{dT}{T - T_S} = \int K\, dt$$

$$\ln|T - T_S| = Kt + C$$

$$T - T_S = e^{Kt+C} = Ce^{Kt}$$

$$T = Ce^{Kt} + T_S$$

$$T_0 = T_S + C \quad \text{so } C = T_0 - T_S$$

$$T = (T_0 - T_S)e^{Kt} + T_S$$

> Newton's Law of Cooling Model: $T = (T_0 - T_S)e^{Kt} + T_S$

Example 3

A hot entree is served at 200°F. After one minute in a 70°F room the entree's temperature has dropped to 160°F. How much time will it take for the entree to drop to 120°F? (round to the nearest tenth)

We know: $T_s = 70°F$; $T_0 = 200°F$; $T(1) = 160°$

$$T = (T_0 - T_s)e^{Kt} + T_s \quad \text{Newton's Law of Cooling}$$

$$T = (200°F - 70°F)e^{Kt} + 70$$

$$T = 130e^{Kt} + 70$$

$$160 = 130e^K + 70 \quad \text{Use initial condition: when } t = 1, T = 160.$$

$$90 = 130e^K$$

$$\frac{90}{130} = e^K$$

$$.6923 = e^K$$

$$K = \ln(.6923)$$

$$K = -.3677$$

$$120 = 130e^{-.3677t} + 70 \quad \text{Answer the question.}$$

$$50 = 130e^{-.3677t}$$

$$\frac{50}{130} = e^{-.3677t}$$

$$\ln(.3846) = -.3667t$$

$$\frac{-.9555}{-.3667} = t$$

$$t \approx 2.6 \text{ min}$$

It will take approximately 2.6 minutes for the entree to drop in temperature to 120°F.

Example 4

A bank account earns interest continuously at a rate of 4% of the current balance per year. A $1,000 initial deposit is made, and there is no other activity (deposits or withdrawals).

a. Write the differential equation for the balance.

$$\frac{dB}{dt} = .04B \quad \text{where "B" is the balance at time "t" in years}$$

b. Solve the differential equation.

$$B_0 = 1,000$$

$$\frac{dB}{B} = .04dt$$
$$\int \frac{dB}{B} = \int .04 \, dt$$
$$\ln(B) = .04t + C$$
$$B = e^{.04t+C}$$
$$B = Ce^{.04t}$$
$$B = 1,000e^{.04t} \quad\quad C = B_0 = 1,000$$

c. What would be the balance in 10 years?

$$B = 1,000e^{.04(10)}$$
$$\approx \$1,491.82$$

d. How many years would it take for the balance to reach $2,500?

$$2,500 = 1,000e^{.04t}$$
$$2.5 = e^{.04t}$$
$$\ln(2.5) = .04t$$
$$\frac{\ln(2.5)}{.04} = t$$
$$22.9 \text{ yr} \approx t$$

Example 5

A rock is dropped from rest from a window 20 meters (m) above the ground. The acceleration of the rock is -16 m/sec^2.

a. Find the velocity of the rock as a function of time t.

$$\frac{dv}{dt} = -16 \frac{m}{sec^2}$$

The derivative of velocity is acceleration.
Integrate to find velocity.

$$dv = -16 dt$$
$$\int dv = \int -16 \, dt$$
$$v = -16t + C$$

Initial velocity $v_0 = C = 0$.

$$v = -16t.$$

b. Find the distance function x(t).

$$\frac{dx}{dt} = -16t$$
$$dx = -16t \, dt$$

The derivative of distance is velocity.
Integrate velocity to find distance.

$$\int dx = \int -16t \, dt$$
$$x = -16 \frac{t^2}{2} + C$$
$$x = -8t^2 + C$$

x(0) is the initial distance from the ground, which was 20 m.

$$20 = C$$
$$x = -8t^2 + 20$$

c. The time when the rock hits the ground and the velocity of the rock upon impact.

The rock will hit the ground when x = 0.

$$-8t^2 + 20 = 0$$
$$t^2 = 2.5$$
$$t \approx 1.6 \text{ seconds}$$

$$v(1.6) = -16(1.6) \approx -25.6 \frac{m}{\text{sec}}$$

The negative sign indicates direction.

The rock will hit the ground at an approximate velocity of 25.6 m/sec at approximately 1.6 seconds.

Calculus Placement Test Solutions

I. **1.**

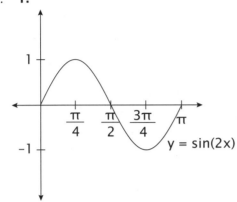

2.

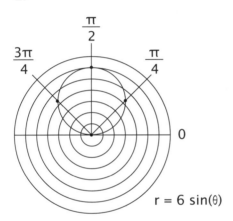

3.

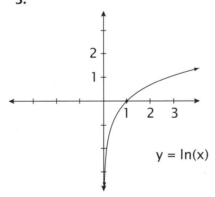

$f(x) = \sec(x) + 1$

II. **1.** $\sum_{i=-1}^{2} 2i^3 = 2(-1)^3 + 2(0)^3 + 2(1)^3 + 2(2)^3 = 16$

2. $\tan(45°) + \sin(225°) + \cos\left(\dfrac{\pi}{3}\right)$

$= 1 - \dfrac{\sqrt{2}}{2} + \dfrac{1}{2} = \dfrac{3 - \sqrt{2}}{2}$

3. $\lim\limits_{x \to \infty} \dfrac{2}{x+1} = 0$

4. $\lim\limits_{x \to 1} \dfrac{x^2 + 3x - 4}{x - 1} = \lim\limits_{x \to 1} \dfrac{(x-1)(x+4)}{(x-1)} = 5$

5. $\log_2 8 = 3$ because $2^3 = 8$

III. $3, -1, \dfrac{1}{3}, -\dfrac{1}{9}$

IV. **1.** $\sqrt{x+1} < 2$
$x + 1 < 4$
$x < 3$

2. $\ln(x - 1) = 3$
$x - 1 = e^3$
$x = e^3 + 1$

CALCULUS PLACEMENT TEST

3. $e^{2x} - 4e^x = -3$
$e^{2x} - 4e^x + 3 = 0$
$y^2 - 4y + 3 = 0$ Let $y = e^x$
$(y-3)(y-1) = 0$
$y = 3, \ y = 1$
$e^x = 3, \ e^x = 1$

so $x = \ln(3)$, and $x = \ln(1)$
$x = 0, \ln(3)$

V. 1. $\dfrac{\cos^2(\theta)}{\sin(\theta)} = \sin(\theta)\ \cos^2(\theta) + \dfrac{\cos^4(\theta)}{\sin(\theta)}$

$\dfrac{\cos^2(\theta)}{\sin(\theta)} = \dfrac{\sin^2(\theta)\ \cos^2(\theta) + \cos^4(\theta)}{\sin(\theta)}$

$\dfrac{\cos^2(\theta)}{\sin(\theta)} = \dfrac{\cos^2\theta(\sin^2(\theta) + \cos^2(\theta))}{\sin(\theta)}$

$\dfrac{\cos^2(\theta)}{\sin(\theta)} = \dfrac{\cos^2(\theta)}{\sin(\theta)}$

VI. Solve for the unknown sides and angles in the triangle below.

$c^2 = a^2 + b^2 - 2ab\ \cos C$
$c^2 = 15^2 + 10^2 - 2(15)(10)\cos 50°$
$c^2 = 225 + 100 - 300(.6428)$
$c^2 = 132.2$
$c = 11.5'$

$\dfrac{a}{\sin A} = \dfrac{c}{\sin C}$

$\dfrac{15}{\sin A} = \dfrac{11.5}{\sin 50°}$

$\sin A = \dfrac{15 \sin 50°}{11.5}$

$\sin A = \dfrac{15(.7660)}{11.5} = .9992$

$A = 87.7°$
$B = 180° - (50° + 87.7°) = 42.3°$

VII. $Q(t) = 5e^{-.005(200)}$
$= 5e^{-1}$
$= 1.8$ grams

VIII. $f(x) = e^x$
$g(x) = \ln(x) + 2$

$f(g(x)) = f(\ln(x) + 2)$
$= e^{\ln(x) + 2}$
$= e^{\ln(x)} \cdot e^2$
$= xe^2$

Student Solutions

Lesson Practice 1A

1. $2 < x \le 4$

   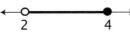

2. $|x| < 3$

   $x < 3 \quad -x < 3$
   $\quad\quad\quad\quad x > -3$

   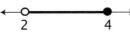

3. $|x - 1| \ge 3$

   $x - 1 \ge 3 \quad -(x-1) \ge 3$
   $x \ge 4 \quad\quad -x + 1 \ge 3$
   $\quad\quad\quad\quad\quad -x \ge 2$
   $\quad\quad\quad\quad\quad x \le -2$

4. $|x^3| < 8$

   $x^3 < 8 \quad -x^3 < 8$
   $x < 2 \quad\quad x^3 > -8$
   $\quad\quad\quad\quad x > -2$

5. $y = |x| + 2$

6. $y = -3x^3$

   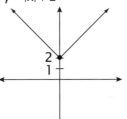
   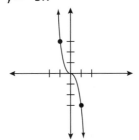

7. $y = -\frac{1}{4}x^2 + 3$

   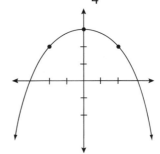

8. $y = 2\sqrt{x} - 2$

   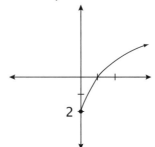

9. $y = \begin{cases} \dfrac{1}{x} & \text{for } x > -1 \\ x^2 & \text{for } x \le -1 \end{cases}$

   discontinuity at $(-1, 1)$
   vertical asymptote at $x = 0$
   It is not a removable discontinuity.

   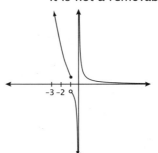

10. $y = \begin{cases} |x| \text{ for } x > -2 \\ \frac{1}{2}x^2 \text{ for } x \leq -2 \end{cases}$

no discontinuities or vertical asymptotes

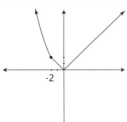

11. $y = \dfrac{x^2 - 1}{x - 1}$

$y = \dfrac{(x-1)(x+1)}{x-1}$

$y = x + 1 \quad x \neq 1$

removable discontinuity at $(1, 2)$
no vertical asymptotes

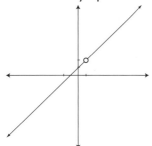

12. $y = \dfrac{2}{x}$

$xy = 2 \quad x \neq 1$

vertical asymptote at $x = 0$
non-removable discontinuity at $x = 0$

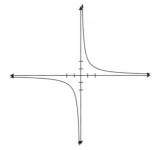

Lesson Practice 1B

1. $-3 \leq x < 2$

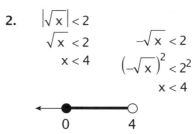

2. $|\sqrt{x}| < 2$

$\sqrt{x} < 2 \qquad -\sqrt{x} < 2$

$x < 4 \qquad (-\sqrt{x})^2 < 2^2$

$\qquad\qquad\qquad x < 4$

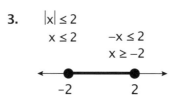

$\sqrt{x}$ must be positive so x is greater than 0.

$0 \leq x < 4$

3. $|x| \leq 2$

$x \leq 2 \qquad -x \leq 2$

$\qquad\qquad x \geq -2$

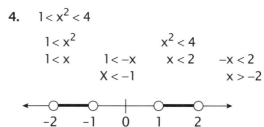

4. $1 < x^2 < 4$

$1 < x^2 \qquad\qquad x^2 < 4$

$1 < x \quad 1 < -x \quad x < 2 \quad -x < 2$

$\qquad\quad X < -1 \qquad\qquad\quad x > -2$

5. $y = \dfrac{2}{x}$

$xy = 2 \quad x \neq 0$

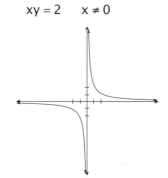

6. $y = 1 - 2x^2$

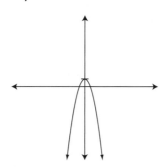

7. $y = |2x| + 1$

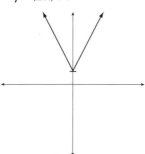

8. $y = \sqrt{x-1} - 2$

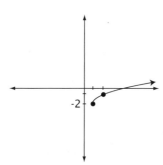

9. $xy = 1$

non-removable discontinuity at $x = 0$
vertical asymptote at $x = 0$

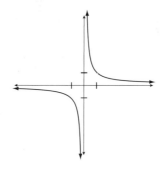

10. $y = \begin{cases} \dfrac{1}{x^3} & \text{for } x > 0 \\ |x| & \text{for } x \leq 0 \end{cases}$

non-removable discontinuity at $x = 0$
vertical asymptote at $x = 0$

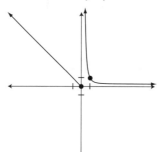

11. $y = \dfrac{x^2 - 4}{x + 2}$

$y = \dfrac{(x-2)(x+2)}{x+2}$

$y = x - 2$

removeable discontinuity at $(-2, -4)$
no vertical asymptotes

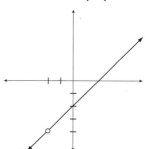

12. $y = \begin{cases} \sqrt{x-2} & \text{for } x > 2 \\ |x-2| & \text{for } x \leq 2 \end{cases}$

no discontinuities or vertical asymptotes

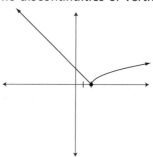

Lesson Practice 1C

1. $|x-1|=5$

   $x-1=5 \qquad -(x-1)=5$
   $x=6 \qquad\quad -x+1=5$
   $\qquad\qquad\qquad -x=4$
   $\qquad\qquad\qquad x=-4$

   $x=-4, 6$

2. $|3-2x|=3$

   $3-2x=3 \qquad -(3-2x)=3$
   $-2x=0 \qquad\quad -3+2x=3$
   $x=0 \qquad\qquad 2x=6$
   $\qquad\qquad\qquad x=3$

   $x=0, 3$

3. $|x-1|\le 3$

   $x-1\le 3 \qquad -(x-1)\le 3$
   $x\le 4 \qquad\quad -x+1\le 3$
   $\qquad\qquad\qquad -x\le 2$
   $\qquad\qquad\qquad x\ge -2$

   $-2\le x\le 4$

4. $|2x-5|>0$

   The absolute value of any non-0 number is greater than 0, so the solution is all x except those values of x that make $|2x-5|$ equal to 0:

   $|2x-5|\ne 0$
   $2x-5\ne 0$
   $2x\ne 5$
   $x\ne 2.5$

5. $(-2,-1)$
   See graph

6. $(2, 1)$

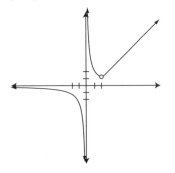

7. $(-2, 0)$

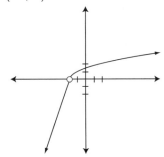

8. y has a discontinuity which is not removable because it is not just a hole.

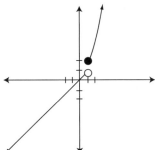

9. A secant line intersects in exactly two points while a tangent line intersects in exactly one point.

10. $y = x^2 + 2$
    domain: all real numbers
    range: all real numbers ≥ 2

11. Graph 1 has a vertical asymptote at $x = 0$.
    Graphs 2 and 4 have a range of all real numbers.
    Graph 5 has a domain of $x \ge 0$.

Lesson Practice 1D

1. $|2-x|=3$

   $2-x=3 \qquad -(2-x)=3$
   $-x=1 \qquad\quad -2+x=3$
   $x=-1 \qquad\quad x=5$

   $x=-1, 5$

2. $|3x+1|=7$

$3x+1=7 \quad\quad -(3x+1)=7$
$3x=6 \quad\quad -3x-1=7$
$x=2 \quad\quad -3x=8$
$ x=-\dfrac{8}{3}$

$x=2,-\dfrac{8}{3}$

3. $|x-2|>3$

$x-2>3 \quad\quad -(x-2)>3$
$x>5 \quad\quad -x+2>3$
$-x>1$
$x<-1$

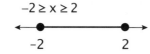

4. $|x^2|\le 4$

$x^2 \le 4$

If x is positive, $x \le 2$

If x is negative, $-x \le 2$
$ x \ge -2$

$-2 \ge x \ge 2$

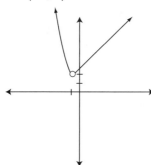

5. $(-3, 4)$ see graph
6. $(-1, 2)$

7. $(1, 1)$

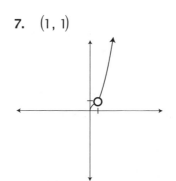

8. y has a discontinuity when x is equal to 2. It is not removable, because an infinite number of points would be needed to fill the gap.

9. An asymptote is a straight line that is closely approached, but never met, by the curve.

10. domain: all real numbers
range: all real numbers ≤ 3

11. domain: all real numbers
range: all real numbers ≤ 2

12. domain: all real numbers ≥ 1
range: all real numbers

Lesson Practice 2A

1. $3x+y=6$
$y=-3x+6$
Neither x nor y are squared, so this is a line. Slope is -3 and y intercept is 6.

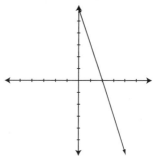

2. $y = 2x^2 + 4x - 3$
$y = 2(x^2 + 2x + 1) - 3 - 2$
$y = 2(x+1)^2 - 5$

One squared term in x plus a 4x term means we have a parabola in the "smile" shape translated away from the origin.
vertex: $(-1, -5)$
axis of symmetry: $x = -1$

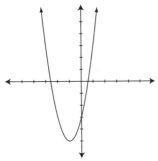

3. $4x^2 + 9y^2 - 8x + 36y + 4 = 0$
$4x^2 - 8x + 9y^2 + 36y = -4$
$4(x^2 - 2x + 1) + 9(y^2 + 4y + 4) = -4 + 4 + 36$
$4(x-1)^2 + 9(y+2)^2 = 36$
$\dfrac{(x-1)^2}{9} + \dfrac{(y+2)^2}{4} = 1$

Two squared terms with different coefficients of the same sign means that we have an ellipse.
center: $(1, -2)$
extremities: $(4, -2); (-2, -2); (1, 0); (1, -4)$

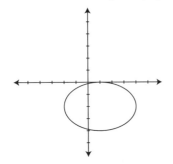

4. $x^2 + (y-1)^2 = 9$

Two squared terms with equal coefficients means we have a circle. The center is at $(0, 1)$ and the radius is $\sqrt{9}$, or 3.

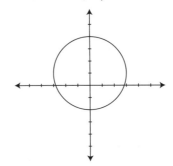

5. $x = -2y$

Neither x nor y is squared, so this is a line. Changing to slope-intercept form we get:
$y = -\dfrac{1}{2}x$

Slope is $-\dfrac{1}{2}$ and y intercept is 0.

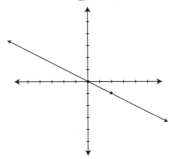

LESSON PRACTICE 2A - LESSON PRACTICE 2B

6. $x = -(y-2)^2$

   Parabola.

   The standard form is $x = a(y-k)^2 + h$. This parabola has a backwards "c" shape because a is negative. The axis of symmetry is $y = 2$.

   vertex: $(0, 2)$

   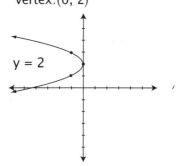

7. $x^2 - 4x + y^2 - 6y + 12 = 0$

   Circle. Gathering terms and completing the square, we get:

   $x^2 - 4x + 4 + y^2 - 6y + 9 = -12 + 4 + 9$

   $(x-2)^2 + (y-3)^2 = 1$

   center: $(2, 3)$

   radius: 1

   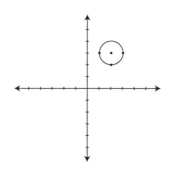

8. $y = 2x^2 + 8x + 7$

   Parabola. Standard form is $y = a(x-h)^2 + k$. Completing the square we get

   $y = 2(x^2 + 4x) + 7$

   $y = 2(x^2 + 4x + 4) + 7 - 8$

   $y = 2(x+2)^2 - 1$

   Because a is positive, this is a "smile" parabola.

   vertex: $(-2, -1)$

   axis of symmetry: $x = -2$

   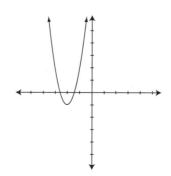

Lesson Practice 2B

1. $3x^2 = 9 - 3y^2$

   Circle. This equation becomes $3x^2 + 3y^2 = 9$ which reduces to $x^2 + y^2 = 3$.

   center: $(0, 0)$.

   radius: $\sqrt{3} \sim 1.7$

   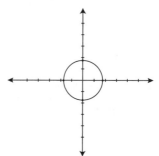

2. $x^2 + y^2 + 2x + 2y = -1$

same coefficients for x^2 and y^2 means we have a circle

$x^2 + 2x + 1 + y^2 + 2y + 1 = -1 + 1 + 1$

$(x+1)^2 + (y+1)^2 = 1$

center: $(-1, -1)$
radius: 1

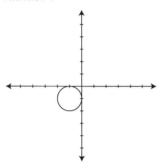

3. $x^2 - 2x + 8y = 7$

$x^2 - 2x + 1 = -8y + 7 + 1$

$(x-1)^2 = -8y + 8$

$y = -\frac{1}{8}(x-1)^2 + 1$

parabola, "frown";
vertex: $(1, 1)$
axis of symmetry: $x = 1$

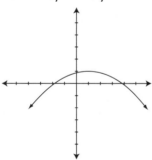

4. $4y^2 + 8y + x^2 = 12$

$x^2 + 4(y^2 + 2y + 1) = 12 + 4$

$\frac{x^2}{16} + \frac{(y+1)^2}{4} = 1$

ellipse, center: $(0, -1)$
extremities: $(4, -1); (-4, -1); (0, 1); (0, -3)$

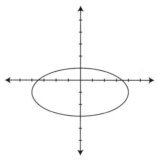

5. $2x + y = -2$

Line. Dividing by -2 we get

$\frac{x}{-1} + \frac{y}{-2} = 1$

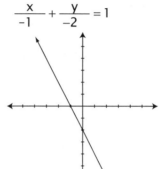

6. $2x^2 + y = 3$

Parabola. $y = -2x^2 + 3$. This is a "frown" parabola with a vertex of $(0, 3)$. The axis of symmetry is the y-axis.

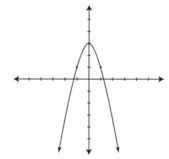

7. $4x^2 + 9y^2 = 36$
Ellipse. Dividing by 36 we get:
$$\frac{x^2}{9} + \frac{y^2}{4} = 1$$
The standard formula is:
$$\frac{(x-h)^2}{a^2} + \frac{(y-k)^2}{b^2} = 1$$
center: $(0, 0)$
extremities: $(\pm 3, 0);\ (0, \pm 2)$

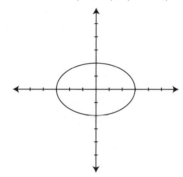

8. $x = 3y^2 - 1$
Parabola. This is a "c" parabola.
vertex: $(-1, 0)$. Axis of symmetry is the x-axis.

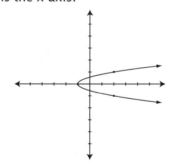

Lesson Practice 2C

1. $12 - 3x^2 = 3y^2$
$3x^2 + 3y^2 = 12$
$x^2 + y^2 = 4$

circle, center: $(0, 0)$; radius: 2

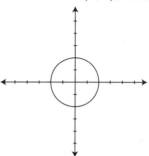

2. $x + 3 = 2y^2$
$x = 2y^2 - 3$

parabola, "c"; vertex: $(-3, 0)$
axis of symmetry: x-axis

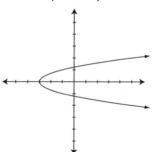

3. $2x - 5y = 10$
$$\frac{x}{5} + \frac{y}{-2} = 1$$
line

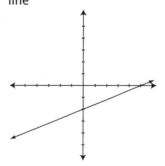

4.
$$x^2 + 4y^2 + 4x = 0$$
$$x^2 + 4x + 4y^2 = 0$$
$$x^2 + 4x + 4 + 4y^2 = 4$$
$$(x+2)^2 + 4y^2 = 4$$
$$\frac{(x+2)^2}{4} + \frac{y^2}{1} = 1$$
ellipse, center: $(-2, 0)$
extremities: $(0, 0); (-4, 0); (-2, 1); (-2, -1)$

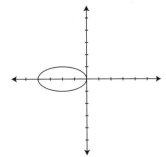

5. $x^2 + 2x + y^2 - 4y = -1$
Circle. Completing the square we get:
$$x^2 + 2x + 1 + y^2 - 4y + 4 = -1 + 1 + 4$$
$$(x+1)^2 + (y-2)^2 = 4$$
center: $(-1, 2)$, radius: 2

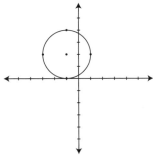

6. $2x = 3y$ Line.
In slope-intercept form we have $y = \frac{2}{3}x$
Slope is $\frac{2}{3}$ and y intercept is 0.

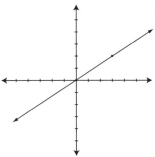

7. $y = -3x^2 - 6x - 1$
parabola: $y = -3(x^2 + 2x) - 1$
$y = -3(x^2 + 2x + 1) - 1 + 3$
$y = -3(x+1)^2 + 2$
vertex: $(-1, 2)$
axis of symmetry: $x = -1$
This is a "frown" parabola.

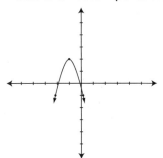

8. $9(x-2)^2 + 4(y+2)^2 = 36$
Ellipse. Dividing by 36 we get:
$$\frac{(x-2)^2}{4} + \frac{(y+2)^2}{9} = 1$$
center: $(2, -2)$
extreme points: $(0, -2); (2, 1)$
$(4, -2); (2, -5)$

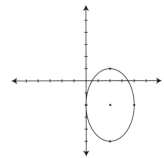

Lesson Practice 2D

1. $6x - 3 = 0$
   $6x = 3$
   $x = \dfrac{1}{2}$: line

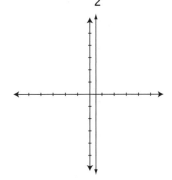

2. $y^2 + 2y + 2 + x = 0$
   Parabola.
   Solving for x we get $x = -y^2 - 2y - 2$
   $x = -(y^2 + 2y) - 2$
   $x = -(y^2 + 2y + 1) - 2 + 1$
   $x = -(y+1)^2 - 1$
   vertex: $(-1, -1)$
   This is a backwards "c" parabola.
   axis of symmetry: $y = -1$

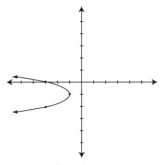

3. $x^2 + y^2 + 2x - 6y = 6$
   $x^2 + 2x + y^2 - 6y = 6$
   $x^2 + 2x + 1 + y^2 - 6y + 9 = 6 + 1 + 9$
   $(x+1)^2 + (y-3)^2 = 16$
   circle, center: $(-1, 3)$; radius: 4

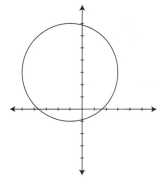

4. $2y^2 + 4y = x - 2$
   $2(y^2 + 2y + 1) = x - 2 + 2$
   $2(y+1)^2 = x$
   $x = 2(y+1)^2$
   parabola, "c"
   vertex: $(0, -1)$
   axis of symmetry: $y = -1$

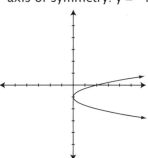

5. $\dfrac{1}{3}x + 3y - 3 = 0$
   Line. Multiplying by 3 we get:
   $x + 9y - 9 = 0$
   $x + 9y = 9$
   $\dfrac{x}{9} + \dfrac{y}{1} = 1$

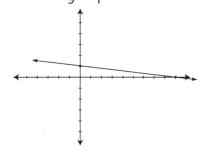

LESSON PRACTICE 2D - LESSON PRACTICE 3A

6. $2x^2 + 4x + 3y^2 - 12y = -8$
   Ellipse. Completing the square, we get:
   $$2(x^2 + 2x) + 3(y^2 - 4y) = -8$$
   $$2(x^2 + 2x + 1) + 3(y^2 - 4y + 4) = -8 + 2 + 12$$
   $$2(x+1)^2 + 3(y-2)^2 = 6$$
   $$\frac{(x+1)^2}{3} + \frac{(y-2)^2}{2} = 1$$
   center: $(-1, +2)$
   extreme points: $(-1+\sqrt{3}, 2)$
   $(-1-\sqrt{3}, 2)$ $(-1, 2+\sqrt{2})$
   $(-1, 2-\sqrt{2})$

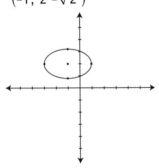

8. $3x^2 + 3y^2 = 6x$
   Circle, Completing the square, we get:
   $$3x^2 - 6x + 3y^2 = 0$$
   $$3(x^2 - 2x) + 3y^2 = 0$$
   $$3(x^2 - 2x + 1) + 3y^2 = 3$$
   $$3(x-1)^2 + 3y^2 = 3$$
   $$\frac{(x-1)^2}{1} + \frac{y^2}{1} = 1$$
   center: $(1, 0)$
   radius: 1

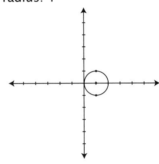

7. $x^2 + y^2 + 8y = 0$
   Circle, Completing the square, we get:
   $$x^2 + y^2 + 8y + 16 = 16$$
   $$x^2 + (y+4)^2 = 16$$
   center $(0, -4)$
   radius = 4

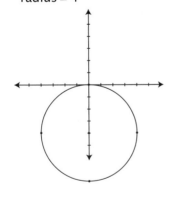

Lesson Practice 3A

1. $\frac{y^2}{9} - \frac{x^2}{16} = 1$
   Two squared terms of opposite signs means this is a hyperbola. It is an up/down hyperbola because the y^2 term is positive.
   The asymptotes are $y = \pm\frac{3}{4}x$.

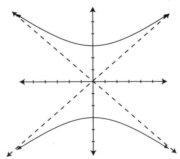

2. $xy = 7$

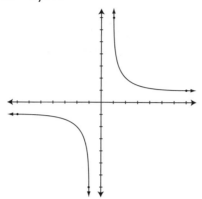

3. $16x^2 + 25y^2 = 400$: ellipse

$\dfrac{x^2}{25} + \dfrac{y^2}{16} = 1$

$x^2 + y^2 = 25$: circle

$(-16)(x^2 + y^2 = 25) \Rightarrow \begin{aligned} 16x^2 + 25y^2 &= 400 \\ -16x^2 - 16y^2 &= -400 \\ \hline 9y^2 &= 0 \\ y &= 0 \end{aligned}$

$16x^2 + 25y^2 = 400$
$16x^2 + 25(0)^2 = 400$
$16x^2 = 400$
$x^2 = 25$
$x = \pm 5$

intersection: $(5, 0); (-5, 0)$

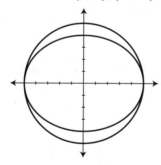

4. $y = x^2 - 2x + 2$: parabola, "smile"
$y - 2x = -2$: line
$y = 2x - 2$

$y = x^2 - 2x + 2$
$(2x - 2) = x^2 - 2x + 2$
$0 = x^2 - 4x + 4$
$0 = (x - 2)(x - 2)$
$x - 2 = 0$
$x = 2$

$y = 2x - 2$
$y = 2(2) - 2$
$y = 4 - 2$
$y = 2$

intersection: $(2, 2)$

$y = x^2 - 2x + 2$
$y = x^2 - 2x + 1 + 1$
$y = (x - 1)^2 + 1$

vertex: $(1, 1)$

This is an example of a tangent line to a curve.

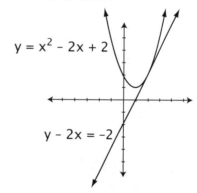

5. $y = x^2 + 1$: parabola, "smile"
   $y - x = 1$: line
   $y = x + 1$

   $$y = x^2 + 1$$
   $$(x + 1) = x^2 + 1$$
   $$0 = x^2 - x$$
   $$0 = (x)(x - 1)$$
   $x = 0$ $x - 1 = 0$
                $x = 1$

   intersection: $(0, 1); (1, 2)$

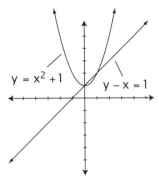

6. Graph $y < x^2 + 2$

   First graph $y = x^2 + 2$ but use dotted lines
   Try $(0, 0)$ $0 < 0^2 + 2$
                    $0 < 2$
   This is true, so shade toward $(0, 0)$.

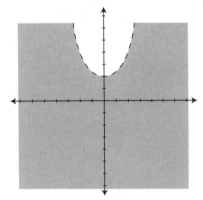

Lesson Practice 3B

1. $xy = -2$
   hyperbola in quadrants II and IV
   asymptotes are x- and y-axis

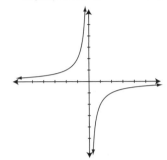

2. $x^2 - y^2 = 16$
   hyperbola, right-left
   $$\frac{x^2}{16} - \frac{y^2}{16} = 1$$
   asymptotes: $y = \pm \frac{4}{4} x$ or $y = \pm x$

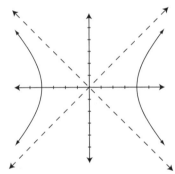

3. First graph $x^2 + y^2 = 4$
   Try $(0, 0)$
   $0^2 + 0^2 = 4$
   $0 \leq 4$ true so shade toward $(0, 0)$

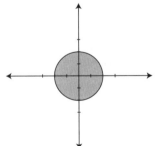

4. $4y = 3x$: line
    $y = \frac{3}{4}x$
    $x^2 + y^2 = 25$: circle
    $$x^2 + \left(\frac{3}{4}x\right)^2 = 25$$
    $$x^2 + \frac{9}{16}x^2 = 25$$
    $$16\left(x^2 + \frac{9}{16}x^2\right) = 16(25)$$
    $$16x^2 + 9x^2 = 400$$
    $$25x^2 = 400$$
    $$x^2 = 16$$
    $$x = \pm 4$$
    $$y = \frac{3}{4}x$$
    $$y = \frac{3}{4}(\pm 4)$$
    $$y = \pm 3$$
    intersection: $(4, 3)$; $(-4, -3)$

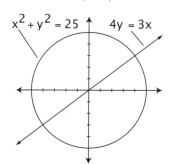

5. $xy = 6$: hyperbola
    $y = -x - 5$: line
    $$x(-x - 5) = 6$$
    $$-x^2 - 5x = 6$$
    $$0 = x^2 + 5x + 6$$
    $$(x+2)(x+3) = 0$$
    $x + 2 = 0 \qquad x + 3 = 0$
    $x = -2 \qquad x = -3$
    $xy = 6 \qquad xy = 6$
    $(-2)y = 6 \qquad (-3)y = 6$
    $y = -3 \qquad y = -2$

intersection: $(-2, -3)$; $(-3, -2)$
Even though the graph appears to meet in many places, they only intersect at 2 points. Between -2 and -3, the hyperbola crosses the line.

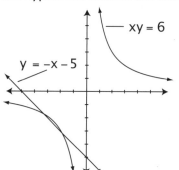

6. $\frac{x^2}{4} - \frac{y^2}{9} = 1$: hyperbola
    $x^2 + y^2 = 4$: circle
    $$36\left(\frac{x^2}{4} - \frac{y^2}{9}\right) = 36(1)$$
    $$9x^2 - 4y^2 = 36$$
    $4(x^2 + y^2 = 4) \Rightarrow \quad 4x^2 + 4y^2 = 16$
    $\qquad\qquad\qquad\qquad 9x^2 - 4y^2 = 36$
    $\qquad\qquad\qquad\qquad \overline{13x^2 = 52}$
    $\qquad\qquad\qquad\qquad\quad x^2 = 4$
    $\qquad\qquad\qquad\qquad\quad x = \pm 2$
    $$x^2 + y^2 = 4$$
    $$(\pm 2)^2 + y^2 = 4$$
    $$4 + y^2 = 4$$
    $$y^2 = 0$$
    $$y = 0$$
    intersection: $(2, 0)$; $(-2, 0)$

There are two tangents at the points $(2, 0)$ and $(-2, 0)$.

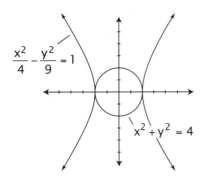

Lesson Practice 3C

1. $y = \dfrac{6}{x}$

   $xy = 6$

   hyperbola, quadrants I and III

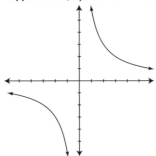

2. $y^2 - x^2 = 4$ $\qquad \dfrac{y^2}{4} - \dfrac{x^2}{4} = 1$

   This is an up-down hyperbola because the y^2 term is positive. The asymptotes are $y = \pm x$.

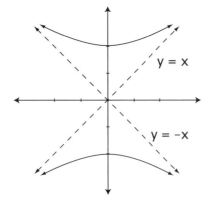

3. Graph $xy = 1$ with dotted lines

   Try $(0, 0)$ $\qquad 0 > 1$ is false

   Try $(3, 3)$ $\qquad 9 > 1$ is true

   Shade towards $(3, 3)$. This graph will also work for $(-3, -3)$.

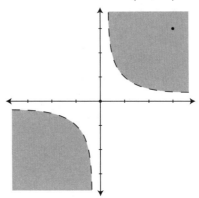

4. $x^2 + y^2 = 4$: circle

   $y = x + 1$: line

   $$x^2 + (x+1)^2 = 4$$
   $$x^2 + x^2 + 2x + 1 = 4$$
   $$2x^2 + 2x - 3 = 0$$

   $$x = \dfrac{-(2) \pm \sqrt{(2)^2 - 4(2)(-3)}}{2(2)}$$

   $$= \dfrac{-2 \pm \sqrt{4 - (-24)}}{4}$$

   $$= \dfrac{-2 \pm \sqrt{28}}{4}$$

   $$= \dfrac{-2 \pm 2\sqrt{7}}{4}$$

   $$= \dfrac{-1 \pm \sqrt{7}}{2}$$

   $y = x + 1$

   $y = \dfrac{-1 \pm \sqrt{7}}{2} + 1$

   $y = +\dfrac{1}{2} \pm \dfrac{\sqrt{7}}{2} = \dfrac{1 \pm \sqrt{7}}{2}$

intersection:

$\left(\dfrac{-1+\sqrt{7}}{2}, \dfrac{1}{2}+\dfrac{\sqrt{7}}{2}\right)$;

$\left(\dfrac{-1-\sqrt{7}}{2}, \dfrac{1}{2}-\dfrac{\sqrt{7}}{2}\right)$

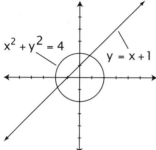

5. $xy = 5$: hyperbola, quadrants I and III
    $y = -x$: line

    $xy = 5$
    $x(-x) = 5$
    $-x^2 = 5$
    $x^2 = -5$
    $x = \sqrt{-5}$
    no real solution

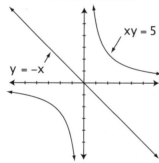

6. $\dfrac{x^2}{4} - y^2 = 1$: hyperbola, left-right

    asymptotes: $y = \pm \dfrac{1}{2}x$

    $y = 1$: line

    $\dfrac{x^2}{4} - (1)^2 = 1$

    $\dfrac{x^2}{4} = 2$

    $x^2 = 8$

    $x = \pm\sqrt{8}$

    $x = \pm 2\sqrt{2}$

    intersection: $(2\sqrt{2}, 1); (-2\sqrt{2}, 1)$

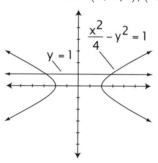

Lesson Practice 3D

1. $16 + x^2 = 4y^2$
    $x^2 - 4y^2 = -16$: hyperbola, up/down
    $4y^2 - x^2 = 16$
    $\dfrac{y^2}{4} - \dfrac{x^2}{16} = 1$

    asymptotes: $y = \pm \dfrac{2}{4}x$ or $y = \pm \dfrac{1}{2}x$

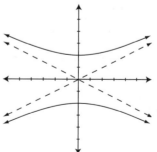

2. $\dfrac{y^2}{4} - \dfrac{x^2}{9} = 1$

hyperbola, up/down asymptotes: $y = \pm \dfrac{2}{3}x$

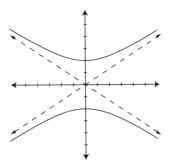

3. $x^2 - 2x + 1 + y^2 = 3 + 1$

$(x-1)^2 + y^2 = 4$

$\dfrac{(x-1)^2}{4} + \dfrac{y^2}{4} = 1$

center: $(1, 0)$

This is a circle.

Try $(0, 0)$. $0 < 3$

true so shade toward $(0, 0)$

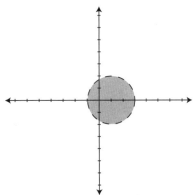

4. $y = x^2 - 2x - 3$

$y = x^2 - 2x + 1 - 3 - 1$

$y = (x-1)^2 - 4$

parabola, "smile"; vertex: $(1, -4)$

$y = -x^2 + 2x - 5$

$y = -(x^2 - 2x + 1) - 5 + 1$

$y = -(x-1)^2 - 4$

parabola, "frown"; vertex: $(1, -4)$

$y = -x^2 + 2x - 5 \Rightarrow$

$x^2 - 2x - 3 = -x^2 + 2x - 5$

$2x^2 - 4x + 2 = 0$

$2(x^2 - 2x + 1) = 0$

$x^2 - 2x + 1 = 0$

$(x-1)^2 = 0$

$x - 1 = 0$

$x = 1$

$y = x^2 - 2x - 3$

$y = (1)^2 - 2(1) - 3$

$y = 1 - 2 - 3$

$y = -4$

intersection: $(1, -4)$

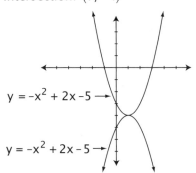

5. $x^2 - y^2 = 1$: hyperbola, left-right

$\dfrac{x^2}{1} - \dfrac{y^2}{1} = 1$

Asymptotes: $y = \pm x$

$x^2 + y^2 = 5$: circle

$\begin{aligned} x^2 - y^2 &= 1 \\ x^2 + y^2 &= 5 \\ \hline 2x^2 &= 6 \end{aligned}$

$x^2 = 3$

$x = \pm \sqrt{3}$

$$x^2 - y^2 = 1$$
$$\left(\sqrt{3}\right)^2 - y^2 = 1$$
$$3 - y^2 = 1$$
$$-y^2 = -2$$
$$y^2 = 2$$
$$y = \pm\sqrt{2}$$

intersection: $\left(\sqrt{3}, \sqrt{2}\right); \left(-\sqrt{3}, \sqrt{2}\right)$
$\left(\sqrt{3}, -\sqrt{2}\right); \left(-\sqrt{3}, -\sqrt{2}\right)$

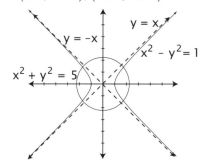

6. $x^2 + y^2 = 4$: circle
$4x^2 + y^2 = 4$: ellipse

$$\begin{aligned} x^2 + y^2 &= 4 \\ -4x^2 - y^2 &= -4 \\ \hline -3x^2 &= 0 \\ x^2 &= 0 \\ x &= 0 \end{aligned}$$

$$x^2 + y^2 = 4$$
$$(0)^2 + y^2 = 4$$
$$y^2 = 4$$
$$y = \pm 2$$

intersection: $(0, 2); (0, -2)$

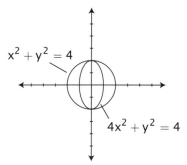

Lesson Practice 4A

1. $f(x) = x^3 - 5x^2 - 4x + 20$

$$\begin{aligned} f(0) &= (0)^3 - 5(0)^2 - 4(0) + 20 \\ &= 0 - 0 - 0 + 20 \\ &= 20 \end{aligned}$$

$$\begin{aligned} f(1) &= (1)^3 - 5(1)^2 - 4(1) + 20 \\ &= 1 - 5 - 4 + 20 \\ &= 12 \end{aligned}$$

$$\begin{aligned} f(3) &= (3)^3 - 5(3)^2 - 4(3) + 20 \\ &= 27 - 5(9) - 12 + 20 \\ &= 27 - 45 - 12 + 20 \\ &= -10 \end{aligned}$$

$$\begin{aligned} f(5) &= (5)^3 - 5(5)^2 - 4(5) + 20 \\ &= 125 - 5(25) - 20 + 20 \\ &= 125 - 125 - 20 + 20 \\ &= 0 \end{aligned}$$

2. $A(r) = \pi r^2$

$$\begin{aligned} A(0) &= \pi(0)^2 \\ &= 0 \end{aligned}$$

$$\begin{aligned} A(2) &= \pi(2)^2 \\ &= 4\pi \end{aligned}$$

$$\begin{aligned} A(3) &= \pi(3)^2 \\ &= 9\pi \end{aligned}$$

LESSON PRACTICE 4A - LESSON PRACTICE 4A

3. $f(x) = 4 - 2x^2 + x^4$

$f(0) = 4 - 2(0)^2 + (0)^4$
$= 4 - 0 + 0$
$= 4$

$f(1) = 4 - 2(1)^2 + (1)^4$
$= 4 - 2(1) + 1$
$= 4 - 2 + 1$
$= 3$

$f(-1) = 4 - 2(-1)^2 + (-1)^4$
$= 4 - 2(1) + 1$
$= 4 - 2 + 1 = 3$

$f(2) = 4 - 2(2)^2 + (2)^4$
$= 4 - 2(4) + 16$
$= 4 - 8 + 16$
$= 12$

$f(-2) = 4 - 2(-2)^2 + (-2)^4$
$= 4 - 2(4) + 16$
$= 4 - 8 + 16$
$= 12$

4. $G(z) = z^2(z^2 - 4)$

$G(0) = (0)^2((0)^2 - 4)$
$= 0(0 - 4)$
$= 0$

$G(1) = (1)^2((1)^2 - 4)$
$= 1(1 - 4)$
$= 1(-3)$
$= -3$

$G(2) = (2)^2((2)^2 - 4)$
$= 4(4 - 4)$
$= (4)(0)$
$= 0$

$G(-2) = (-2)^2((-2)^2 - 4)$
$= 4(4 - 4)$
$= 4(0)$
$= 0$

5. $F(y) = 2^y$

$F(0) = 2^{(0)} = 1$

$F\left(\frac{3}{2}\right) = 2^{\left(\frac{3}{2}\right)} = 2^{(3)\left(\frac{1}{2}\right)} = 8^{\frac{1}{2}}$
$= \sqrt{8} = 2\sqrt{2}$

$F(-1) = 2^{(-1)} = \frac{1}{2^1} = \frac{1}{2}$

$F(y - 2) = 2^{(y-2)} = \frac{2^y}{2^2} = \frac{2^y}{4}$

6. $f(x) = x + 2$

$f(x + h) = (x + h) + 2$
$= x + h + 2$

$f(x + h) - f(x) = (x + h) + 2 - (x + 2)$
$= x + h + 2 - X - 2$
$= h$

7. $f(x) = \frac{1}{x}; \; f(x+h) = \frac{1}{x+h}$

$f(x+h) - f(x)$

$= \frac{1}{x+h} - \frac{1}{x}$

$= \frac{1(x)}{(x+h)(x)} - \frac{1(x+h)}{(x)(x+h)}$

$= \frac{x - (x+h)}{x^2 + xh}$

$= \frac{x - x - h}{x^2 + xh}$

$= \frac{-h}{x^2 + xh}$

8. $f(y) = y^2 - 2y + 6$

$f(y + h)$

$= (y + h)^2 - 2(y + h) + 6$
$= (y + h)(y + h) - 2y - 2h + 6$
$= y^2 + yh + yh + h^2 - 2y - 2h + 6$
$= y^2 + 2yh + h^2 - 2y - 2h + 6$
$= y^2 - 2y + 6 + 2yh - 2h + h^2$
$= y^2 - 2y + 6 + 2(y - 1)h + h^2$

9. $D = RT$
    $(1) = RT$
    $\dfrac{1}{R} = T$
    $T(R) = \dfrac{1}{R}$

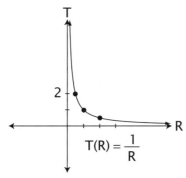
$T(R) = \dfrac{1}{R}$

10. $A = S^2$
    $S = \sqrt{A}$
    $S(A) = \sqrt{A}$

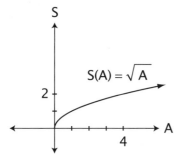

11. $V = S^3$
    $S = \sqrt[3]{V}$
    $S(V) = \sqrt[3]{V}$

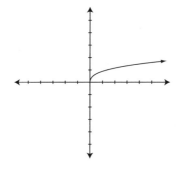

12. $A = S^2$
    $A(S) = S^2$

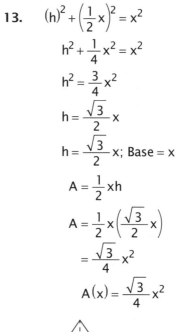

13. $(h)^2 + \left(\dfrac{1}{2}x\right)^2 = x^2$
    $h^2 + \dfrac{1}{4}x^2 = x^2$
    $h^2 = \dfrac{3}{4}x^2$
    $h = \dfrac{\sqrt{3}}{2}x$
    $h = \dfrac{\sqrt{3}}{2}x$; Base $= x$
    $A = \dfrac{1}{2}xh$
    $A = \dfrac{1}{2}x\left(\dfrac{\sqrt{3}}{2}x\right)$
    $ = \dfrac{\sqrt{3}}{4}x^2$
    $A(x) = \dfrac{\sqrt{3}}{4}x^2$

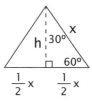

Lesson Practice 4B

1. $f(x) = x^3 - x^2 + 2x + 3$

   $f(0) = (0)^3 - (0)^2 + 2(0) + 3$
   $= 0 - 0 + 0 + 3$
   $= 3$

   $f(1) = (1)^3 - (1)^2 + 2(1) + 3$
   $= 1 - 1 + 2 + 3$
   $= 5$

   $f(2) = (2)^3 - (2)^2 + 2(2) + 3$
   $= 8 - 4 + 4 + 3$
   $= 11$

2. $f(x) = x^2(x^2 + 4)$

   $f(0) = (0)^2((0)^2 + 4)$
   $= 0(0 + 4)$
   $= 0$

   $f(1) = (1)^2((1)^2 + 4)$
   $= 1(1 + 4)$
   $= 1(5)$
   $= 5$

   $f(2) = (2)^2((2)^2 + 4)$
   $= 4(4 + 4)$
   $= 4(8)$
   $= 32$

   $f(-2) = (-2)^2((-2)^2 + 4)$
   $= 4(4 + 4)$
   $= 4(8)$
   $= 32$

3. $m(x) = \dfrac{x^2 - 2x}{x^2 + 1}$

   $m(0) = \dfrac{(0)^2 - 2(0)}{(0)^2 + 1}$
   $= \dfrac{0 - 0}{0 + 1}$
   $= \dfrac{0}{1}$
   $= 0$

   $m(1) = \dfrac{(1)^2 - 2(1)}{(1)^2 + 1}$
   $= \dfrac{1 - 2}{1 + 1}$
   $= \dfrac{-1}{2}$ or $-\dfrac{1}{2}$

   $m(-1) = \dfrac{(-1)^2 - 2(-1)}{(-1)^2 + 1}$
   $= \dfrac{1 + 2}{1 + 1}$
   $= \dfrac{3}{2}$

   $m(2) = \dfrac{(2)^2 - 2(2)}{(2)^2 + 1}$
   $= \dfrac{4 - 4}{4 + 1}$
   $= \dfrac{0}{5}$
   $= 0$

4. $f(x) = x^2 - 2x + 5$

   $f(1) = (1)^2 - 2(1) + 5$
   $= 1 - 2 + 5$
   $= 4$

   $f(b) = (b)^2 - 2(b) + 5$
   $= b^2 - 2b + 5$

   $f(b + h) = (b + h)^2 - 2(b + h) + 5$
   $= (b + h)(b + h) - 2b - 2h + 5$
   $= b^2 + 2bh + h^2 - 2b - 2h + 5$

5. $f(x) = |2 - x|$

   $f(0) = |2 - (0)|$
   $= |2|$
   $= 2$

   $f(3) = |2 - (3)|$
   $= |-1|$
   $= 1$

$f(-2) = |2-(-2)|$
$= |4|$
$= 4$

6. $f(x) = |x| - |x+1|$
$f(0) = |(0)| - |(0)+1|$
$= 0 - |1|$
$= 0 - 1$
$= -1$
$f(-1) = |(-1)| - |(-1)+1|$
$= 1 - |0|$
$= 1 - 0$
$= 1$
$f(-2) = |(-2)| - |(-2)+1|$
$= 2 - |-1|$
$= 2 - 1$
$= 1$

7. $b(x) = 3 - x;\ j(x) = -2x$
$b(j(x)) = 3 - (-2x)$
$= 3 + 2x$
$j(b(x)) = -2(3-x)$
$= -6 + 2x$
$b(j(0)) = 3 - 2(0)$
$= 3$
$j(b(0)) = -2(3-0)$
$= -2(3)$
$= -6$

b and j are not inverse functions
If b and j were inverse functions,
then $b(j(x)) = j(b(x)) = x$.

8. $f(x) = x - 2;\ g(x) = 3x^2$
$f(g(x)) = f(3x^2) = 3x^2 - 2$
$g(f(x)) = 3(x-2)^2$
$= 3(x^2 - 4x + 4)$
$= 3x^2 - 12x + 12$

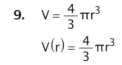
$f(g(1)) = 3 - 2 = 1$
$g(f(1)) = 3(1-2)^2 = 3$
No. $f(g(1)) \neq g(f(1))$

9. $V = \frac{4}{3}\pi r^3$
$V(r) = \frac{4}{3}\pi r^3$

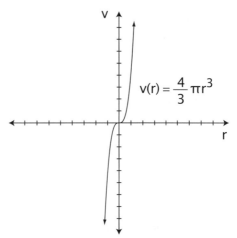

10. $P = 4 \cdot s$
$s = \frac{1}{4}P$
$s(P) = \frac{1}{4}P$

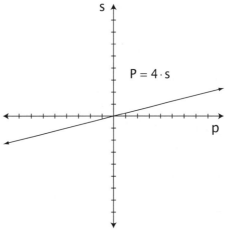

LESSON PRACTICE 4B - LESSON PRACTICE 4C

11.
1. $y = \frac{1}{3}x + 5$
2. $x = \frac{1}{3}y + 5$ switch variables
3. $(x-5)3 = y$
4. $f^{-1}(x) = 3(x-5)$

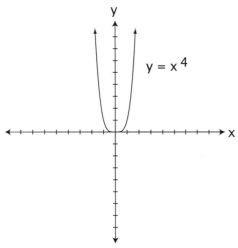

$y = x^4$

12. No, because $y = x^4$ fails the horizontal line test.

13.
1. $y = 4 - x^3$
2. $x = 4 - y^3$ switch variables
3. $y = \sqrt[3]{4-x}$
4. $f^{-1}(x) = \sqrt[3]{4-x}$

Lesson Practice 4C

1. $R(t) = 3^{t-1}$

$R(0) = 3^{(0)-1}$
$= 3^{-1}$
$= \frac{1}{3}$

$R(1) = 3^{(1)-1}$
$= 3^0$
$= 1$

$R(2) = 3^{(2)-1}$
$= 3^1$
$= 3$

$R(1-t) = 3^{(1-t)-1}$
$= 3^{-t}$
$= \frac{1}{3^t}$

2. $f(y) = y^2 - 2y + 6$

$f(-1) = (-1)^2 - 2(-1) + 6$
$= 1 + 2 + 6$
$= 9$

$f(2y) = (2y)^2 - 2(2y) + 6$
$= 4y^2 - 4y + 6$

$f(y+h) = (y+h)^2 - 2(y+h) + 6$
$= y^2 + 2yh + h^2 - 2y - 2h + 6$

3. $P(z) = 4^z$

$P(2) = 4^{(2)} = 16$

$P(-2) = 4^{(-2)} = \frac{1}{4^2} = \frac{1}{16}$

$3P(z) = 3 \cdot 4^z$

$P(z+1) = 4^{z+1}$

4. $3P(z) = 3 \cdot 4^z$

$P(z+1) - P(z) = 4^{z+1} - 4^z$

$3 \cdot 4^z = 4^{z+1} - 4^z$
$3 \cdot 4^z = (4^z)(4) - 4^z$
$3 \cdot 4^z = 4^z(4-1)$
$3 = 4 - 1$
$3 = 3$

LESSON PRACTICE 4C - LESSON PRACTICE 4C

5. $f(x) = \dfrac{1}{\sqrt{2x}}$

   A. No, we use only positive square roots in this course.
   B. No, division by 0 is undefined.
   C. $x > 0$
   D. $f\left(\dfrac{1}{2}\right) = \dfrac{1}{\sqrt{2\left(\dfrac{1}{2}\right)}} = \dfrac{1}{\sqrt{1}} = 1$

6. $f(x) = \begin{cases} 3-x \text{ if } x \leq 1 \\ 2x \text{ if } x > 1 \end{cases}$

   $f(0) = 3 - (0)$
   $\quad\quad = 3$
   $f(1) = 3 - (1)$
   $\quad\quad = 2$
   $f(4) = 2(4)$
   $\quad\quad = 8$

7. $f(x) = \sqrt{x}\,;\, g(x) = x - 1$

   A. $f(g(x)) = f(x-1) = \sqrt{x-1}$
   B. $x \geq 1$
   C. $f(x) \geq 0$
   D. $f(g(2)) = f(2-1) = \sqrt{2-1} = \sqrt{1} = 1$

8. $f(x) = \dfrac{1}{x-2}\,;\, g(x) = \sqrt{x}$

   A. $f(g(x)) = f(\sqrt{x}) = \dfrac{1}{\sqrt{x}-2}$
   B. $g(f(x)) = g\left(\dfrac{1}{x-2}\right) = \sqrt{\dfrac{1}{x-2}}$
   C. $f(g(9)) = f(\sqrt{9}) = f(3) = \dfrac{1}{(3)-2} = 1$
   D. $x > 2$

9. $f(x) = \dfrac{1}{4}x + 3$

   $y = \dfrac{1}{4}x + 3$
   $x = \dfrac{1}{4}y + 3$ (switch variables)
   $x - 3 = \dfrac{1}{4}y$
   $4(x-3) = y$
   $f^{-1}(x) = 4(x-3)$
   $f^{-1}(x) = 4x - 12$

10. $f^{-1}(f(x)) = f^{-1}\left(\dfrac{1}{4}x + 3\right)$
    $\quad\quad\quad\quad = 4\left(\dfrac{1}{4}x + 3\right) - 12$
    $\quad\quad\quad\quad = x + 12 - 12$
    $\quad\quad\quad\quad = x$

11. $f(x) = (x-1)^2$

    $y = (x-1)^2$
    $x = (y-1)^2$ (switch variables)
    $\pm\sqrt{x} = y - 1$
    $\pm\sqrt{x} + 1 = y$
    $f^{-1}(x) = 1 \pm \sqrt{x}$

12. No. The original function does not pass the horizontal line test.

13. Problem 1:
    domain: all real numbers
    range: $R(t) > 0$

Lesson Practice 4D

1. $H(r) = r^2 - 3$
   $H(-1) = (-1)^2 - 3$
   $\quad = 1 - 3$
   $\quad = -2$

   $H(1) = (1)^2 - 3$
   $\quad = 1 - 3$
   $\quad = -2$

   $H(\sqrt{3}) = (\sqrt{3})^2 - 3$
   $\quad = 3 - 3$
   $\quad = 0$

2. $W(t) = 3^t - 2$
   $W(0) = 3^{(0)} - 2$
   $\quad = 1 - 2$
   $\quad = -1$

   $W(-1) = 3^{(-1)} - 2$
   $\quad = \frac{1}{3} - \frac{6}{3}$
   $\quad = -\frac{5}{3}$ or $-1\frac{2}{3}$

   $W(2) = 3^{(2)} - 2$
   $\quad = 9 - 2$
   $\quad = 7$

3. $f(x) = \frac{1}{\sqrt{x-1}}$
   $f(2) = \frac{1}{\sqrt{(2)-1}}$
   $\quad = \frac{1}{\sqrt{1}}$
   $\quad = \frac{1}{1}$
   $\quad = 1$

   $f(5) = \frac{1}{\sqrt{(5)-1}}$
   $\quad = \frac{1}{\sqrt{4}}$
   $\quad = \frac{1}{2}$

   $f(10) = \frac{1}{\sqrt{(10)-1}}$
   $\quad = \frac{1}{\sqrt{9}}$
   $\quad = \frac{1}{3}$

4. $F(S) = 3 - 2S^2$
   $F(0) = 3 - 2(0)^2$
   $\quad = 3 - 0$
   $\quad = 3$

   $F(1) = 3 - 2(1)^2$
   $\quad = 3 - 2(1)$
   $\quad = 3 - 2$
   $\quad = 1$

   $F(-1) = 3 - 2(-1)^2$
   $\quad = 3 - 2(1)$
   $\quad = 3 - 2$
   $\quad = 1$

5. $f(x) = 2^{\sqrt{x}}$

   A. No, we use only positive square roots in this course.

   B. Yes, $2^0 = 1$

   C. $x \geq 0$

   D. $f(1) = 2^{\sqrt{(1)}}$
   $\quad = 2^1$
   $\quad = 2$

6. $f(x) = \begin{cases} 4 - x^2 & \text{if } x \leq 0 \\ \sqrt{x} & \text{if } x > 0 \end{cases}$

   $f(-1) = 4 - (-1)^2$
   $\quad = 4 - 1$
   $\quad = 3$

   $f(0) = 4 - (0)^2$
   $\quad = 4 - 0$
   $\quad = 4$

   $f(1) = \sqrt{(1)}$
   $\quad = 1$

7. $f(x) = x+1$; $g(x) = 2^x$
   A. $f(g(x)) = f(2^x) = 2^x + 1$
   B. all real numbers
   C. all $y > 1$
   D. $f(g(3)) = 2^3 + 1 = 9$

8. $f(x) = x^2 - 1$; $g(x) = \frac{x}{2}$
   A. $f(g(x)) = f\left(\frac{x}{2}\right) = \left(\frac{x}{2}\right)^2 - 1 = \frac{x^2}{4} - 1$
   B. $g(f(x)) = g(x^2 - 1) = \frac{x^2 - 1}{2}$
   C. $f(g(3)) = \frac{(3)^2}{4} - 1 = \frac{9}{4} - \frac{4}{4} = \frac{5}{4}$
   D. all real numbers

9. $f(x) = 4 - x$
   $y = 4 - x$
   $x = 4 - y$ (switch variables)
   $x - 4 = -y$
   $-x + 4 = y$
   $y = 4 - x$
   $f^{-1}(x) = 4 - x$

10. $f^{-1}(f(x)) = f^{-1}(4 - x)$
    $= 4 - (4 - x)$
    $= 4 - 4 + x$
    $= x$

11. $f(x) = x^3 + 1$
    $y = x^3 + 1$
    $x = y^3 + 1$ (switch variables)
    $x - 1 = y^3$
    $y = \sqrt[3]{x - 1}$
    $f^{-1}(x) = \sqrt[3]{x - 1}$

12. Yes. $f(x) = x^3 + 1$
    passes the horizontal line test.

13. Problem 3:
    domain: $x > 1$
    range: $f(x) > 0$

Lesson Practice 5A

1. period = π

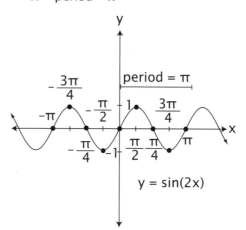

2. period = 2π

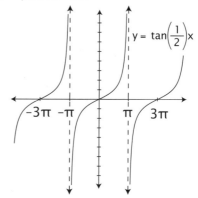

3. $y = \cos(3x) + 1$

x	y
0	2
$\frac{\pi}{6}$	1
$\frac{\pi}{3}$	0
$\frac{\pi}{2}$	1

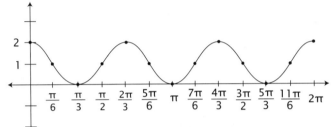

4. $y = 2 - \sin(x)$

x	y
0	2
$\frac{\pi}{2}$	1
π	2
$\frac{3\pi}{2}$	3
2π	2

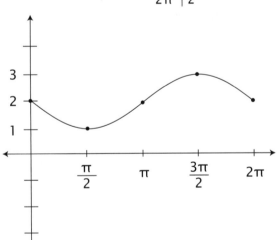

5. A. $\sin\left(\frac{3\pi}{4}\right) = \sin\left(\frac{540°}{4}\right) = \sin(135°) = \frac{\sqrt{2}}{2}$

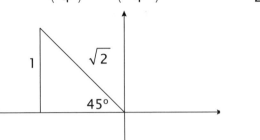

B. $\cos\left(\frac{\pi}{3}\right) = \cos\left(\frac{180°}{3}\right) = \cos(60°) = \frac{1}{2}$

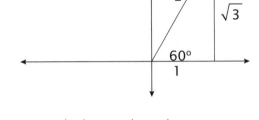

C. $\sec\left(\frac{\pi}{6}\right) = \sec\left(\frac{180°}{6}\right) = \sec(30°)$

$= \frac{1}{\cos(30°)} = \frac{1}{\frac{\sqrt{3}}{2}}$

$= \frac{2}{\sqrt{3}} = \frac{2\sqrt{3}}{3}$

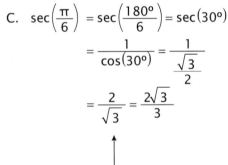

D. $\csc\left(\dfrac{\pi}{2}\right) = \csc\left(\dfrac{180°}{2}\right) = \csc(90°)$
$= \dfrac{1}{\sin(90°)} = \dfrac{1}{1} = 1$

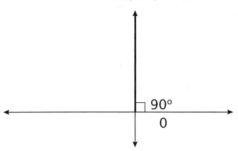

E. $\tan\left(\dfrac{7\pi}{6}\right) = \tan\left(\dfrac{7(180°)}{6}\right) = \tan 7(30°)$
$= \tan(210°) = \dfrac{-1}{-\sqrt{3}} = \dfrac{1}{\sqrt{3}} = \dfrac{\sqrt{3}}{3}$

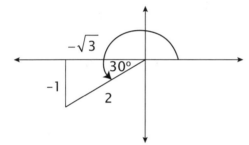

F. $\cot\left(\dfrac{5\pi}{4}\right) = \cot\dfrac{5(180°)}{4} = \cot 5(45°) =$
$\cot(225°) = \dfrac{1}{\tan(225°)} = \dfrac{1}{\frac{-1}{-1}} = \dfrac{1}{1} = 1$

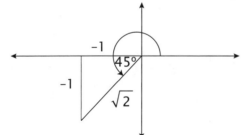

6. $\cos(2x) = 1$ for x in [0, π]
$\cos(\theta) = 1$ when θ = 0, 2π, 4π etc.
$2x = 0$ $2x = 2\pi$
$x = 0$ $x = \pi$

Answers $x = 0, \pi$

Lesson Practice 5B

1. period = 4π

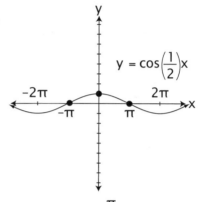

2. period = $\dfrac{\pi}{4}$

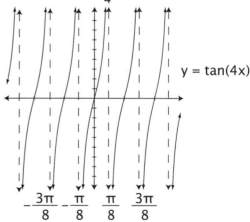

3. $y = \tan(x) + 2$

x	y
0	2
$\frac{\pi}{4}$	3
$-\frac{\pi}{4}$	1

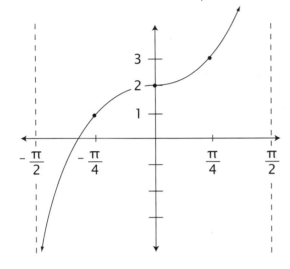

4. $y = 3 - \sin(2x)$

x	y
0	3
$\frac{\pi}{4}$	2
$\frac{\pi}{2}$	3
$\frac{3\pi}{4}$	4

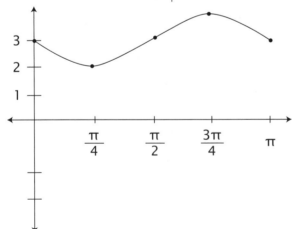

5. A. $\cos\left(\frac{3\pi}{4}\right) = \cos(135°) = -\frac{\sqrt{2}}{2}$
   Quadrant II

   B. $\sec\left(\frac{\pi}{3}\right) = \sec(60°) = \frac{2}{1}$
   Quadrant I

   C. $\csc(\pi) = \csc(180°) = \frac{1}{0}$
   undefined

   D. $\sin\left(\frac{7\pi}{6}\right) = \sin(210°) = -\frac{1}{2}$
   Quadrant III

   E. $\tan\left(\frac{\pi}{3}\right) = \tan(60°) = \frac{\sqrt{3}}{1} = \sqrt{3}$
   Quadrant I

   F. $\cot\left(\frac{\pi}{2}\right) = \cot(90°) = \frac{0}{1} = 0$
   Quadrant I

6. $\cos(2x - \pi) = 0$ for x in $\left[0, \frac{3\pi}{2}\right]$

   $\cos(\theta) = 0$ when $\theta = \frac{\pi}{2}, \frac{3\pi}{2}, \frac{5\pi}{2}$, etc

   $2x - \pi = \frac{\pi}{2}$ $2x - \pi = \frac{3\pi}{2}$

   $2x = \frac{\pi}{2} + \pi$ $2x = \frac{3\pi}{2} + \pi$

   $2x = \frac{3\pi}{2}$ $2x = \frac{5\pi}{2}$

   $x = \frac{3\pi}{4}$ $x = \frac{5\pi}{4}$

   $x = \frac{\pi}{4}, \frac{3\pi}{4}, \frac{5\pi}{4}$ $2x - \pi = -\frac{\pi}{2}$

   $2x = \pi - \frac{\pi}{2}$

   $2x = \frac{\pi}{2}$

   $x = \frac{\pi}{4}$

Lesson Practice 5C

1. The frequency is 2, because the graph has 2 cycles from 0 to 2π.
2. 2
3.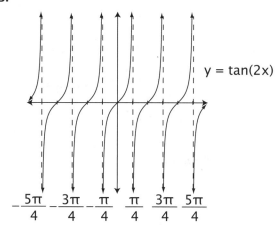

4.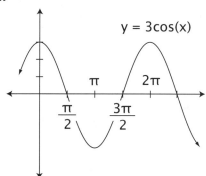

5. $\cot\left(\dfrac{3\pi}{4}\right) = \cot(135°) = -1$
   cot is negative in quadrant II

6. $\csc\left(\dfrac{5\pi}{6}\right) = \csc(150°) = 2$
   csc is positive in quadrant II

7. $\sin(4x) = 0$ x in $[0, 2\pi]$
   $\sin(\theta)$ is 0 when $\theta = 0, \pi, 2\pi, 3\pi, 4\pi$ etc.

   $4x = 0$ $4x = \pi$ $4x = 2\pi$ $4x = 3\pi$
   $x = 0$ $x = \dfrac{\pi}{4}$ $x = \dfrac{\pi}{2}$ $x = \dfrac{3\pi}{4}$

   $4x = 4\pi$ $4x = 5\pi$ ... $4x = 8\pi$
   $x = \pi$ $x = \dfrac{5\pi}{4}$ ... $x = 2\pi$

   Answers $x = 0, \dfrac{\pi}{4}, \dfrac{\pi}{2}, \dfrac{3\pi}{4}, \pi,$
   $\dfrac{5\pi}{4}, \dfrac{3\pi}{2}, \dfrac{7\pi}{4}, 2\pi$

8. $\tan(2x + 1) = 0$ for x in $[0, \pi]$

   $\tan(\theta)$ is 0 when $\theta = 0$
   $2x + 1 = 0$
   $2x = -1$
   $x = -\dfrac{1}{2}$

   $\tan(\theta) = 0$ when $\theta = \pi$ $\tan(\theta) = 0$ when $\theta = 2\pi$
   $2x + 1 = \pi$ $2x + 1 = 2\pi$
   $2x = \pi - 1$ $2x = 2\pi - 1$
   $x = \dfrac{\pi - 1}{2}$ $x = \dfrac{2\pi - 1}{2}$

   possible answers $x = -\dfrac{1}{2}, \dfrac{\pi - 1}{2}, \dfrac{2\pi - 1}{2}$

   Since $x = -\dfrac{1}{2}$ is not in our interval $[0, \pi]$ it is not a true solution.

   Answers: $x = \dfrac{\pi - 1}{2}$ and $\dfrac{2\pi - 1}{2}$

Lesson Practice 5D

1. $f(x) = 2 + 4 \sin 2(x + \pi)$
   amplitude = 4
   vertical shift = 2
   period = $\frac{2\pi}{2} = \pi$
   phase shift = $-\pi$

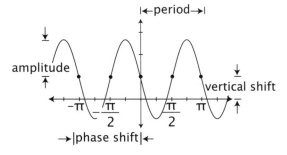

2. $f(x) = 1 - 3 \sin \frac{1}{2}(x - \pi)$
   amplitude = 3
   vertical shift = 1
   phase shift = π
   period = $\frac{2\pi}{\frac{1}{2}} = 4\pi$

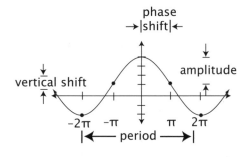

3. $f(x) = 3 + 2 \cos 4\left(x - \frac{\pi}{2}\right)$
   amplitude = 2
   vertical shift = 3
   phase shift = $\frac{\pi}{2}$
   period = $\frac{2\pi}{4} = \frac{\pi}{2}$

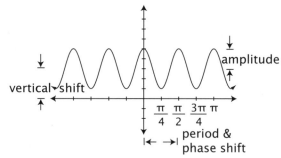

4. $f(x) = 2 - 4\cos 2(x + \pi)$
   amplitude = 4
   vertical shift = 2
   phase shift = $-\pi$
   period = $\frac{2\pi}{2} = \pi$

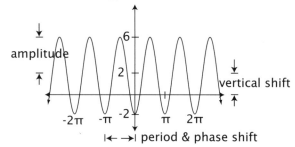

5. $\csc\left(\frac{2\pi}{3}\right) = \csc(120°) = \frac{2}{\sqrt{3}} = \frac{2\sqrt{3}}{3}$

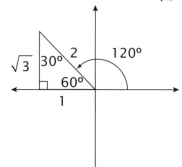

6. $\tan\left(\frac{3\pi}{4}\right) = \tan(135°) = -1$

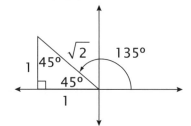

7. $\cos(2x) = 0$ for x in $[0, \pi]$

$\cos(\theta) = 0$ when $\theta = \dfrac{\pi}{2}, \dfrac{3\pi}{2}, \dfrac{5\pi}{2}$ etc

$2x = \dfrac{\pi}{2}$ $2x = \dfrac{3\pi}{2}$ $2x = \dfrac{5\pi}{2}$

$x = \dfrac{\pi}{4}$ $x = \dfrac{3\pi}{4}$ $x = \dfrac{5\pi}{4}$

$x = \dfrac{\pi}{4}$ and $\dfrac{3\pi}{4}$

8. $\tan\left(\dfrac{1}{2}x\right) = 0$ $\left[0, \dfrac{\pi}{2}\right]$

$\tan(\theta) = 0$ when $\theta = 0, \pi, 2\pi, 3\pi$ etc.

$\dfrac{1}{2}x = 0$ $\dfrac{1}{2}x = \pi$

$x = 0$ $x = 2\pi$

$x = 0$ is the only answer in $\left[0, \dfrac{\pi}{2}\right]$

Lesson Practice 6A

1. $y = \dfrac{e^x}{3}$

$x = \dfrac{e^y}{3}$ (Switch variables)

$3x = e^y$

$\ln(3x) = \ln(e^y)$

$\ln(3x) = y$

$f^{-1}(x) = \ln(3x)$

2. $y = 2e^x$

$x = 2e^y$ (reverse variables)

$\ln(x) = \ln(2e^y)$

$\ln(x) = \ln 2 + y$

$\ln(x) - \ln(2) = y$

$f^{-1}(x) = \ln(x) - \ln(2)$

Note: This problem is the same as example 4 in the instruction manual but solved differently. Both solutions are correct.

$\ln(x) - \ln(2) = \ln\left(\dfrac{x}{2}\right)$

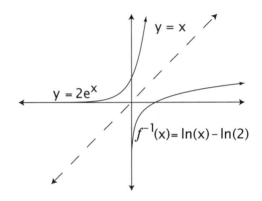

3. A. $e^{2x+1} = 1$

$\ln(e^{2x+1}) = \ln(1)$

$2x + 1 = 0$

$2x = -1$

$x = -\dfrac{1}{2}$

B. $2e^{3x} = e^0$

$\ln(2e^{3x}) = \ln(e^0)$

$\ln(2) + 3x = 0$

$3x = -\ln(2)$

$x = -\dfrac{\ln(2)}{3}$

C. $\quad 0 = \ln(2x+5)$
$\quad e^0 = e^{\ln(2x+5)}$
$\quad 1 = 2x+5$
$\quad -4 = 2x$
$\quad x = \dfrac{-4}{2} = -2$

D. $\quad \ln(x) + \ln(5) = 6$
$\quad \ln(5x) = 6$
$\quad e^{\ln(5x)} = e^6$
$\quad 5x = e^6$
$\quad x = \dfrac{e^6}{5}$ or $\dfrac{1}{5}e^6$

4. A. $\quad e^{2x} - 5e^x = -6 \quad$ Let $y = e^x$.
$\quad y^2 - 5y + 6 = 0$
$\quad (y-2)(y-3) = 0$

$y - 2 = 0 \qquad\qquad y - 3 = 0$
$y = 2 \qquad\qquad\qquad y = 3$
$e^x = 2 \qquad\qquad\quad e^x = 3$
$\ln(e^x) = \ln(2) \quad\quad \ln(e^x) = \ln(3)$
$x = \ln(2) \qquad\qquad x = \ln(3)$

B. $\quad 2e^{2x} + 7e^x = 4 \quad$ Let $y = e^x$.
$\quad 2y^2 + 7y - 4 = 0$
$\quad (2y - 1)(y + 4) = 0$

$2y - 1 = 0 \qquad\qquad y + 4 = 0$
$2y = 1 \qquad\qquad\qquad y = -4$
$y = \dfrac{1}{2} \qquad\qquad\qquad e^x = -4$
$e^x = \dfrac{1}{2} \qquad\qquad \ln(e^x) = \ln(-4)$
$\ln(e^x) = \ln\left(\dfrac{1}{2}\right) \quad$ error:
$x = \ln(1) - \ln(2) \quad$ no solution
$x = 0 - \ln(2)$
$x = -\ln(2)$

Lesson Practice 6B

1. $\quad y = e^{x+1}$
$\quad x = e^{y+1}$ (reverse variables)
$\quad \ln(x) = \ln(e^{y+1})$
$\quad \ln(x) = y + 1$
$\quad \ln(x) - 1 = y$
$\quad f^{-1}(x) = \ln(x) - 1$

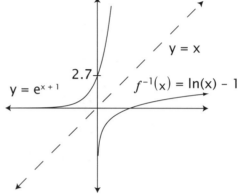

2. $\quad y = e^{\frac{x}{2}}$
$\quad x = e^{\frac{y}{2}}$ (reverse variables)
$\quad \ln(x) = \ln\left(e^{\frac{y}{2}}\right)$
$\quad \ln(x) = \dfrac{y}{2}$
$\quad 2\ln(x) = y$
$\quad f^{-1}(x) = 2\ln(x)$

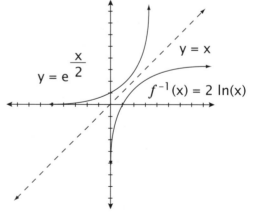

3. A. $e^{x+\ln(3)} = 2$
    $\ln(e^{x+\ln(3)}) = \ln(2)$
    $x + \ln(3) = \ln(2)$
    $x = \ln(2) - \ln(3)$
    $x = \ln\left(\frac{2}{3}\right)$

    B. $e^{x+1} = e^{2x-2}$
    $\ln(e^{x+1}) = \ln(e^{2x-2})$
    $x + 1 = 2x - 2$
    $-x + 1 = -2$
    $-x = -3$
    $x = 3$

    C. $\ln(x^2 + 3x + 5) = \ln(1 - x)$
    $x^2 + 3x + 5 = 1 - x$
    $x^2 + 4x + 4 = 0$
    $(x + 2)(x + 2) = 0$
    $x + 2 = 0$
    $x = -2$

    D. $\ln\left(\frac{x}{2}\right) = 3$
    $e^{\ln\left(\frac{x}{2}\right)} = e^3$
    $\frac{x}{2} = e^3$
    $x = 2e^3$

4. A. $2\ln^2(x) + 3 = 7\ln(x)$ Let $y = \ln(x)$
    $2y^2 + 3 = 7y$
    $2y^2 - 7y + 3 = 0$
    $(2y - 1)(y - 3) = 0$

    $2y - 1 = 0$ $y - 3 = 0$
    $2Y = 1$ $y = 3$
    $y = \frac{1}{2}$ $\ln(x) = 3$
                         $e^{\ln(x)} = e^3$
    $\ln(x) = \frac{1}{2}$ $x = e^3$
    $e^{\ln(x)} = e^{\frac{1}{2}}$
    $x = \sqrt{e}$

    B. $e^{2x} = 2e^x$ Let $y = e^x$
    $y^2 = 2y$
    $y^2 - 2y = 0$
    $(y)(y - 2) = 0$

    $y = 0$ $y - 2 = 0$
    $e^x = 0$ $y = 2$
    $\ln(e^x) = \ln(0)$ $e^x = 2$
    $x = \ln(0)$ $\ln(e^x) = \ln(2)$
    undefined $x = \ln(2)$

Lesson Practice 6C

1. $y = 2x^2$
    $x = 2y^2$ (reverse variables)
    $\frac{x}{2} = y^2$
    $\pm\sqrt{\frac{x}{2}} = y$
    $f^{-1}(x) = \pm\sqrt{\frac{x}{2}}$

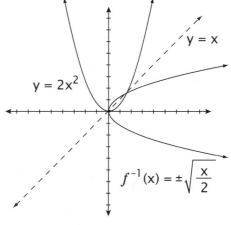

LESSON PRACTICE 6C - LESSON PRACTICE 6D

2. $e^{4x} = e$
$\ln(e^{4x}) = \ln(e)$
$4x = 1$
$x = \dfrac{1}{4}$

3. $\ln(3x - 1) = 1$
$e^{\ln(3x-1)} = e^1$
$3x - 1 = e$
$3x = e + 1$
$x = \dfrac{e+1}{3}$

4. $e^{2x} - 7e^x + 10 = 0$ let $y = e^x$
$y^2 - 7y + 10 = 0$
$(y - 5)(y - 2) = 0$

$y - 5 = 0$ $y - 2 = 0$
$y = 5$ $y = 2$
$e^x = 5$ $e^x = 2$
$\ln(e^x) = \ln(5)$ $\ln(e^x) = \ln(2)$
$x = \ln(5)$ $x = \ln(2)$

5. $\ln^2(x) = 2\ln(x)$ let $y = \ln(x)$
$y^2 = 2y$
$y^2 - 2y = 0$
$(y)(y - 2) = 0$

$y = 0$ $y - 2 = 0$
$\ln(x) = 0$ $y = 2$
$e^{\ln(x)} = e^0$ $\ln(x) = 2$
$x = e^0$ $e^{\ln(x)} = e^2$
$x = 1$ $x = e^2$

6. $e^{2x} - 3e^x + 2 = 0$
$(e^x - 1)(e^x - 2) = 0$

$e^x - 1 = 0$ $e^x - 2 = 0$
$e^x = 1$ $e^x = 2$
$x = \ln(1)$ $x = \ln(2)$
$x = 0, \ln(2)$

Lesson Practice 6D

1. $e^{2x+2} = 5$
$\ln(e^{2x+2}) = \ln(5)$
$2x + 2 = \ln(5)$
$2x = \ln(5) - 2$
$x = \dfrac{\ln(5) - 2}{2}$

2. $2e^{2x} + 5e^x = 3$ let $y = e^x$
$2y^2 + 5y = 3$
$2y^2 + 5y - 3 = 0$
$(2y - 1)(y + 3) = 0$

$2y - 1 = 0$ $y + 3 = 0$
$2y = 1$ $y = -3$
$y = \dfrac{1}{2}$ $e^x = -3$
$e^x = \dfrac{1}{2}$ $x = \ln(-3)$
$x = \ln\left(\dfrac{1}{2}\right)$ undefined
$x = \ln 1 - \ln(2)$
$x = 0 - \ln(2)$
$x = -\ln(2)$

3. $\ln(x + 2) = 2$
$e^{\ln(x+2)} = e^2$
$x + 2 = e^2$
$x = e^2 - 2$

4. $\ln(x + 1) + \ln(4) = 3$
$\ln(4(x + 1)) = 3$
$\ln(4x + 4) = 3$
$e^{\ln(4x+4)} = e^3$
$4x + 4 = e^3$
$4x = e^3 - 4$
$x = \dfrac{e^3 - 4}{4}$

5. $\ln(2x - 4) = 2$
$2x - 4 = e^2$
$2x = e^2 + 4$
$x = \dfrac{e^2 + 4}{2}$

6.

x	y
0	1
1	e^3
-1	e^{-3}
$\frac{1}{3}$	e

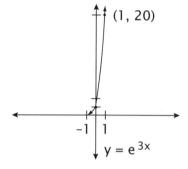

Finding the inverse

$y = e^{3x}$
$x = e^{3y}$
$\ln(x) = 3y$
$y = \frac{\ln(x)}{3}$
$f^{-1}(x) = \frac{\ln(x)}{3}$

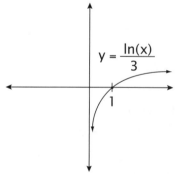

Lesson Practice 7A

1. $\lim\limits_{x \to 3}(x^2 - x + 3) =$
   $\lim\limits_{x \to 3} x^2 - \lim\limits_{x \to 3} x + \lim\limits_{x \to 3} 3 =$
   $9 - 3 + 3 = 9$

2. $\lim\limits_{x \to 2}(7 - 2x - x^2) = 7 - 4 - 4 = -1$

3. $\lim\limits_{t \to 0} \frac{t^2 - 3t + 3}{2t^2 + 2} = \frac{0 - 0 + 3}{0 + 2} = \frac{3}{2}$

4. $\lim\limits_{z \to -2} \frac{z+1}{z^2 - 2} = \frac{-2+1}{4-2} = \frac{-1}{2}$

5. $\lim\limits_{\theta \to \frac{\pi}{4}} \frac{\tan(3\theta)}{\sin(\theta)\cos(\theta)} =$

   $\lim\limits_{\theta \to \frac{\pi}{4}} \frac{\tan\left(\frac{3\pi}{4}\right)}{\sin\left(\frac{\pi}{4}\right)\cos\left(\frac{\pi}{4}\right)} =$

   $\frac{-1}{\frac{\sqrt{2}}{2} \cdot \frac{\sqrt{2}}{2}} = \frac{-1}{\frac{2}{4}} = -2$

6. $\lim\limits_{x \to 3} \frac{x^2 - 9}{x^2 - 3x}$

   When $x = 3$, the function is undefined. Factoring we get,

   $\lim\limits_{x \to 3} \frac{(x+3)\cancel{(x-3)}}{x\cancel{(x-3)}} =$

   $\lim\limits_{x \to 3} \frac{(x+3)}{x} = \frac{6}{3} = 2$

7. $\lim\limits_{x \to -3} \frac{x^2 + 5x + 6}{x + 3} =$

   $\lim\limits_{x \to -3} \frac{(x+2)\cancel{(x+3)}}{\cancel{(x+3)}} = -1$

LESSON PRACTICE 7A - LESSON PRACTICE 7C

8. $\lim_{x \to -1} \frac{x^2 + 5x + 6}{(x+1)} =$

   $\lim_{x \to -1} \frac{(x+2)(x+3)}{(x+1)} =$

   limit does not exist

9. $\lim_{\theta \to 0} (\sin(\theta) + 2\cos(\theta)) =$

   $0 + 2(1) = 2$

10. $\lim_{\theta \to 0} (4\sec(\theta) - \sin(\theta) + 3) =$

    $4(1) - 0 + 3 = 7$

7. $\lim_{t \to -2} \frac{t^2 + 3t + 2}{t^2 + t - 2}$

   $\lim_{t \to -2} \frac{(t+2)(t+1)}{(t+2)(t-1)} = \lim_{t \to -2} \frac{t+1}{t-1}$

   $= \frac{-2+1}{-2-1} = \frac{-1}{-3} = \frac{1}{3}$

8. $\lim_{\theta \to 0} \left(\frac{\sin^2(\theta) + 1}{\cos(\theta) + 1} \right) = \frac{0+1}{1+1} = \frac{1}{2}$

9. $\lim_{\theta \to 0} \left(\frac{\csc(\theta) - \cos(\theta)}{3} \right)$

   does not exist because $\csc(\theta)$ when $\theta = 0$ is undefined.

10. $\lim_{x \to e} \frac{\ln^2(x) - 1}{\ln(x) - 1}$

    $\lim_{x \to e} \frac{\ln^2(x) - 1}{\ln(x) - 1} =$

    $\lim_{x \to e} \frac{(\ln(x) - 1)(\ln(x) + 1)}{(\ln(x) - 1)} =$

    $\ln(e) + 1 = 2$

Lesson Practice 7B

1. $\lim_{x \to 2} (x^2 - 5) = \lim_{x \to 2} x^2 - \lim_{x \to 2} 5$

   $= 4 - 5 = -1$

2. $\lim_{x \to -1} \frac{x^2 + 2}{2x^2 - 3} = \frac{(-1)^2 + 2}{2(-1)^2 - 3}$

   $= \frac{3}{2-3} = -3$

3. $\lim_{\theta \to \frac{\pi}{2}} (-3\cos(\theta)) = -3(0) = 0$

4. $\lim_{x \to 1} \frac{\ln(x) + 2}{3} = \frac{\ln(1) + 2}{3} = \frac{2}{3}$

5. $\lim_{t \to 1} \frac{e^{2t} + e^2}{4} = \frac{e^{2(1)} + e^2}{4}$

   $= \frac{2e^2}{4} = \frac{1}{2}e^2$

6. $\lim_{x \to -2} \frac{x^2 + 2x}{x + 2} = \lim_{x \to -2} \frac{x(x+2)}{(x+2)}$

   $= \lim_{x \to -2} x = -2$

Lesson Practice 7C

1. $\lim_{x \to a} f(x) \cdot g(x) = 5(-3) = -15$

2. $\lim_{x \to a} 2f(x) + 3g(x) = 2(5) + 3(-3) = 1$

3. $\lim_{x \to 1} \frac{x^2 + 3x}{x + 1} = \frac{1^2 + 3(1)}{1 + 1} = \frac{4}{2} = 2$

4. $\lim_{x \to -3} \frac{2(x+3)}{x^2+4x+3} =$

$\lim_{x \to -3} \frac{2(\cancel{x+3})}{(\cancel{x+3})(x+1)} =$

$\lim_{x \to -3} \frac{2}{x+1} = -1$

5. $\lim_{x \to 1} \frac{x^3-1}{x-1} =$

$\lim_{x \to 1} \frac{(\cancel{x-1})(x^2+x+1)}{(\cancel{x-1})} =$

$\lim_{x \to 1} (x^2+x+1) = 3$

6. $\lim_{x \to 4} \frac{x+4}{x^2-16} = \lim_{x \to 4} \frac{(\cancel{x+4})}{(\cancel{x+4})(x-4)}$

$= \lim_{x \to 4} \frac{1}{x-4}$

which does not exist

7. $\lim_{x \to 0} 4 \tan(x) = 4 \cdot 0 = 0$

8. $\lim_{x \to \pi} \frac{\sin\left(\frac{x}{2}\right)}{x} = \frac{\sin\left(\frac{\pi}{2}\right)}{\pi} = \frac{1}{\pi}$

9. $\lim_{x \to \frac{3\pi}{2}} \frac{\sin^2(x) + 4\sin(x) + 3}{\sin(x) + 1}$

(now let $u = \sin(x)$)

$\lim_{x \to \frac{3\pi}{2}} \frac{u^2+4u+3}{u+1} =$

$\lim_{x \to \frac{3\pi}{2}} \frac{(u+3)(\cancel{u+1})}{(\cancel{u+1})} = \lim_{x \to \frac{3\pi}{2}} u+3$

(replace u with sin(x))

$\lim_{x \to \frac{3\pi}{2}} (\sin(x)+3) = -1+3 = 2$

10. $\lim_{x \to 1} \frac{\ln^2(x) + 5\ln(x)}{\ln(x)}$

(let $u = \ln(x)$)

$\lim_{x \to 1} \frac{u^2+5u}{u} = \lim_{x \to 1} \frac{\cancel{u}(u+5)}{\cancel{u}}$

(replace u with ln(x))

$\lim_{x \to 1}(\ln(x) + 5) = 5$

11. $\lim_{x \to 1} \frac{e^{2x}-1}{e^x-1} =$ (let $u = e^x$)

$\lim_{x \to 1} \frac{u^2-1}{u-1} = \lim_{x \to 1} \frac{(u+1)(\cancel{u-1})}{(\cancel{u-1})} =$

$\lim_{x \to 1} u+1 = \lim_{x \to 1} e^x + 1 = e+1$

Systematic Review 7D

1. $\lim_{x \to a} \frac{g(x)}{f(x)} = \frac{-4}{-2} = 2$

2. $\lim_{x \to a} [3g(x) - 2f(x)] = 3(-4) - 2(-2)$
$= -12 + 4 = -8$

3. $\lim_{x \to \frac{1}{2}} (2x-1) = 2 \cdot \frac{1}{2} - 1 = 0$

4. $\lim_{x \to -3} 3x^2 = 3(-3)^2 = 27$

5. $\lim_{x \to 0} \frac{4x^6 - 3x^3}{5x^4 - x^3} = \lim_{x \to 0} \frac{x^3(4x^3 - 3)}{x^3(5x^4 - 1)}$
$= \lim_{x \to 0} \frac{4x^3 - 3}{5x^4 - 1} = 3$

6. $\lim_{x \to -2} |x| = |-2| = 2$
from the graph, the limit is 2

7. $\lim_{x \to -1} |x + 4| = |-1 + 4| = 3$

8. $\lim_{x \to 0} \frac{3 \sec(x)}{\cos(x)} = \frac{3(1)}{1} = 3$

9. $\lim_{x \to 0} x \cos(x) = 0 \cdot 1 = 0$

10. $\lim_{x \to -5} \frac{x^2 + 6x + 5}{x + 5}$
$= \lim_{x \to -5} \frac{(x+5)(x+1)}{(x+5)}$
$= -5 + 1 = -4$

11. $\lim_{x \to 1} \frac{2e^{2x} - e^x - 1}{e^x - 1}$ (let $u = e^x$)
$= \lim_{x \to 1} \frac{2u^2 - u - 1}{u - 1}$
$= \lim_{x \to 1} \frac{(2u + 1)(u - 1)}{(u - 1)}$
$= \lim_{x \to 1} 2e^x + 1 = 2e + 1$

Lesson Practice 8A

1. $\lim_{x \to \infty} \sqrt{x} = \sqrt{\infty} = \infty$

2. $\lim_{x \to \infty} \frac{1 - x}{1 - 2x}$
Dividing by the highest power of x, we get:
$\lim_{x \to \infty} \frac{\frac{1}{x} - 1}{\frac{1}{x} - 2} = \frac{1}{2}$

3. $\lim_{x \to \infty} 10^x = 10^\infty = \infty$

4. $\lim_{x \to \infty} \sin(x)$
This limit does not exist. It takes on all values between −1 and 1.

5. $\lim_{x \to \infty} x^{\frac{3}{2}} = \infty^{\frac{3}{2}} = \infty$

6. $\lim_{x \to \infty} e^x = e^\infty = \infty$

7. $\lim_{x \to \infty} \frac{x^2 + 2}{3x^2 - 5}$
Dividing by the highest power of x, we get:
$\lim_{x \to \infty} \frac{1 + \frac{2}{x^2}}{3 - \frac{5}{x^2}} = \frac{1}{3}$

8. $\lim_{x \to \infty} \frac{2x^2 + 2}{3x^2 + x - 7}$
Dividing by the highest power of x, we get:
$\lim_{x \to \infty} \frac{2 + \frac{2}{x^2}}{3 + \frac{1}{x} - \frac{7}{x^2}} = \frac{2}{3}$

9. $\lim_{x \to 2} \frac{1}{x-2} = $ DNE

Vertical asymptote at $x = 2$

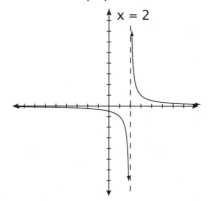

10. $\lim_{x \to 0} 2^{\frac{1}{x}} = $ DNE

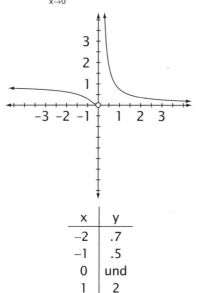

x	y
−2	.7
−1	.5
0	und
1	2
2	1.4

11. $x - 1 = 0$
    $x = 1$ is a discontinuity

12. $x^2 - 4 = 0$
    $(x-2)(x+2) = 0$
    $x = \pm 2$ are discontinuities

13. $x^2 + 5x = 0$
    $x(x+5) = 0$
    $x = 0, -5$ are discontinuities

14. $x^2 + 2x + 5 = 0$
    There are no real factors for this polynomial, so there are no discontinuities

15. There are no discontinuities for $f(\theta) = \sin(\theta)$

16. Yes, it passes the vertical line test.

17. Yes, at $x = -2, 0, 2, 3$

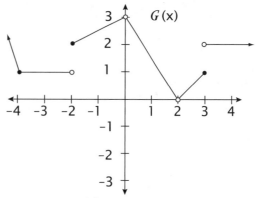

18. $\lim_{x \to -2} G(x) = $ DNE

19. $\lim_{x \to 0} G(x) = 3$

20. $\lim_{x \to 2} G(x) = 0$

21. $\lim_{x \to 3^+} G(x) = 2$

22. $\lim_{x \to 3^-} G(x) = 1$

23. $\lim_{x \to 3} G(x) = $ DNE

24. Yes, all the discontinuities are removable on [−1, 2].

25. There is a removable discontinuity at $x = 0$ and $x = 2$.
    The fixes would be $(2, 0)$ and $(0, 3)$.

Lesson Practice 8B

1. $\lim\limits_{x \to \infty} 4^{-x} = 4^{-\infty} = \dfrac{1}{4^{\infty}} = 0$

2. $\lim\limits_{x \to \infty} \dfrac{2x+3}{4x^2-7}$

   Dividing by the highest power of x, we get

   $\lim\limits_{x \to \infty} \dfrac{\frac{2}{x}+\frac{3}{x^2}}{4-\frac{7}{x^2}} = \dfrac{0}{4} = 0$

3. $\lim\limits_{x \to \infty} \dfrac{x}{3} = \dfrac{\infty}{3} = \infty$

4. $\lim\limits_{x \to 0} \dfrac{\tan^2(x)}{\sec(x)-1} = \lim\limits_{x \to 0} \dfrac{\sec^2(x)-1}{\sec(x)-1} =$

   $\lim\limits_{x \to 0} \dfrac{(\sec(x)-1)(\sec(x)+1)}{(\sec(x)-1)} =$

   $\lim\limits_{x \to 0} (\sec(x)+1) = 2$

5. $\lim\limits_{x \to 3} \dfrac{x^3-3x^2}{x-3} =$

   $\lim\limits_{x \to 3} \dfrac{x^2(x-3)}{x-3} =$

   $\lim\limits_{x \to 3} x^2 = 9$

6. $\lim\limits_{x \to 5} \dfrac{5-x}{x^2-4x-5} =$

   $\lim\limits_{x \to 5} \dfrac{5-x}{(x+1)(x-5)} =$

   $\lim\limits_{x \to 5} \dfrac{-(x-5)}{(x+1)(x-5)} =$

   $\lim\limits_{x \to 5} \dfrac{-1}{x+1} = \dfrac{-1}{6}$

7. $\lim\limits_{x \to \frac{\pi}{2}} \dfrac{\sin(2x)}{2\cos(x)} =$

   $\lim\limits_{x \to \frac{\pi}{2}} \dfrac{2\sin(x)\cos(x)}{2\cos(x)} =$

   $\lim\limits_{x \to \frac{\pi}{2}} \sin(x) = 1$

8. $\lim\limits_{x \to 4} \dfrac{x^2-3x-4}{\sqrt{x}-2} =$

   $\lim\limits_{x \to 4} \dfrac{(x^2-3x-4)}{(\sqrt{x}-2)} \cdot \dfrac{(\sqrt{x}+2)}{(\sqrt{x}+2)} =$

   $\lim\limits_{x \to 4} \dfrac{(x^2-3x-4)(\sqrt{x}+2)}{x-4} =$

   $\lim\limits_{x \to 4} \dfrac{(x-4)(x+1)(\sqrt{x}+2)}{(x-4)} =$

   $5 \cdot 4 = 20$

9. $\lim\limits_{x \to 1} \dfrac{1}{x^2-1} = \text{DNE}$

   $f(x) = \dfrac{1}{x^2-1}$

   vertical asymptotes at 1 and -1

   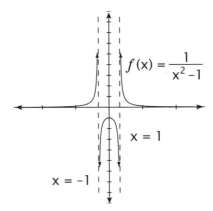

10. $x = -\dfrac{3}{2}$ has a discontinuity

    $2x+3 = 0$

    $x = -\dfrac{3}{2}$

11. $x^2 + 2x = 0$
$x(x+2) = 0$
$x = 0, -2$ are discontinuities

12. $x^2 - 1 = 0$
$(x+1)(x-1) = 0$
$x = \pm 1$ are discontinuities

13. $2x^2 - 5x - 3 = 0$
$(2x+1)(x-3) = 0$
$x = -\frac{1}{2}, 3$ are discontinuities

14. $f(\theta) = \tan(\theta)$, $0 \leq \theta \leq 2\pi$
discontinuities at $x = \frac{\pi}{2}$
and $x = \frac{3\pi}{2}$
vertical asymptotes at $\frac{\pi}{2}, \frac{3\pi}{2}$

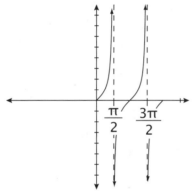

Lesson Practice 8C

1. $f(x) = \dfrac{x^2}{x^2 - 9}$

$\lim\limits_{x \to \infty^-} \dfrac{x^2}{x^2 - 9} = 1$

$\lim\limits_{x \to \infty^+} \dfrac{x^2}{x^2 - 9} = 1$

Therefore $y = 1$ is a vertical asymptote
$x = \pm 3$ makes the denominator $= 0$, so there are two horizontal asymptotes, namely $x = 3$ and $x = -3$.

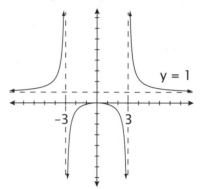

Vertical asymptotes
$x = 3$, $x = -3$
Horizontal asymptotes
$y = 1$

2. If $f(x) = \dfrac{3x-1}{|x|}$ find all the vertical and or horizontal asymptotes. Graph the function and evaluate.

$\lim\limits_{x \to 0} \dfrac{3x-1}{|x|}$

These are two cases to consider

i. $\lim\limits_{x \to \infty} \dfrac{3x-1}{x}$

ii. $\lim\limits_{x \to -\infty} \dfrac{3x-1}{-x}$

Case 1: $\lim\limits_{x \to \infty} \dfrac{3x-1}{x} = \lim\limits_{x \to \infty} \dfrac{3 - \frac{1}{x}}{1} = 3$

Case 2: $\lim\limits_{x \to -\infty} \dfrac{3x-1}{x} = \lim\limits_{x \to -\infty} \dfrac{3 - \frac{1}{x}}{-1} = -3$

Therefore there are two horizontal asymptotes at $y = 3$ and $y = -3$. Because $x = 0$ makes the denominator $= 0$, there is a vertical asymptote at $x = 0$.

LESSON PRACTICE 8C - LESSON PRACTICE 8D

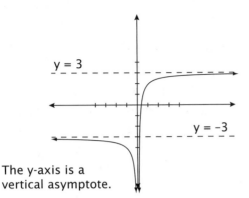

The y-axis is a vertical asymptote.

3. $\lim\limits_{x \to 2} \dfrac{x-2}{x+2} = \dfrac{2-2}{2+2} = \dfrac{0}{4} = 0$

4. $\lim\limits_{x \to \infty} x^{-3} = \lim\limits_{x \to \infty} \dfrac{1}{x^3} = 0$

5. $\lim\limits_{x \to \infty} \dfrac{3x}{7x+2} = \lim\limits_{x \to \infty} \dfrac{3}{7+\frac{2}{x}} = \dfrac{3}{7}$

6. $\lim\limits_{x \to 3} \dfrac{x-3}{x^2+2} = \dfrac{3-3}{3^2+2} = 0$

7. $\lim\limits_{x \to \infty} (e^{-x} + 2) =$

   $\lim\limits_{x \to \infty} \dfrac{1}{e^x} + 2 = 2$

8. $\lim\limits_{x \to 4} \dfrac{x-4}{\sqrt{x}-2} =$

   $\lim\limits_{x \to 4} \dfrac{x-4}{\sqrt{x}-2} \cdot \dfrac{\sqrt{x}+2}{\sqrt{x}+2} =$

   $\lim\limits_{x \to 4} \dfrac{(x-4)(\sqrt{x}+2)}{(x-4)} =$

   $\lim\limits_{x \to 4} \sqrt{x} + 2 = \sqrt{4} + 2 = 4$

9. $\lim\limits_{x \to \frac{\pi}{2}} \dfrac{3\tan^2(x)}{\sec(x)-1} =$

   $\lim\limits_{x \to \frac{\pi}{2}} \dfrac{3(\sec^2(x)-1)}{(\sec(x)-1)} =$

   $\lim\limits_{x \to \frac{\pi}{2}} \dfrac{3(\sec(x)-1)(\sec(x)+1)}{(\sec(x)-1)} =$

   $\lim\limits_{x \to \frac{\pi}{2}} 3(\sec(x)+1) =$ does not exist

10. a. 1
    b. 2
    c. does not exist
    d. 3
    e. 3
    f. 3

Lesson Practice 8D

1. $f(x) = \dfrac{4x+3}{2x+4}$

   $f(x) = \dfrac{4x+3}{2(x+2)}$

   When $x = -2$, the denominator $= 0$
   So there is a vertical asymptote at $x = -2$.

   $\lim\limits_{x \to \infty^-} \dfrac{4x+3}{2x+4} =$

   $\lim\limits_{x \to \infty^-} \dfrac{4+\frac{3}{x}}{2+\frac{4}{x}} = 2$

   Therefore there is a horizontal asymptote at $y = 2$.

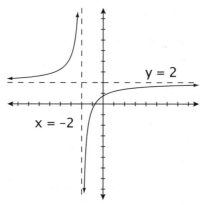

x	y
0	$\frac{3}{4}$
1	$\frac{7}{6}$
-1	$-\frac{1}{2}$
-3	$\frac{-9}{-2} = 4\frac{1}{2}$
-4	$\frac{-13}{-4} = 3\frac{1}{4}$

2. $f(x) = \dfrac{x^2 - 6x + 5}{x^2 - 4x - 5} = \dfrac{(x-1)(x-5)}{(x+1)(x-5)}$

There is a hole when $x = 5$ and a vertical asymptote at $x = -1$.

$\lim\limits_{x \to \infty} \dfrac{x^2 - 6x + 5}{x^2 - 4x - 5} = \dfrac{1 - \frac{6}{x} + \frac{5}{x^2}}{1 - \frac{4}{x} - \frac{5}{x^2}} = 1$

There is a horizontal asymptote at $y = 1$.

$\lim\limits_{x \to \infty} \dfrac{x^2 - 6x + 5}{x^2 - 4x - 5} = 1$

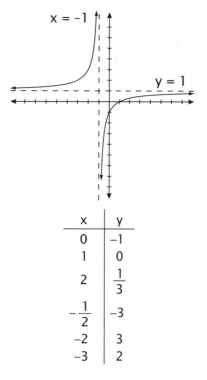

x	y
0	-1
1	0
2	$\frac{1}{3}$
$-\frac{1}{2}$	-3
-2	3
-3	2

3. $\lim\limits_{x \to \infty} \dfrac{ax+2}{3-8x}$

Will determine the horizontal asymptotes if any exist.

$= \lim\limits_{x \to \infty} \dfrac{a + \frac{2}{x}}{\frac{3}{x} - 8} = \dfrac{a}{-8}$

$\dfrac{a}{-8} = \dfrac{-5}{4}$ When $a = 10$

4. If there is a hole at $x = 3$ then $(x-3)$ is a factor of the denominator. If there is a vertical asymptote at $x = -3$, then $(x+3)$ is a factor of the denominator also. Therefore, $(x+3)(x-3) = x^2 - 9$ and $C = 9$.

5. $\lim_{x \to \infty} \cos(x)$ = does not exist

Cosine never approaches a single value. It oscillates between 1 and –1.

6. $\lim_{x \to \infty} \dfrac{2-x}{8-3x} =$

$\lim_{x \to \infty} \dfrac{\frac{2}{x}-1}{\frac{8}{x}-3} = \dfrac{1}{3}$

7. $\lim_{x \to 0} \dfrac{\sin^2(x)}{\sin(2x)} =$

$\lim_{x \to 0} \dfrac{(\cancel{\sin(x)})(\sin(x))}{2(\cancel{\sin(x)})(\cos(x))} =$

$\lim_{x \to 0} \dfrac{\sin(x)}{2\cos(x)} = 0$

8. $\lim_{x \to \infty} \dfrac{2x^2}{x^2+1} =$

$\lim_{x \to \infty} \dfrac{2}{1+\frac{1}{x^2}} = 2$

9. $\lim_{x \to 0} \dfrac{\sin^2(x)}{1-\cos(x)} =$

$\lim_{x \to 0} \dfrac{1-\cos^2(x)}{1-\cos(x)} =$

$\lim_{x \to 0} \dfrac{(\cancel{1-\cos(x)})(1+\cos(x))}{(\cancel{1-\cos(x)})} =$

$\lim_{x \to 0} 1+\cos(x) = 2$

10. a. 2
    b. 2
    c. 2
    d. 0

Lesson Practice 9A

1. find $\dfrac{dy}{dx}$

$f(x) = 4x - 7$

$f(x+h) = 4(x+h) - 7$
$ 4x + 4h - 7$

$f(x+h) - f(x) = 4x + 4h - 7 - (4x - 7) = 4h$

$\lim_{h \to 0} \dfrac{f(x+h) - f(x)}{h} =$

$\lim_{h \to 0} \dfrac{4h}{h} =$

$\lim_{h \to 0} 4 = 4$

2. $f(x) = x^2 - 2$

$f(x+h) = (x+h)^2 - 2$
$ = x^2 + 2xh + h^2 - 2$

$\lim_{h \to 0} \dfrac{f(x+h) - f(x)}{h} =$

$\lim_{h \to 0} \dfrac{(x^2 + 2xh + h^2 - 2) - (x^2 - 2)}{h} =$

$\lim_{h \to 0} \dfrac{2xh + h^2}{h} =$

$\lim_{h \to 0} 2x + h = 2x$

3. $f(x) = 2x^2 + 3x + 1$

$$f(x+h) = 2(x+h)^2 + 3(x+h) + 1$$
$$= 2(x^2 + 2xh + h^2) + 3x + 3h + 1$$
$$= 2x^2 + 4xh + 2h^2 + 3x + 3h + 1$$

$$\lim_{h \to 0} \frac{f(x+h) - f(x)}{h} =$$
$$\lim_{h \to 0} \frac{2x^2 + 4xh + 2h^2 + 3x + 3h + 1 - (2x^2 + 3x + 1)}{h} =$$
$$\lim_{h \to 0} \frac{4xh + 2h^2 + 3h}{h} =$$
$$\lim_{h \to 0} 4x + 2h + 3 = 4x + 3$$

4. $f(x) = 2x^3$

$$\lim_{h \to 0} \frac{f(x+h) - f(x)}{h} =$$
$$\lim_{h \to 0} \frac{2(x+h)^3 - 2x^3}{h} =$$
$$\lim_{h \to 0} \frac{2(x^3 + 3x^2h + 3xh^2 + h^3) - 2x^3}{h} =$$
$$\lim_{h \to 0} \frac{6x^2h + 6xh^2 + 2h^3}{h} =$$
$$\lim_{h \to 0} 6x^2 + 6xh + 2h^2 = 6x^2$$

5. $f(x) = 3x - x^2$

$$\lim_{h \to 0} \frac{f(x+h) - f(x)}{h} =$$
$$\lim_{h \to 0} \frac{3(x+h) - (x+h)^2 - 3x + x^2}{h} =$$
$$\lim_{h \to 0} \frac{3x + 3h - x^2 - 2xh - h^2 - 3x + x^2}{h} =$$
$$\lim_{h \to 0} \frac{3h - 2xh - h^2}{h} =$$
$$\lim_{h \to 0} 3 - 2x - h = 3 - 2x$$

6. $f(x) = \dfrac{1}{3x}$

$$\lim_{h \to 0} \frac{f(x+h) - f(x)}{h} =$$
$$\lim_{h \to 0} \frac{\frac{1}{3(x+h)} - \frac{1}{3x}}{h} =$$
$$\lim_{h \to 0} \frac{\frac{x - (x+h)}{3(x)(x+h)}}{h} =$$
$$\lim_{h \to 0} \frac{\frac{-h}{3(x)(x+h)}}{h} =$$
$$\lim_{h \to 0} \frac{-1}{3x(x+h)} = \frac{-1}{3x^2}$$

7. $f(x) = \dfrac{2}{x+2}$

$$\lim_{h \to 0} \frac{f(x+h) - f(x)}{h} =$$
$$\lim_{h \to 0} \frac{\frac{2}{x+h+2} - \frac{2}{x+2}}{h} =$$
$$\lim_{h \to 0} \frac{\frac{2(x+2) - 2(x+h+2)}{(x+2)(x+h+2)}}{h} =$$
$$\lim_{h \to 0} \frac{\frac{2x + 4 - 2x - 2h - 4}{(x+2)(x+h+2)}}{h} =$$
$$\lim_{h \to 0} \frac{\frac{-2h}{(x+2)(x+h+2)}}{h} =$$
$$\lim_{h \to 0} \frac{-2}{(x+2)(x+h+2)} = \frac{-2}{(x+2)^2}$$

8. $f(x) = -3\sqrt{x}$

$$\lim_{h \to 0} \frac{f(x+h) - f(x)}{h} =$$

$$\lim_{h \to 0} \frac{(-3\sqrt{x+h}) - (-3\sqrt{x})}{h} =$$

$$\lim_{h \to 0} \frac{-3\sqrt{x+h} + 3\sqrt{x}}{h} =$$

$$\lim_{h \to 0} \frac{-3\sqrt{x+h} + 3\sqrt{x}}{h} \cdot \frac{-3\sqrt{x+h} - 3\sqrt{x}}{-3\sqrt{x+h} - 3\sqrt{x}} =$$

$$\lim_{h \to 0} \frac{9(x+h) - 9x}{h \cdot (-3\sqrt{x+h} - 3\sqrt{x})} =$$

$$\lim_{h \to 0} \frac{9h}{h(-3\sqrt{x+h} - 3\sqrt{x})} =$$

$$\lim_{h \to 0} \frac{9}{-3\sqrt{x+h} - 3\sqrt{x}} =$$

$$\frac{9}{-3\sqrt{x} - 3\sqrt{x}} =$$

$$\frac{9}{-6\sqrt{x}} = \frac{-3}{2\sqrt{x}}$$

Lesson Practice 9B

1. $f(x) = 6 - 2x$
$f(x+h) = 6 - 2(x+h)$
$= 6 - 2x - 2h$

$$\lim_{h \to 0} \frac{f(x+h) - f(x)}{h} =$$

$$\lim_{h \to 0} \frac{6 - 2x - 2h - (6 - 2x)}{h} =$$

$$\lim_{h \to 0} \frac{-2h}{h} =$$

$$\lim_{h \to 0} -2 = -2$$

2. $f(x) = 3x^2 + 7$

$$\lim_{h \to 0} \frac{f(x+h) - f(x)}{h} =$$

$$\lim_{h \to 0} \frac{[3(x+h)^2 + 7] - (3x^2 + 7)}{h} =$$

$$\lim_{h \to 0} \frac{3(x^2 + 2xh + h^2) + 7 - 3x^2 - 7}{h} =$$

$$\lim_{h \to 0} \frac{6xh + 3h^2}{h} =$$

$$\lim_{h \to 0} 6x + 3h = 6x$$

3. $f(x) = 4x^2 - x + 2$

$$\lim_{h \to 0} \frac{f(x+h) - f(x)}{h} =$$

$$\lim_{h \to 0} \frac{[4(x+h)^2 - (x+h) + 2] - [4x^2 - x + 2]}{h} =$$

$$\lim_{h \to 0} \frac{4(x^2 + 2xh + h^2) - x - h + 2 - 4x^2 + x - 2}{h} =$$

$$\lim_{h \to 0} \frac{4x^2 + 8xh + 4h^2 - h - 4x^2}{h} =$$

$$\lim_{h \to 0} \frac{8xh + 4h^2 - h}{h} =$$

$$\lim_{h \to 0} 8x + 4h - 1 = 8x - 1$$

4. $f(x) = 1 - 2x^3$

$$\lim_{h \to 0} \frac{f(x+h) - f(x)}{h} =$$

$$\lim_{h \to 0} \frac{[1 - 2(x+h)^3] - [1 - 2x^3]}{h} =$$

$$\lim_{h \to 0} \frac{1 - 2(x^3 + 3x^2h + 3xh^2 + h^3) - 1 + 2x^3}{h} =$$

$$\lim_{h \to 0} \frac{-2x^3 - 6x^2h - 6xh^2 - 2h^3 + 2x^3}{h} =$$

$$\lim_{h \to 0} \frac{-6x^2h - 6xh^2 - 2h^3}{h} =$$

$$\lim_{h \to 0} -6x^2 - 6xh - 2h^2 = -6x^2$$

5. $f(x) = 2\sqrt{x}$

$$\lim_{h \to 0} \frac{f(x+h) - f(x)}{h} =$$

$$\lim_{h \to 0} \frac{2\sqrt{x+h} - 2\sqrt{x}}{h} =$$

$$\lim_{h \to 0} \frac{2\sqrt{x+h} - 2\sqrt{x}}{h} \cdot \left(\frac{2\sqrt{x+h} + 2\sqrt{x}}{2\sqrt{x+h} + 2\sqrt{x}}\right) =$$

$$\lim_{h \to 0} \frac{4(x+h) - 4x}{h(2\sqrt{x+h} + 2\sqrt{x})} =$$

$$\lim_{h \to 0} \frac{4x + 4h - 4x}{h(2\sqrt{x+h} + 2\sqrt{x})} =$$

$$\lim_{h \to 0} \frac{4h}{h(2\sqrt{x+h} + 2\sqrt{x})} =$$

$$\lim_{h \to 0} \frac{4h}{h(2\sqrt{x+h} + 2\sqrt{x})} =$$

$$\lim_{h \to 0} \frac{2}{\sqrt{x+h} + \sqrt{x}} =$$

$$\frac{2}{2\sqrt{x}} = \frac{1}{\sqrt{x}}$$

6. $f(x) = \frac{3}{x}$

$$\lim_{h \to 0} \frac{f(x+h) - f(x)}{h} =$$

$$\lim_{h \to 0} \frac{\frac{3}{x+h} - \frac{3}{x}}{h} =$$

$$\lim_{h \to 0} \frac{\frac{3x - 3(x+h)}{x(x+h)}}{h} =$$

$$\lim_{h \to 0} \frac{\frac{3x - 3x - 3h}{x(x+h)}}{h} =$$

$$\lim_{h \to 0} \frac{\frac{-3h}{x(x+h)}}{h} =$$

$$\lim_{h \to 0} \frac{-3}{x(x+h)} = \frac{-3}{x^2}$$

7. $f(x) = \frac{-2}{1-x}$

$$\lim_{h \to 0} \frac{f(x+h) - f(x)}{h} =$$

$$\lim_{h \to 0} \frac{\frac{-2}{1-(x+h)} - \frac{-2}{1-x}}{h} =$$

$$\lim_{h \to 0} \frac{\frac{-2}{1-x-h} + \frac{2}{1-x}}{h} =$$

$$\lim_{h \to 0} \frac{\frac{-2(1-x) + 2(1-x-h)}{(1-x-h)(1-x)}}{h} =$$

$$\lim_{h \to 0} \frac{-2}{(1-x-h)(1-x)} = \frac{-2}{(1-x)^2}$$

LESSON PRACTICE 9B - LESSON PRACTICE 9C

8. $f(x) = \dfrac{1}{x} + 2x$

$$\lim_{h \to 0} \frac{f(x+h) - f(x)}{h} =$$

$$\lim_{h \to 0} \frac{\left[\dfrac{1}{x+h} + 2(x+h)\right] - \left(\dfrac{1}{x} + 2x\right)}{h} =$$

$$\lim_{h \to 0} \frac{\dfrac{1}{x+h} + 2x + 2h - \dfrac{1}{x} - 2x}{h} =$$

$$\lim_{h \to 0} \frac{\dfrac{1}{x+h} - \dfrac{1}{x} + 2h}{h} =$$

$$\lim_{h \to 0} \left(\dfrac{\dfrac{1}{x+h} - \dfrac{1}{x}}{h}\right) + \lim_{h \to 0} \frac{2h}{h} =$$

$$\lim_{h \to 0} \left(\dfrac{\dfrac{x - (x+h)}{x(x+h)}}{h}\right) + 2 =$$

$$\lim_{h \to 0} \dfrac{\dfrac{x - x - h}{x(x+h)}}{h} + 2 =$$

$$\lim_{h \to 0} \dfrac{\dfrac{-h}{x(x+h)}}{h} + 2 =$$

$$\lim_{h \to 0} \dfrac{-1}{x(x+h)} + 2 = \dfrac{-1}{x^2} + 2$$

Lesson Practice 9C

1. $f(x) = 3x + 2$

$$\lim_{h \to 0} \frac{f(x+h) - f(x)}{h} =$$

$$\lim_{h \to 0} \frac{3(x+h) + 2 - (3x+2)}{h} =$$

$$\lim_{h \to 0} \frac{3x + 3h + 2 - 3x - 2}{h} =$$

$$\lim_{h \to 0} \frac{3h}{h} = 3$$

2. $f(x) = 5 - 2x^2$

$$\lim_{h \to 0} \frac{f(x+h) - f(x)}{h} =$$

$$\lim_{h \to 0} \frac{\left[5 - 2(x+h)^2\right] - (5 - 2x^2)}{h} =$$

$$\lim_{h \to 0} \frac{5 - 2x^2 - 4xh - 2h^2 - 5 + 2x^2}{h} =$$

$$\lim_{h \to 0} -4x - 2h = -4x$$

3. $f(x) = \dfrac{2}{(x+1)}$

$$\lim_{h \to 0} \frac{\dfrac{2}{x+h+1} - \dfrac{2}{x+1}}{h} =$$

$$\lim_{h \to 0} \frac{\dfrac{2(x+1) - 2(x+h+1)}{(x+1)(x+h+1)}}{h} =$$

$$\lim_{h \to 0} \frac{\dfrac{2x + 2 - 2x - 2h - 2}{(x+1)(x+h+1)}}{h} =$$

$$\lim_{h \to 0} \frac{\dfrac{-2h}{(x+1)(x+h+1)}}{h} =$$

$$\lim_{h \to 0} \frac{-2}{(x+1)(x+h+1)} = \frac{-2}{(x+1)^2}$$

4. $$\lim_{x \to 3} \frac{2x^2 - 18}{x^2 - 3x} =$$

$$\lim_{x \to 3} \frac{2(x^2 - 9)}{x(x^2 - 3)} = 0$$

$$\lim_{x \to 3} \frac{2(x-3)(x+3)}{x(x-3)} =$$

$$\lim_{x \to 3} \frac{2(x+3)}{x} = \frac{12}{3} = 4$$

5. $$\lim_{x \to \infty} \frac{2x + 7}{4 - 10x} =$$

$$\lim_{x \to \infty} \frac{2 + \dfrac{7}{x}}{\dfrac{4}{x} - 10} = -\frac{1}{5}$$

6. $\lim\limits_{x \to \infty} 3^x = 3^{-\infty} = \dfrac{1}{3^\infty} = 0$

7. $d = \sqrt{(5-2)^2 + (3-1)^2}$
   $= \sqrt{9+4}$
   $= \sqrt{13}$

8. $d = \sqrt{(-2+1)^2 + (-2-3)^2}$
   $= \sqrt{1+25}$
   $= \sqrt{26}$

9. $f(x) = x - [x]$ $[-2,2]$

x	y
-2	0
-1.5	.5
-1.2	.8
-1	0
-.5	.5
0	0
.2	.2
.8	.8
1	0
2	0

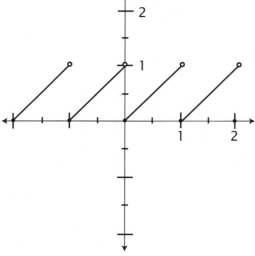

10. $f(x) = 3[x]$ $[-2,2]$

x	y
-2	-6
-1.9	-6
-1.2	-6
-1	-3
0	0
1	3
1.1	3
1.9	3
2	6

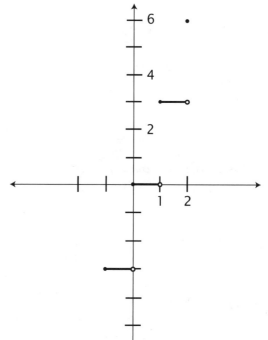

Lesson Practice 9D

1. $f(x) = -2x$

$$\lim_{h \to 0} \frac{f(x+h) - f(x)}{h} =$$

$$\lim_{h \to 0} \frac{-2(x+h) - (-2x)}{h} =$$

$$\lim_{h \to 0} \frac{-2x - 2h + 2x}{h} =$$

$$\lim_{h \to 0} \frac{-2h}{h} = -2 = -2$$

2. $f(x) = 3x^2 - 2x + 4$

$$\lim_{h \to 0} \frac{f(x+h) - f(x)}{h} =$$

$$\lim_{h \to 0} \frac{[3(x+h)^2 - 2(x+h) + 4] - (3x^2 - 2x + 4)}{h} =$$

$$\lim_{h \to 0} \frac{3x^2 + 6xh + 3h^2 - 2x - 2h + 4 - 3x^2 + 2x - 4}{h} =$$

$$\lim_{h \to 0} \frac{6xh + 3h^2 - 2h}{h} =$$

$$\lim_{h \to 0} 6x + 3h - 2 = 6x - 2$$

3. $f(x) = 2\sqrt{x}$

$$\lim_{h \to 0} \frac{f(x+h) - f(x)}{h} =$$

$$\lim_{h \to 0} \frac{2\sqrt{x+h} - 2\sqrt{x}}{h} =$$

$$\lim_{h \to 0} \frac{(2\sqrt{x+h} - 2\sqrt{x})}{h} \cdot \frac{(2\sqrt{x+h} + 2\sqrt{x})}{(2\sqrt{x+h} + 2\sqrt{x})} =$$

$$\lim_{h \to 0} \frac{4(x+h) - 4x}{h(2\sqrt{x+h} + 2\sqrt{x})} =$$

$$\lim_{h \to 0} \frac{4h}{h(2\sqrt{x+h} + 2\sqrt{x})} =$$

$$\frac{4}{4\sqrt{x}} = \frac{1}{\sqrt{x}}$$

4. $\lim_{x \to 2} \frac{3x^2}{x-2} = $ does not exist

The denominator will always be zero, so no limit exists at $x = 2$.

5. $\lim_{x \to \infty} \frac{3}{\sqrt{x}} = 0$

Because the denominator becomes infinitely large.

6. $\lim_{x \to 0} \frac{1 - \cos^2(x)}{\tan^2(x)} =$

$$\lim_{x \to 0} \frac{\sin^2(x)}{\frac{\sin^2(x)}{\cos^2(x)}} =$$

$$\lim_{x \to 0} \cos^2(x) = 1$$

7. $d = \sqrt{(-5-3)^2 + (-2-0)^2}$
$= \sqrt{64 + 4} = \sqrt{68} = 2\sqrt{17}$

8. $d = \sqrt{(-7+3)^2 + (1+2)^2}$
$= \sqrt{16 + 9} = \sqrt{25} = 5$

9. $f(x) = [x] - 1$ $[-2, 2]$

x	$f(x)$
-2	-3
-1.5	-3
-1.2	-3
-1	-2
0	-1
.8	-1
1	0
2	1

LESSON PRACTICE 9D - LESSON PRACTICE 10A

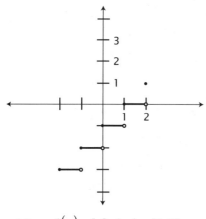

10. $f(x) = [x] - |x|$ $[0,3]$

x	$f(x)$
0	0
.5	−.5
.7	−.7
1	0
1.2	−.2
1.8	−.8
2	0
3	0

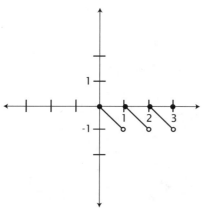

Lesson Practice 10A

1. $y = 6x - 3$

$$y' = \frac{d}{dx}(6x) - \frac{d}{dx}(3)$$
$$= 6 \cdot \frac{d}{dx}(x)$$
$$= 6$$

2. $y = 5x^2 - 4x + 1$

$$y' = \frac{d}{dx}(5x^2) - \frac{d}{dx}(4x) + \frac{d}{dx}(1)$$
$$= 5\frac{d}{dx}(x^2) - 4\frac{d}{dx}(x) + 0$$
$$= 5(2x) - 4(1)$$
$$= 10x - 4$$

3. $y = (4 - 3t)^2$

$$y' = 2(4 - 3t) \cdot \frac{d}{dt}(4 - 3t)$$
$$= 2(4 - 3t) \cdot \left[\frac{d}{dt}4 - \frac{d}{dt}(3t)\right]$$
$$= 2(4 - 3t) \cdot \left[0 - 3\frac{d}{dt}(t)\right]$$
$$= 2(4 - 3t) \cdot (-3)$$
$$= -6(4 - 3t)$$

Hopefully now you can differentiate constants and lines quickly. The solutions from here on will make that assumption.

4. $y = (5 + 2t)^3$

$$y' = 3(5 + 2t)^2 \cdot \frac{d}{dt}(5 + 2t)$$
$$= 3(5 + 2t)^2 \cdot [0 + 2]$$
$$= 6(5 + 2t)^2$$

5. $y = 2z^6$

$$y' = 2 \cdot \frac{d}{dz}(z^6)$$
$$= 2 \cdot (6z^5) = 12z^5$$

LESSON PRACTICE 10A - LESSON PRACTICE 10B

6. $y = \dfrac{2}{3x+2}$

$y' = \dfrac{(3x+2) \cdot \frac{d}{dx}(2) - 2 \cdot \frac{d}{dx}(3x-2)}{(3x+2)^2}$

$= \dfrac{-2\left[\frac{d}{dx}(3x) - \frac{d}{dx}(2)\right]}{(3x+2)^2}$

$= \dfrac{-2(3)}{(3x+2)^2} = \dfrac{-6}{(3x+2)^2}$

OR: $y = 2(3x+2)^{-1}$
using the product rule

$y' = 2 \cdot \dfrac{d}{dx}(3x+2)^{-1}$

$= 2\left[-(3x+2)^{-2} \dfrac{d}{dx}(3x+2)\right]$

$= 2\left[-(3x+2)^{-2}(3)\right]$

$= \dfrac{-6}{(3x+2)^2}$

7. $y = \sqrt{7-3x^2} \qquad y = (7-3x^2)^{\frac{1}{2}}$

$y' = \dfrac{1}{2}(7-3x^2)^{-\frac{1}{2}} \cdot \dfrac{d}{dx}(7-3x^2)$

$= \dfrac{1}{2}(7-3x^2)^{-\frac{1}{2}}\left[\dfrac{d}{dx}(7) - \dfrac{d}{dx}(3x^2)\right]$

$= \dfrac{1}{2}(7-3x^2)^{-\frac{1}{2}}(-6x)$

$= \dfrac{-3x}{\sqrt{7-3x^2}}$

8. $z = \dfrac{1}{\sqrt{x}} \qquad z = x^{-\frac{1}{2}}$

$z'(x) = -\dfrac{1}{2}x^{-\frac{3}{2}} = \dfrac{-1}{2x^{\frac{3}{2}}}$

9. $y = \sqrt[3]{5x^2+2} = (5x^2+2)^{\frac{1}{3}}$

$y' = \dfrac{1}{3}(5x^2+2)^{-\frac{2}{3}} \cdot \dfrac{d}{dx}(5x^2+2)$

$= \dfrac{1}{3}(5x^2+2)^{-\frac{2}{3}}(10x)$

$= \dfrac{10x}{3(5x^2+2)^{\frac{2}{3}}}$

10. $y = \dfrac{\sqrt{x}}{2} - \dfrac{2}{\sqrt{x}}$

$y = \dfrac{1}{2}x^{\frac{1}{2}} - 2x^{-\frac{1}{2}}$

$y' = \dfrac{1}{2} \cdot \dfrac{d}{dx}\left(x^{\frac{1}{2}}\right) - 2 \cdot \dfrac{d}{dx}\left(x^{-\frac{1}{2}}\right)$

$= \dfrac{1}{2}\left(\dfrac{1}{2}x^{-\frac{1}{2}}\right) - 2\left(-\dfrac{1}{2}x^{-\frac{3}{2}}\right)$

$= \dfrac{1}{4\sqrt{x}} + \dfrac{1}{x^{\frac{3}{2}}}$

Lesson Practice 10B

1. $y = -2x$

$y' = \dfrac{d}{dx}(-2x) = -2\dfrac{d}{dx}(x) = -2$

2. $y = 2 - 3x^2$

$y' = \dfrac{d}{dx}(2) - \dfrac{d}{dx}(3x^2)$

$= 0 - 3\dfrac{d}{dx}(x^2)$

$= -3(2x) = -6x$

3. $y = (2x+3)^3$

$y' = 3(2x+3)^2 \cdot \dfrac{d}{dx}(2x+3)$

$= 3(2x+3)^2 \cdot 2$

$= 6(2x+3)^2$

4. $f(x) = -\dfrac{2}{x^2}$

$f(x) = -2(x^{-2})$

$f'(x) = -2 \cdot \dfrac{d}{dx}(x^{-2})$

$ = -2(-2x^{-3}) = \dfrac{4}{x^3}$

5. $f(x) = x^2(x-1)$

$f'(x) = x^2 \cdot \dfrac{d}{dx}(x-1) + (x-1) \cdot \dfrac{d}{dx}(x^2)$

$ = x^2(1) + (x-1)(2x)$

$ = x^2 + 2x^2 - 2x$

$ = 3x^2 - 2x$

6. $y = 3x^7 - 2x^5$

$y' = 21x^6 - 10x^4$

7. $y = \sqrt{5x^2 - 3x}$

$y' = \dfrac{1}{2}(5x^2 - 3x)^{-\frac{1}{2}}\left[\dfrac{d}{dx}(5x^2 - 3x)\right]$

$ = \dfrac{1}{2}(5x^2 - 3x)^{-\frac{1}{2}}\left[\dfrac{d}{dx}(5x^2) - \dfrac{d}{dx}(3x)\right]$

$ = \dfrac{1}{2}(5x^2 - 3x)^{-\frac{1}{2}}\left[5 \cdot \dfrac{d}{dx}(x^2) - 3\right]$

$ = \dfrac{1}{2}(5x^2 - 3x)^{-\frac{1}{2}}[5(2x) - 3]$

$ = \dfrac{1}{2}(5x^2 - 3x)^{-\frac{1}{2}}(10x - 3)$

$ = \dfrac{10x - 3}{2(5x^2 - 3x)^{\frac{1}{2}}} = \dfrac{10x - 3}{2\sqrt{5x^2 - 3x}}$

8. $y = \sqrt{2x}\,(4 - 2x)$

$y' = \sqrt{2x} \cdot \dfrac{d}{dx}(4 - 2x) + (4 - 2x) \cdot \dfrac{d}{dx}(2x)^{\frac{1}{2}}$

$ = \sqrt{2x}\left[\dfrac{d}{dx}(4) - \dfrac{d}{dx}(2x)\right]$

$ + (4 - 2x) \cdot \dfrac{1}{2}(2x)^{-\frac{1}{2}} \cdot \dfrac{d}{dx}(2x)$

$ = \sqrt{2x}\,(-2) + (4 - 2x) \cdot \dfrac{1}{2}(2x)^{-\frac{1}{2}}(2)$

$ = -2\sqrt{2x} + \dfrac{(4 - 2x)}{\sqrt{2x}}$

9. $y = (5x - 8x^2)^3$

$y' = 3(5x - 8x^2)^2 \cdot \dfrac{d}{dx}(5x - 8x^2)$

$y' = 3(5x - 8x^2)^2 \cdot \dfrac{d}{dx}(5x - 8x^2)$

$ = 3(5x - 8x^2)^2 \cdot \left[\dfrac{d}{dx}(5x) - \dfrac{d}{dx}(8x^2)\right]$

$ = 3(5x - 8x^2)^2\,[5 - 16x]$

10. $y = (2x + 1)^{\frac{5}{2}}$

$y' = \dfrac{5}{2}(2x + 1)^{\frac{3}{2}} \cdot \dfrac{d}{dx}(2x + 1)$

$ = 5(2x + 1)^{\frac{3}{2}}$

Lesson Practice 10C

1. $y = 3 - 5x$

$$\frac{dy}{dx} = \lim_{h \to 0} \frac{f(x+h) - f(x)}{h} =$$

$$\lim_{h \to 0} \frac{[3 - 5(x+h)] - [3 - 5x]}{h} =$$

$$\lim_{h \to 0} \frac{3 - 5x - 5h - 3 + 5x}{h} =$$

$$\lim_{h \to 0} \frac{-5h}{h} =$$

$$\lim_{h \to 0} -5 = -5$$

$y = 3 - 5x$;

$$\frac{dy}{dx} = \frac{d}{dx}(3) - \frac{d}{dx}(5x) = 0 - 5 = -5$$

2. $y = 2x^2 - x$

$$\frac{dy}{dx} = \lim_{h \to 0} \frac{f(x+h) - f(x)}{h} =$$

$$\lim_{h \to 0} \frac{[2(x+h)^2 - (x+h)] - (2x^2 - x)}{h} =$$

$$\lim_{h \to 0} \frac{2x^2 + 4xh + 2h^2 - x - h - 2x^2 + x}{h} =$$

$$\lim_{h \to 0} \frac{4xh + 2h^2 - h}{h} =$$

$$\lim_{h \to 0} 4x + 2h - 1 = 4x - 1$$

$y = 2x^2 - x$;

$$\frac{dy}{dx} = \frac{d}{dx}(2x^2) - \frac{d}{dx}(x)$$

$$\frac{dy}{dx} = 2\frac{d}{dx}(x^2) - 1 = 2(2x) - 1 = 4x - 1$$

3. $y = 1 - 4x - x^2$

Method 1:

$$y = -(x^2 + 4x) + 1$$
$$-4 + y = -(x^2 + 4x + 4) + 1$$
$$y - 4 = -(x + 2)^2 + 1$$
$$y = -(x + 2)^2 + 5$$

Vertex $(-2, 5)$

Method 2:

$$\frac{dy}{dx} = -4 - 2x$$
$$-4 - 2x = 0$$
$$4 = -2x$$
$$x = -2$$

Substituting, we get $(-2, 5)$ for the vertex.

4. $y = 3x^2 + 6x + 4$

Method 1:

$$3 + y = 3(x^2 + 2x + 1) + 4$$
$$y = 3(x + 1)^2 + 1$$

vertex $(-1, 1)$

$$\frac{dy}{dx} = 6x + 6$$
$$6x + 6 = 0$$
$$6x = -6$$
$$x = -1$$

Substituting, we get $(-1, 1)$ for our vertex.

LESSON PRACTICE 10C

5. $y = x^2 + 12x - 3$

$\frac{dy}{dx} = 2x + 12$

Tangents parallel to the x-axis have a slope of 0.

$2x + 12 = 0$
$2x = -12$
$x = -6$

$(-6)^2 + 12(-6) - 3 = 36 - 72 - 3 = -39$
$(-6, -39)$

6. $y = \frac{1}{3}x^3 - 3x^2 + 5x$

$\frac{dy}{dx} = x^2 - 6x + 5$

$x^2 - 6x + 5 = 0$
$(x - 5)(x - 1) = 0$
$x = 1, 5$

$y(1) = \frac{1}{3} - 3 + 5 = 2\frac{1}{3}$

$y(5) = \frac{1}{3}(5)^3 - 3(5)^2 + 5(5)$

$= \frac{125}{3} - 75 + 25$

$= 41\frac{2}{3} - 50$

$= -8\frac{1}{3}$

Points where tangent is parallel to the x-axis are:

$\left(1, 2\frac{1}{3}\right)$ $\left(5, -8\frac{1}{3}\right)$

7. $y = \frac{x^2}{2 - x^2}$

$\frac{dy}{dx} = \frac{(2-x^2)(2x) - x^2(-2x)}{(2-x^2)^2}$

$= \frac{4x - 2x^3 + 2x^3}{(2-x^2)^2} = \frac{4x}{(2-x^2)^2}$

$\frac{4x}{(2-x^2)^2} = 0$ when $x = 0$

$y(0) = \frac{0}{2} = 0$

Point (0,0) is where the tangent line is parallel to the x-axis.

8. $\left(\frac{0+2}{2}, \frac{1+5}{2}\right)$ midpoint is (1,3)

9. $\left(\frac{-3+(-5)}{2}, \frac{-1+6}{2}\right) = \left(-4, \frac{5}{2}\right)$

10. $d = 5$, then $5 = \sqrt{(4-a)^2 + (2-5)^2}$

$5 = \sqrt{(4-a)^2 + 9}$
$25 = (4-a)^2 + 9$
$16 = (4-a)^2$
$\pm 4 = 4 - a$
$a = 0, 8$

LESSON PRACTICE 10D - LESSON PRACTICE 10D

Lesson Practice 10D

1. $y = 7 - 8x^3$

$$\lim_{h \to 0} \frac{f(x+h) - f(x)}{h} =$$

$$\lim_{h \to 0} \frac{[7 - 8(x+h)^3] - [7 - 8x^3]}{h} =$$

$$\lim_{h \to 0} \frac{[7 - 8[x^3 + 3x^2h + 3xh^2 + h^3] - 7 + 8x^3]}{h} =$$

$$\lim_{h \to 0} \frac{-24x^2h - 24xh^2 - 8h^3}{h} =$$

$$\lim_{h \to 0} (-24x^2 - 24xh - 8h^2) = -24x^2$$

$y = 7 - 8x^3$

$\frac{dy}{dx} = -8\left(\frac{d}{dx}x^3\right) = -24x^2$

2. $y = \frac{2x}{1 - 2x}$

$$\lim_{h \to 0} \frac{\frac{2x + 2h}{1 - 2x - 2h} - \frac{2x}{1 - 2x}}{h} =$$

$$\lim_{h \to 0} \frac{\frac{(2x + 2h)(1 - 2x) - 2x(1 - 2x - 2h)}{(1 - 2x - 2h)(1 - 2x)}}{h} =$$

$$\lim_{h \to 0} \frac{\frac{2x - 4x^2 + 2h - 4hx - 2x + 4x^2 + 4xh}{(1 - 2x - 2h)(1 - 2x)}}{h} =$$

$$\lim_{h \to 0} \frac{\frac{2h}{(1 - 2x - h)(1 - 2x)}}{h} =$$

$$\lim_{h \to 0} \frac{2}{(1 - 2x - h)(1 - 2x)} = \frac{2}{(1 - 2x)^2}$$

$y = \frac{2x}{1 - 2x}$

$\frac{dy}{dx} = \frac{(1 - 2x)(2) - 2x(-2)}{(1 - 2x)^2}$

$= \frac{2 - 4x + 4x}{(1 - 2x)^2} = \frac{2}{(1 - 2x)^2}$

3. $y = -(x^2 + 6x) - 12$
$-9 + y = -(x^2 + 6x + 9) - 12$
$y = -(x + 3)^2 - 3$
vertex $(-3, -3)$

$y = -x^2 - 6x - 12$
$\frac{dy}{dx} = -2x - 6$
$-2x - 6 = 0$
$6 = -2x$
$x = -3$

$y(-3) = (-3)^2 - 6(-3) - 12 = -3$
vertex $(-3, -3)$

4. $y = 3x^2 + 6x$
$3 + y = 3(x^2 + 2x + 1)$
$y = 3(x + 1)^2 - 3$
vertex $(-1, -3)$

$\frac{dy}{dx} = 6x + 6$
$6x + 6 = 0$
$6x = -6$
$x = -1$

$y(-1) = 3(-1)^2 + 6(-1)$
$= -3$
vertex $(-1, -3)$

5.
$y = (2x+1)(x^2)$
$y = 2x^3 + x^2$
$\dfrac{dy}{dx} = 6x^2 + 2x$

$6x^2 + 2x = 0$
$2x(3x+1) = 0$
$x = 0, -\dfrac{1}{3}$

$y(0) = 0$
$y\left(-\dfrac{1}{3}\right) = \left[2\left(-\dfrac{1}{3}\right)+1\right]\left(-\dfrac{1}{3}\right)^2$
$= \left(-\dfrac{2}{3}+1\right)\left(\dfrac{1}{9}\right)$
$= \left(\dfrac{1}{3}\right)\left(\dfrac{1}{9}\right) = \left(\dfrac{1}{27}\right)$

$(0, 0) \qquad \left(-\dfrac{1}{3}, \dfrac{1}{27}\right)$

6. $y = \sqrt{x} = x^{\frac{1}{2}}$
$\dfrac{dy}{dx} = \dfrac{1}{2}x^{-\frac{1}{2}}$
$= \dfrac{1}{2\sqrt{x}}$

$\dfrac{1}{2\sqrt{x}} = 0$ nowhere.

There is no place where the tanget is equal to zero.

7. $y = (2+x^2)^3$
$\dfrac{dy}{dx} = 3(2+x^2)^2 \cdot \dfrac{d}{dx}(2+x^2)$

$3(2+x^2)^2 \cdot (2x) = 6x(2+x^2)^2$
The derivative is equal to zero when $x = 0$.
$y(0) = 8$
$(0, 8)$ is where the tangent is parallel to the x-axis.

8. $(7, -2)\ (-1, 0)$
$\left(\dfrac{7-1}{2}, \dfrac{-2+0}{2}\right) = (3, -1)$

9. $(-4, 1)\ (-3, 6)$
$\left(\dfrac{-4-3}{2}, \dfrac{1+6}{2}\right) = \left(-\dfrac{7}{2}, \dfrac{7}{2}\right)$

10. $(1, -2)\ (8, a)\ d = 5\sqrt{2}$
$5\sqrt{2} = \sqrt{(a+2)^2 + (8-1)^2}$
$5\sqrt{2} = \sqrt{(a+2)^2 + 49}$
$50 = (a+2)^2 + 49$

$1 = (a+2)^2$
$\pm 1 = a+2$
$a = -1\text{ or }-3$

Lesson Practice 11A

1. $y = 5u^2$ and $u = 1-x$
$\dfrac{dy}{dx} = \dfrac{dy}{du} \cdot \dfrac{du}{dx}$
$= 10u \cdot (-1)$
$= 10(1-x)(-1)$
$= 10(x-1)$
Check: $y = 5(1-x)^2$
$= 5(1-2x+x^2)$
$= 5 - 10x + 5x^2$
$y' = -10 + 10x$
$= 10(x-1)$

2. $y = 2 - 3u\quad u = 1 - 2x$
$\dfrac{dy}{dx} = \dfrac{dy}{du} \cdot \dfrac{du}{dx}$
$= (-3)(-2) = 6$
check: $y = 2 - 3(1-2x)$
$= 2 - 3 + 6x$
$y' = 6$

LESSON PRACTICE 11A - LESSON PRACTICE 11A

3. $y = \sqrt{u} \qquad u = x^2 + 1$

    $\dfrac{dy}{dx} = \dfrac{dy}{du} \cdot \dfrac{du}{dx}$

    $= \dfrac{1}{2} u^{-\frac{1}{2}} (2x)$

    $= \dfrac{x}{\sqrt{u}} = \dfrac{x}{\sqrt{x^2+1}}$

    check: $y = \sqrt{x^2+1}$

    $\dfrac{dy}{dx} = \dfrac{1}{2}(x^2+1)^{-\frac{1}{2}}(2x)$

    $= \dfrac{x}{\sqrt{x^2+1}}$

4. $y = 2v^3 - 1 \qquad v = x^2 + 2$

    $\dfrac{dy}{dx} = \dfrac{dy}{dv} \cdot \dfrac{dv}{dx}$

    $= (6v^2)(2x)$

    $= 12xv^2$

    $= 12x(x^2+2)^2$

5. $y = 2\sqrt{2v+1} \qquad v = 2x^3$

    $\dfrac{dy}{dx} = \dfrac{dy}{dv} \cdot \dfrac{dv}{dx}$

    $y = 2(2v+1)^{\frac{1}{2}} \qquad v = 2x^3$

    $\dfrac{dy}{dv} = 2 \cdot \dfrac{1}{2}(2v+1)^{-\frac{1}{2}} \cdot \dfrac{d}{dv}(2v+1)$

    $\dfrac{dv}{dx} = 6x^2$

    $\dfrac{dy}{dx} = (2v+1)^{-\frac{1}{2}}(2)(6x^2) =$

    $\dfrac{12x^2}{\sqrt{2v+1}} = \left(\dfrac{12x^2}{\sqrt{4x^3+1}}\right)$

6. $y = u^6, \quad u = 1+2\sqrt{x} \qquad u = 1+2x^{\frac{1}{2}}$

    $\dfrac{dy}{du} = 6u^5 \qquad\qquad \dfrac{du}{dx} = x^{-\frac{1}{2}}$

    $\dfrac{dy}{du} \cdot \dfrac{du}{dx} = (6u^5)(x^{-\frac{1}{2}}) =$

    $\dfrac{6u^5}{\sqrt{x}} = \dfrac{6(1+2\sqrt{x})^5}{\sqrt{x}}$

7. $f(x) = 6x^5 - 4x^4 + 3x^2 - 10$

    $f'(x) = 30x^4 - 16x^3 + 6x$

    $f'(0) = 30(0)^4 - 16(0)^3 + 6(0) = 0$

    $f''(x) = 120x^3 - 48x^2 + 6$

    $f''(0) = 120(0)^3 - 48(0)^2 + 6 = 6$

8. $y = 2\sqrt{2x+1}$

    $y' = 2(2x+1)^{-\frac{1}{2}} \cdot \dfrac{1}{2} \cdot \dfrac{d}{dx}(2x+1)$

    $= \dfrac{2}{\sqrt{2x+1}}$

    $y'(0) = \dfrac{2}{\sqrt{1}} = 2$

    $y'' = 2\left[-\dfrac{1}{2}(2x+1)^{-\frac{3}{2}} \cdot \dfrac{d}{dx}(2x+1)\right]$

    $= \dfrac{-2}{(2x+1)^{\frac{3}{2}}}$

    $y''(0) = \dfrac{-2}{(1)^{\frac{3}{2}}} = -2$

9. $y = \sqrt{u} \qquad u = r^2+1 \qquad r = -3x$

    $\dfrac{dy}{dx} = \dfrac{dy}{du} \cdot \dfrac{du}{dr} \cdot \dfrac{dr}{dx}$

    $= \dfrac{1}{2}u^{-\frac{1}{2}}(2r)(-3)$

    $= \dfrac{-3r}{\sqrt{u}} = \dfrac{-3(-3x)}{\sqrt{u}} = \dfrac{9x}{\sqrt{u}}$

    $= \dfrac{9x}{\sqrt{r^2+1}} = \dfrac{9x}{\sqrt{(-3x)^2+1}} = \dfrac{9x}{\sqrt{9x^2+1}}$

10. $y = 2u^3 \quad u = \dfrac{r}{2} \quad r = x^2$

$\dfrac{dy}{dx} = \dfrac{dy}{du} \cdot \dfrac{du}{dr} \cdot \dfrac{dr}{dx}$

$= 6u^2 \left(\dfrac{1}{2}\right)(2x)$

$= 6u^2 x$

$= 6\left(\dfrac{r}{2}\right)^2 x$

$= \dfrac{3r^2 x}{2} = 3(x^2)^2 x = \dfrac{3x^5}{2}$

Lesson Practice 11B

1. $y = -2u^3 \quad u = \dfrac{x}{3}$

$\dfrac{dy}{dx} = \dfrac{dy}{du} \cdot \dfrac{du}{dx}$

$= -6u^2 \left(\dfrac{1}{3}\right)$

$= -2u^2 = -2\left(\dfrac{x}{3}\right)^2 = \dfrac{-2x^2}{9}$

Check: $y = -2\left(\dfrac{x}{3}\right)^3$

$= -\dfrac{2}{27} x^3$

$y' = -\dfrac{2}{9} x^2$

2. $y = 3u + 2 \quad u = -3x^2$

$\dfrac{dy}{dx} = \dfrac{dy}{du} \cdot \dfrac{du}{dx}$

$= (3)(-6x) = -18x$

Check: $y = 3(-3x^2) + 2$

$= -9x^2 + 2$

$y' = -18x$

3. $y = 5u^2 \quad u = \sqrt{x}$

$\dfrac{dy}{dx} = 10u \cdot \left(\dfrac{1}{2} x^{-\frac{1}{2}}\right)$

$= \dfrac{10\sqrt{x} \cdot \dfrac{1}{2}}{\sqrt{x}} = 5$

Check: $y = 5(\sqrt{x})^2 = 5x$

$y' = 5$

4. $y = 2u^5 + 1 \quad u = 3x$

$\dfrac{dy}{dx} = \dfrac{dy}{du} \cdot \dfrac{du}{dx}$

$\dfrac{dy}{dx} = 10u^4 (3)$

$= 30(3x)^4 = 2430 x^4$

5. $y = 1 - 4u^2 \quad u = -x^3$

$\dfrac{dy}{dx} = \dfrac{dy}{du} \cdot \dfrac{du}{dx}$

$= (-8u)(-3x^2)$

$= -8(-x^3)(-3x^2) = -24x^5$

6. $y = \sqrt{2u} \quad u = 8x$

$\dfrac{dy}{dx} = \dfrac{dy}{du} \cdot \dfrac{du}{dx}$

$= \dfrac{1}{2}(2u)^{-\frac{1}{2}} \cdot 2(8)$

$= \dfrac{8}{\sqrt{2u}} = \dfrac{8}{\sqrt{2(8x)}}$

$= \dfrac{8}{\sqrt{16x}} = \dfrac{8}{4\sqrt{x}} = \dfrac{2}{\sqrt{x}}$

7. $f(x) = 2x^5 - 3x^4 + 4x^3 - x^2 + 1$

$f'(x) = 10x^4 - 12x^3 + 12x^2 - 2x$

$f'(0) = 0$

$f''(x) = 40x^3 - 36x^2 + 24x - 2$

$f''(0) = -2$

8. $y = \dfrac{8}{\sqrt{1-8x}} = 8(1-8x)^{-\frac{1}{2}}$

$y' = 8\left[-\dfrac{1}{2}(1-8x)^{-\frac{3}{2}} \cdot (-8)\right]$

$= 32(1-8x)^{-\frac{3}{2}}$

$= \dfrac{32}{(1-8x)^{\frac{3}{2}}}$

$y'(0) = 32$

$y'' = 32\left[-\dfrac{3}{2}(1-8x)^{-\frac{5}{2}}(-8)\right]$

$= 384(1-8x)^{-\frac{5}{2}}$

$= \dfrac{384}{(1-8x)^{\frac{5}{2}}}$

$y''(0) = 384$

9. $y = 2u - 3 \quad u = \sqrt{r} \quad r = 4x$

$\dfrac{dy}{dx} = \dfrac{dy}{du} \cdot \dfrac{du}{dr} \cdot \dfrac{dr}{dx}$

$= (2)\left(\dfrac{1}{2}r^{-\frac{1}{2}}\right)(4)$

$= \dfrac{4}{\sqrt{r}} = \dfrac{4}{\sqrt{4x}} = \dfrac{4}{2\sqrt{x}} = \dfrac{2}{\sqrt{x}}$

10. $y = 2 - 5u \quad u = r^3 \quad r = \sqrt{1-5x}$

$\dfrac{dy}{dx} = \dfrac{dy}{du} \cdot \dfrac{du}{dr} \cdot \dfrac{dr}{dx}$

$= (-5)(3r^2)\left(\dfrac{1}{2}(1-5x)^{-\frac{1}{2}}\right)(-5)$

$= \dfrac{75r^2}{2\sqrt{1-5x}}$

$= \dfrac{75(1-5x)}{2\sqrt{1-5x}} = \dfrac{75\sqrt{1-5x}}{2}$

Lesson Practice 11C

1. At $x = -5$, the graph is a parabola and is differentiable.

2. At $x = -2$, there is a cusp, so no derivative exists.

3. At $x = 1$, there is another cusp, therefore the derivative does not exist.

4. At $x = 4$ $f(x)$ is discontinuous, so there is no derivative there.

5. $f(x) = \dfrac{\sqrt{2x-1}}{x+1}$

$f'(x) = \dfrac{(x+1)\dfrac{1}{2}(2x-1)^{-\frac{1}{2}}(2) - \sqrt{2x-1}\,(1)}{(x+1)^2}$

$f'(x) = \dfrac{(2x-1)^{-\frac{1}{2}}\left[x+1-(2x-1)\right]}{(x+1)^2}$

$= \dfrac{x+1-2x+1}{\sqrt{2x-1}\,(x+1)^2} = \dfrac{-x+2}{(x+1)^2\sqrt{2x-1}}$

This function is not differentiable at $x = -1$ and at $x = \dfrac{1}{2}$.

6. $y = 4u^2 + 1 \quad u = 3x - 2$

$\dfrac{dy}{dx} = \dfrac{dy}{du} \cdot \dfrac{du}{dx}$

$= (8u)(3) = 24u = 24(3x-2)$

$= 72x - 48$

7. $y = 1 - u^2 \quad u = \sqrt{r} \quad r = 2x$

$\dfrac{dy}{dx} = \dfrac{dy}{du} \cdot \dfrac{du}{dr} \cdot \dfrac{dr}{dx}$

$= (-2u)\left(\dfrac{1}{2}r^{-\frac{1}{2}}\right)(2)$

$= \dfrac{-2u}{\sqrt{r}} = \dfrac{-2\sqrt{r}}{\sqrt{r}} = -2$

8. $\lim\limits_{x \to \infty} \dfrac{2}{1 + e^{-.2x}}$

$\lim\limits_{x \to \infty} \dfrac{2}{1 + \dfrac{1}{e^{\infty}}} = 2$

9. $\lim\limits_{x \to \infty} \dfrac{\ln(x)}{1 - \ln(x)} =$

$\lim\limits_{x \to \infty} \dfrac{1}{\dfrac{1}{\ln(x)} - 1} =$ dividing by $\ln(x)$

$= -1$

10. If $\lim_{x \to 2} f(x) = 4$ and $\lim_{x \to 2} g(x) = 3$

then $\lim_{x \to 2} \frac{2[g(x)]^2}{2 + f(x)} = \frac{2(3)^2}{2 + 4} = \frac{18}{6} = 3$

Lesson Practice 11D

1. No. There is a corner at $x = -5$.
2. No. The left and right hand limits disagree.
3. No. There is a discontinuity there.
4. No. There is a cusp at $x = 0$.
5. Yes. This hyperbola is continuous without cusps or corners or discontinuities.
6. $f(x) = -x^3 - 3x^2 + 5$

   $f'(x) = -3x^2 - 6x = 0$

   $-3x(x + 2) = 0$

   $x = 0, -2$

   $f''(x) = -6x - 6 = 0$

   $-6(x + 1) = 0$

   $x = -1$

7. $f(x) = 2\sqrt{2x} = 2(2x)^{\frac{1}{2}}$

   $f'(x) = 2 \cdot \frac{1}{2}(2x)^{-\frac{1}{2}}(2)$

   $= \frac{2}{\sqrt{2x}}$ undefined at $x = 0$

   $f''(x) = -\frac{1}{2}(2)(2x)^{-\frac{3}{2}}(2)$

   $= \frac{-2}{(2x)^{\frac{3}{2}}}$ undefined at $x = 0$

8. $\lim_{\theta \to \frac{\pi}{2}} \frac{2\sin(\theta) + 3}{2\sin(\theta) - 3} = \frac{2\sin\left(\frac{\pi}{2}\right) + 3}{2\sin\left(\frac{\pi}{2}\right) - 3}$

   $= \frac{5}{-1} = -5$

9. $\lim_{\theta \to 0} \frac{\tan(\theta)}{\csc(\theta)} =$

   $\lim_{\theta \to 0} \frac{\sin(\theta)}{\cos(\theta)} \cdot \frac{\sin(\theta)}{1} =$

   $\lim_{\theta \to 0} \frac{\sin^2(\theta)}{\cos(\theta)} = 0$

10. $\lim_{x \to 0} \frac{2e^{2x} - 2e^x}{2e^x - 2} =$

    $\lim_{x \to 0} \frac{e^{2x} - e^x}{e^x - 1} =$

    $\lim_{x \to 0} \frac{e^x(e^x - 1)}{(e^x - 1)} =$

    $\lim_{x \to 0} e^x = 1$

Lesson Practice 12A

1. $f(x) = \cos(3x)$

   $f'(x) = -\sin(3x) \cdot \frac{d}{dx}(3x) = -3\sin(3x)$

2. $f(x) = \sin\left(\frac{x}{2}\right)$

   $f'(x) = \cos\left(\frac{x}{2}\right) \cdot \frac{d}{dx}\left(\frac{x}{2}\right)$

   $= \frac{1}{2}\cos\left(\frac{x}{2}\right)$

3. $f(x) = \tan(2x)$

   $f'(x) = \sec^2(2x) \cdot \frac{d}{dx}(2x)$

   $= 2\sec^2(2x)$

4. $f(x) = \sec(4x)$

   $f'(x) = \sec(4x)\tan(4x) \cdot \frac{d}{dx}(4x)$

   $= 4\sec(4x)\tan(4x)$

5. $f(x) = 2\cot\left(\frac{x}{2}\right)$

   $f'(x) = 2\left(-\csc^2\left(\frac{x}{2}\right)\right) \cdot \frac{d}{dx}\left(\frac{x}{2}\right)$

   $= -2\left(\csc^2\left(\frac{x}{2}\right)\right)\left(\frac{1}{2}\right)$

   $= -\csc^2\left(\frac{x}{2}\right)$

6. $f(x) = x\sin(2x)$ Product formula

$f'(x) = x \cdot \frac{d}{dx}(\sin(2x)) + \sin(2x) \cdot \frac{d}{dx}(x)$

$ = x\cos(2x) \cdot \frac{d}{dx}(2x) + \sin(2x)$

$ = 2x\cos(2x) + \sin(2x)$

7. $f(X) = x^2 \tan(2x)$

$f'(x) = x^2 \cdot \frac{d}{dx}(\tan(2x)) + \tan(2x) \cdot \frac{d}{dx}(x^2)$

$ = x^2 \sec^2(2x) \cdot \frac{d}{dx}(2x) + (\tan(2x))(2x)$

$ = 2x^2 \sec^2(2x) + 2x\tan(2x)$

$ = 2x(x\sec^2(2x) + \tan(2x))$

8. $f(x) = \sin(x)\tan(x)$

$f'(x) = \sin(x)(\sec^2(x)) + \tan(x)(\cos(x))$

$ = \sin(x)(\sec^2(x)) + \sin(x)$

$ = \sin(x)(\sec^2(x) + 1)$

9. $f(x) = \frac{1}{2}\sin^2(x) = \frac{1}{2}(\sin(x))^2$

$f'(x) = \frac{1}{2} \cdot 2(\sin(x)) \cdot \frac{d}{dx}(\sin(x))$

$ = \sin(x)\cos(x)$

10. $f(x) = \sqrt{\cos(2x)} = (\cos(2x))^{\frac{1}{2}}$

$f'(x) = \frac{1}{2}(\cos(2x))^{-\frac{1}{2}} \cdot \frac{d}{dx}(\cos(2x))$

$ = \frac{1}{2}(\cos(2x))^{-\frac{1}{2}} \cdot (-\sin(2x) \cdot 2)$

$ = \frac{-\sin(2x)}{\sqrt{\cos(2x)}}$

11. $f(x) = \sin(2x) + \cos(3x)$

$f'(x) = (\cos(2x))(2) - (\sin(3x))(3)$

$ = 2\cos(2x) - 3\sin(3x)$

12. $f(x) = [\sin(x^2+1)]^3$

$f'(x) = 3[\sin(x^2+1)]^2 \cdot \frac{d}{dx}(\sin(x^2+1))$

$ = 3\sin^2(x^2+1)\cos(x^2+1) \cdot \frac{d}{dx}(x^2+1)$

$ = 6x\sin^2(x^2+1)\cos(x^2+1)$

13. $f(x) = \dfrac{\cot(4x)}{\csc(2x)}$

$f'(x) = \dfrac{\csc(2x) \cdot \frac{d}{dx}(\cot(4x)) - \cot(4x) \cdot \frac{d}{dx}(\csc(2x))}{(\csc 2x)^2}$

$f'(x) = \dfrac{\csc(2x) \cdot (-\csc^2(4x)) \cdot \frac{d}{dx}(4x) - \cot(4x)\left(-\csc(2x)\cot(2x)\frac{d}{dx}(2x)\right)}{(\csc(2x))^2}$

$= \dfrac{-4(\csc(2x))(\csc^2(4x)) + 2(\cot(4x))(\csc(2x))(\cot(2x))}{(\csc(2x))^2}$

$= \dfrac{-2\left[2\csc^2(4x) - (\cot(4x))(\cot(2x))\right]}{\csc(2x)}$

14. $f(x) = (\tan^2(3x))(\sec(3x))$

$f'(x) = (\tan^2(3x)) \cdot \frac{d}{dx}(\sec(3x)) + (\sec(3x)) \cdot \frac{d}{dx}(\tan(3x))^2$

$= (\tan^2(3x)) \cdot (\sec(3x))(\tan(3x)) \cdot \frac{d}{dx}(3x) + (\sec(3x))(2\tan(3x)) \cdot \frac{d}{dx}(\tan(3x))$

$= 3(\tan^3(3x))(\sec(3x)) + 2(\sec(3x))(\tan(3x))(\sec^2(3x)) \cdot 3$

$= 3(\tan^3(3x))(\sec(3x)) + 6(\sec^3(3x))(\tan(3x))$

$= 3(\tan(3x))(\sec(3x))\left[\tan^2(3x) + 2\sec^2(3x)\right]$

Lesson Practice 12B

1. $f(x) = \sin(ax)$

   $f'(x) = \cos(ax) \cdot \frac{d}{dx}(ax)$

   $= a\cos(ax)$

2. $f(x) = 3\cos(2x)$

   $f'(x) = 3 \cdot \frac{d}{dx}(\cos(2x))$

   $= 3 \cdot (-\sin(2x) \cdot 2)$

   $= -6\sin(2x)$

3. $f(x) = 2\csc(3x)$

   $f'(x) = 2 \cdot (-\csc(3x)\cot(3x)) \cdot \frac{d}{dx}(3x)$

   $= -6\csc(3x)\cot(3x)$

4. $f(x) = \tan(x) - x$

   $f'(x) = \sec^2(x) - 1$

   $= \tan^2(x)$

LESSON PRACTICE 12B - LESSON PRACTICE 12B

5. $f(x) = \dfrac{4}{\sqrt{\sec(x)}} = 4(\sec(x))^{-\frac{1}{2}}$

   $f'(x) = 4 \cdot \left[-\dfrac{1}{2}(\sec(x))^{-\frac{3}{2}}\right] \cdot \dfrac{d}{dx}(\sec(x))$

   $= -2(\sec(x))^{-\frac{3}{2}} \cdot \sec(x)\tan(x)$

   $= -2[\sec(x)]^{-\frac{1}{2}} \tan(x)$

   $= \dfrac{-2\tan(x)}{\sqrt{\sec(x)}}$

6. $f(x) = x\cos(x)$

   $f'(x) = x \cdot \dfrac{d}{dx}(\cos(x)) + \cos(x) \cdot \dfrac{d}{dx}(x)$

   $= x(-\sin(x)) + \cos(x)$

   $= \cos(x) - x\sin(x)$

7. $f(x) = \cot^3\left(\dfrac{x}{3}\right) = \left[\cot\left(\dfrac{x}{3}\right)\right]^3$

   $f'(x) = 3\left(\cot\left(\dfrac{x}{3}\right)\right)^2 \cdot \dfrac{d}{dx}\left(\cot\left(\dfrac{x}{3}\right)\right)$

   $= 3\left(\cot\left(\dfrac{x}{3}\right)\right)^2 \cdot \left(-\csc^2\left(\dfrac{x}{3}\right)\right)\left(\dfrac{1}{3}\right)$

   $= -\left(\cot\left(\dfrac{x}{3}\right)\right)^2 \left(\csc^2\left(\dfrac{x}{3}\right)\right)$

8. $f(x) = (1-x)\sin(2x)$

   $f'(x) = (1-x) \cdot \dfrac{d}{dx}(\sin(2x)) + \sin(2x) \cdot \dfrac{d}{dx}(1-x)$

   $= (1-x)\cos(2x) \cdot 2 + \sin(2x)(-1)$

   $= 2(1-x)\cos(2x) - \sin(2x)$

9. $f(x) = 2\sin^2\left(\dfrac{x}{2}\right) = 2\left(\sin\left(\dfrac{x}{2}\right)\right)^2$

   $f'(x) = 2 \cdot 2\left(\sin\left(\dfrac{x}{2}\right)\right) \cdot \dfrac{d}{dx}\left(\sin\left(\dfrac{x}{2}\right)\right)$

   $= 4\left(\sin\left(\dfrac{x}{2}\right)\right)\left(\cos\left(\dfrac{x}{2}\right)\right) \cdot \dfrac{d}{dx}\left(\dfrac{x}{2}\right)$

   $= 4\left(\sin\left(\dfrac{x}{2}\right)\right)\left(\cos\left(\dfrac{x}{2}\right)\right) \cdot \dfrac{1}{2}$

   $= 2\sin\left(\dfrac{x}{2}\right)\cos\left(\dfrac{x}{2}\right)$

   $= \sin(x)$ using double angle formula in reverse: $\sin(2y) = 2\sin(y)\cos(y)$

10. $f(x) = 3\sin(x) - \sin(3x)$

    $f'(x) = 3\cos(x) - 3\cos(3x)$

    $= 3(\cos(x) - \cos(3x))$

11. $f(x) = (x + \cos(x))(\tan(x))$

    $f'(x) = (x + \cos(x)) \cdot \dfrac{d}{dx}(\tan(x)) + \tan(x) \cdot \dfrac{d}{dx}(x + \cos(x))$

    $f'(x) = (x + \cos(x))\sec^2(x) + \tan(x)(1 - \sin(x))$

12. $f(x) = \dfrac{x^2 - 2}{1 + \sin(x)}$

    $f'(x) = \dfrac{(1 + \sin(x))(2x) - (x^2 - 2)(\cos(x))}{(1 + \sin(x))^2}$

13. $f(x) = \sec(x)\tan(x)$

    $f'(x) = \sec(x) \cdot \dfrac{d}{dx}(\tan(x)) + \tan(x) \cdot \dfrac{d}{dx}(\sec(x))$

    $= \sec(x) \cdot \sec^2(x) + \tan(x) \cdot \sec(x)\tan(x)$

    $= \sec^3(x) + \sec(x)\tan^2(x)$

    $= \sec(x)(\sec^2(x) + \tan^2(x))$

14. $f(x) = (\sin(2x)\cos(2x))^2$

$= 2(\sin(2x)\cos(2x)) \cdot \frac{d}{dx}(\sin(2x)\cos(2x))$

$= 2(\sin(2x)\cos(2x))\left[\sin(2x) \cdot \frac{d}{dx}(\cos(2x)) + \cos(2x) \cdot \frac{d}{dx}(\sin(2x))\right]$

$= 2(\sin(2x)\cos(2x))[\sin(2x)(-\sin(2x) \cdot 2) + \cos(2x) \cdot \cos(2x) \cdot 2]$

$= -4(\sin(2x)\cos(2x))[\sin^2(2x) - \cos^2(2x)]$

Lesson Practice 12C

1. $f(x) = \frac{\sin(x)}{x}$

$f'(x) = \frac{x \cdot \frac{d}{dx}(\sin(x)) - \sin(x) \cdot \frac{d}{dx}(x)}{x^2}$

$= \frac{x(\cos(x)) - \sin(x)(1)}{x^2}$

$= \frac{x\cos(x) - \sin(x)}{x^2}$

2. $f(x) = \sqrt[3]{\tan(3x)} = (\tan(3x))^{\frac{1}{3}}$

$f'(x) = \frac{1}{3}(\tan(3x))^{-\frac{2}{3}} \cdot \frac{d}{dx}(\tan(3x))$

$= \frac{1}{3}(\tan(3x))^{-\frac{2}{3}}(\sec^2(3x))3$

$= \frac{\sec^2(3x)}{(\tan(3x))^{\frac{2}{3}}}$

3. $f(x) = \frac{\csc(2x)}{-8x}$

$f'(x) = \frac{-8x \cdot \frac{d}{dx}(\csc(2x)) - \csc(2x) \cdot \frac{d}{dx}(-8x)}{64x^2}$

$= \frac{8x\csc(2x)\cot(2x) \cdot 2 + 8\csc(2x)}{64x^2}$

$= \frac{2x\csc(2x)\cot(2x) + \csc(2x)}{8x^2}$

$= \frac{\csc(2x)[2x\cot(2x) + 1]}{8x^2}$

LESSON PRACTICE 12C - LESSON PRACTICE 12C

4. $f(x) = \tan^2(1-x) = [\tan(1-x)]^2$

$f'(x) = 2[\tan(1-x)] \cdot \dfrac{d}{dx}(\tan(1-x))$

$= 2\tan(1-x)\sec^2(1-x)(-1)$

$= -2[\tan(1-x)\sec^2(1-x)]$

5. $f(x) = \cot(x^2) \cdot \sec(x^2)$ Let $A = x^2$

$f(x) = \cot(A) \cdot \sec(A)$

$f'(x) = \cot(A)\sec(A)\tan(A)\dfrac{dA}{dx} +$

$\quad \sec(A)(-\csc^2(A))\dfrac{dA}{dx}$

$= \sec(A)\dfrac{dA}{dx}(\cot(A)\tan(A) - \csc^2(A))$

$= \sec(A)\dfrac{dA}{dx}(1 - \csc^2(A))$

Replace $A = x^2$, $\dfrac{dA}{dx} = 2x$.

$= 2x\sec(x^2)(1-\csc^2(x^2))$

$1 - \csc^2(u) = -\cot^2(u)$ trig I.D.

$= 2x\sec(x^2)(-\cot^2(x^2)) =$
$-2x\sec(x^2)\cot^2(x^2)$

Another way to do this problem is to rename cot and sec with sin and cos.

$f(x) = \dfrac{\cos(x^2)}{\sin(x^2)} \cdot \dfrac{1}{\cos x^2} = \dfrac{1}{\sin(x^2)} = \csc(x^2)$

$= -2x(\csc(x^2))(\cot(x^2))$

These answers, although they may appear different, are equivalent.

6. $f(x) = 4\cos(2x)$ x in [0, π]

$f'(x) = -4(\sin(2x))(2) = -8\sin(2x)$

$f''(x) = -8(\cos(2x))(2) = -16\cos(2x)$

Find all points for x in [0, π] where the second derivative is equal to zero.

$-16\cos(2x) = 0$

$\cos(2x) = 0$ divide by (-16)

Since the cos is 0 at $\dfrac{\pi}{2}$ then:

$2x = \dfrac{\pi}{2}$

$x = \dfrac{\pi}{4}$ divide by (2)

Since the cos is 0 at $\dfrac{3\pi}{2}$ then:

$2x = \dfrac{3\pi}{2}$

$x = \dfrac{3\pi}{4}$ divide by (2)

Put values $\dfrac{\pi}{4}$ and $\dfrac{3\pi}{4}$ back into original equation.

$4\cos\left(\dfrac{2\pi}{4}\right) = 4\cos\left(\dfrac{\pi}{2}\right) = 4(0) = 0$

$4\cos\left(\dfrac{6\pi}{4}\right) = 4\cos\left(\dfrac{3\pi}{2}\right) = 4(0) = 0$

The points are $\left(\dfrac{\pi}{4}, 0\right), \left(\dfrac{3\pi}{4}, 0\right)$.

7. $f(x) = \sin(2x)$ x in $[0, \pi]$
$f'(x) = (\cos(2x))(2) = 2\cos(2x)$
$f''(x) = 2(-\sin(2x))(2) = -4\sin(2x)$
Find all points for x in $[0, \pi]$ where the second derivative is equal to zero.
$-4\sin(2x) = 0$
$\sin(2x) = 0$ divide by (-4)
Since the sin is 0 at 0 and π:
$2x = 0$ $2x = \pi$ $2x = 2\pi$
$x = 0$ $x = \dfrac{\pi}{2}$ $x = \pi$

Replacing 0, $\dfrac{\pi}{2}$, and π into the original function, yields the following points.

                       Points
$\sin(2(0)) = \sin(0) = 0$ $(0, 0)$
$\sin\left(2\left(\dfrac{\pi}{2}\right)\right) = \sin(\pi) = 0$ $\left(\dfrac{\pi}{2}, 0\right)$
$\sin(2(\pi)) = \sin(2\pi) = 0$ $(\pi, 0)$

8. $f(x) = 1 + \sin(x)$ x in $[0, 2\pi]$
$f'(x) = 0 + \cos(x) = \cos(x)$
$f''(x) = -\sin(x)$
$-\sin(x)$ is 0, when $x = 0, \pi, 2\pi$
Replacing these values into the original equation yields:

                       Points
$1 + \sin(x) = 1 + \sin(0) = 1 + 0 = 1$ $(0, 1)$
$1 + \sin(x) = 1 + \sin(\pi) = 1 + 0 = 1$ $(\pi, 1)$
$1 + \sin(x) = 1 + \sin(2\pi) = 1 + 0 = 1$ $(2\pi, 1)$

9. $\lim\limits_{x \to \infty} \dfrac{5 - 2x^2}{3x + 5x^2} = \dfrac{\dfrac{5}{x^2} - 2}{\dfrac{3}{x} + 5} = -\dfrac{2}{5}$

10. $\lim\limits_{t \to 0} \dfrac{4t^2 + 3t + 2}{t^3 + 2t - 6} = -\dfrac{2}{6} = -\dfrac{1}{3}$

11. $\lim\limits_{y \to \infty} \dfrac{4y^2 - 3}{2y^3 + 3y^2} = \lim\limits_{y \to \infty} \dfrac{\dfrac{4}{y} - \dfrac{3}{y^3}}{2 + \dfrac{3}{y}} = 0$

12. $\lim\limits_{x \to \infty} \dfrac{ax^4 + bx^2 + c}{dx^3 + cx^2 + fx + g}$

$= \lim\limits_{x \to \infty} \dfrac{a + \dfrac{b}{x^2} + \dfrac{c}{x^4}}{\dfrac{d}{x} + \dfrac{c}{x^2} + \dfrac{f}{x^3} + \dfrac{g}{x^4}}$

$= \dfrac{a}{0} =$ undefined

Lesson Practice 12D

1. $f(x) = \dfrac{\sin(x)+1}{1-\cos(x)}$

   $f'(x) = \dfrac{(1-\cos(x))\cdot \dfrac{d}{dx}(\sin(x)+1) - (\sin(x)+1)\cdot \dfrac{d}{dx}(1-\cos(x))}{(1-\cos(x))^2}$

   $= \dfrac{(1-\cos(x))(\cos(x)) - (\sin(x)+1)(\sin(x))}{(1-\cos(x))^2}$

   $= \dfrac{\cos(x) - \cos^2(x) - \sin^2(x) - \sin(x)}{(1-\cos(x))^2}$

   $= \dfrac{-\cos(x) + \cos^2(x) + \sin^2(x) + \sin(x)}{-(1-\cos(x))^2}$ Hint: $\sin^2(x) + \cos^2(x) = 1$

   $= \dfrac{1 - \cos(x) + \sin(x)}{-(1-\cos(x))^2}$

2. $f(x) = (\tan(x))(2x^2)$

   $f'(x) = \tan(x)\cdot \dfrac{d}{dx}(2x^2) + 2x^2\cdot \dfrac{d}{dx}(\tan(x))$

   $= (\tan(x))(4x) + 2x^2 \cdot \sec^2(x)$

   $= 2x(2\tan(x) + x\sec^2(x))$

3. $f(x) = \dfrac{1}{3}\tan^3(x) - \tan(x) + x$

   $f'(x) = \tan^2(x)\cdot \dfrac{d}{dx}(\tan(x)) - \sec^2(x) + 1$

   $= \tan^2(x)(\sec^2(x)) - \sec^2(x) + 1$

   $= \sec^2(x)\left[\tan^2(x) - 1\right] + 1$

   $= (\tan^2(x)+1)[\tan^2(x)-1] + 1$

   $= \tan^4(x) - \tan^2(x) + \tan^2(x) - 1 + 1$

   $= \tan^4(x)$

4. $f(x) = (1-\tan(2x))(\sec(2x))$

   $f'(x) = (1-\tan(2x))\cdot \dfrac{d}{dx}(\sec(2x)) + \sec(2x)\cdot \dfrac{d}{dx}(1-\tan(2x))$

   $= (1-\tan(2x))(\sec(2x)\tan(2x))(2) + \sec(2x)(-\sec^2(2x))(2)$

   $= 2(1-\tan(2x))\sec(2x)\tan(2x) - 2\sec^3(2x)$

   $= 2\sec(2x)[(1-\tan(2x))(\tan(2x)) - \sec^2(2x)]$

5. $f(x) = \sin(2x)\csc(3x)$

   $f'(x) = \sin(2x)\dfrac{d}{dx}(\csc(3x)) + \csc(3x)\dfrac{d}{dx}(\sin(2x))$

   $f'(x) = \sin(2x)(-\csc(3x)\cot(3x))(3) + \csc(3x)(\cos(2x))(2)$

   $f'(x) = -3\sin(2x)\csc(3x)\cot(3x) + 2\csc(3x)\cos(2x)$

6. $f(x) = \cos(4x)$ $\quad$ x in $[0, \pi]$
$f'(x) = -(\sin(4x))(4) = -4\sin(4x)$
$f''(x) = -4(\cos(4x))(4) = -16\cos(4x)$
$\cos(x)$ is zero at $\dfrac{\pi}{2}, \dfrac{3\pi}{2}, \dfrac{5\pi}{2}, \dfrac{7\pi}{2}$
$-16\cos(4x) = 0 \qquad$ divide by (-16)
$\cos(4x) = 0$
$(4x) = \dfrac{\pi}{2} \qquad$ divide by (4)
$x = \dfrac{\pi}{8}$
For $\dfrac{3\pi}{2}$, $x = \dfrac{3\pi}{8}$.
For $\dfrac{5\pi}{2}$, $x = \dfrac{5\pi}{8}$.
For $\dfrac{7\pi}{2}$, $x = \dfrac{7\pi}{8}$.
x is = $\dfrac{\pi}{8}, \dfrac{3\pi}{8}, \dfrac{5\pi}{8}, \dfrac{7\pi}{8}$.
Putting these in the original equation yields the points:
$\cos(4x) = \cos\left(4\left(\dfrac{\pi}{8}\right)\right) = \cos\left(\dfrac{\pi}{2}\right) = 0$
$\left(\dfrac{\pi}{8}, 0\right)$
$\cos(4x) = \cos\left(4\left(\dfrac{3\pi}{8}\right)\right) = \cos\left(\dfrac{3\pi}{2}\right) = 0$
$\left(\dfrac{3\pi}{8}, 0\right)$
$\cos(4x) = \cos\left(4\left(\dfrac{5\pi}{8}\right)\right) = \cos\left(\dfrac{5\pi}{2}\right) = 0$
$\left(\dfrac{5\pi}{8}, 0\right)$
$\cos(4x) = \cos\left(4\left(\dfrac{7\pi}{8}\right)\right) = \cos\left(\dfrac{7\pi}{2}\right) = 0$
$\left(\dfrac{7\pi}{8}, 0\right)$

7. $f(x) = 2\sin(x) + 3 \qquad$ x in $[0, 2\pi]$
$f'(x) = 2\cos(x)$
$f''(x) = -2\sin(x)$
The sin is 0, when $x = 0, \pi, 2\pi$
Replacing these into the original equation yields the points:
$2\sin(x) + 3 = 2\sin(0) + 3 = 0 + 3 = 3$
$(0, 3)$
$2\sin(x) + 3 = 2\sin(\pi) + 3 = 0 + 3 = 3$
$(\pi, 3)$
$2\sin(x) + 3 = 2\sin(2\pi) + 3 = 0 + 3 = 3$
$(2\pi, 3)$

8. $f(x) = \dfrac{1}{\sin(x)}$
$f'(x) = \dfrac{\sin(x(0)) - 1(\cos(x))}{\sin^2(x)}$
$= \dfrac{-\cos(x)}{\sin^2(x)} = 0$
$-\cos(x) = 0 \qquad$ in $[0, 2\pi]$
$x = \dfrac{\pi}{2}, \dfrac{3\pi}{2}$
$f\left(\dfrac{\pi}{2}\right) = \dfrac{1}{\sin\left(\dfrac{\pi}{2}\right)} = 1$
$f\left(\dfrac{3\pi}{2}\right) = \dfrac{1}{\sin\left(\dfrac{3\pi}{2}\right)} = -1$
$\left(\dfrac{\pi}{2}, 1\right) \left(\dfrac{3\pi}{2}, -1\right)$

9. $\lim\limits_{x \to \infty} \dfrac{\dfrac{6}{x^2} - 3}{\dfrac{2}{x^2} + 1} = -3$

10. $\lim\limits_{x \to \infty} \dfrac{1 + \dfrac{x^4}{x^5}}{A - \dfrac{3}{x^5}} = \dfrac{1}{A}$

11. $\lim\limits_{x \to 0} \dfrac{2(0)^3 - 3(0)^2 + 7}{(0)^2 + 1} = 7$

12. $\lim\limits_{x \to 2\pi} \dfrac{\sin(x)}{\tan(x)} =$

$\lim\limits_{x \to 2\pi} \dfrac{\sin(x)}{\frac{\sin(x)}{\cos(x)}} =$

$\lim\limits_{x \to 2\pi} \cos(x) =$

$\cos(2\pi) = 1$

Lesson Practice 13A

1. $f(x) = 4e^x$
$f'(x) = 4 \dfrac{d}{dx}(e^x)$
$f'(x) = 4e^x$

2. $f(x) = e^{\sqrt{x}}$
$f'(x) = e^{\sqrt{x}} \cdot \dfrac{d}{dx}\sqrt{x}$
$f'(x) = e^{\sqrt{x}} \cdot \dfrac{1}{2}x^{-\frac{1}{2}}$
$f'(x) = \dfrac{e^{\sqrt{x}}}{2\sqrt{x}}$

3. $f(x) = e^{3x^2}$
$f'(x) = e^{3x^2} \dfrac{d}{dx}(3x^2)$
$f'(x) = 6x\, e^{3x^2}$

4. $f(x) = \dfrac{4}{e^x}$
$f'(x) = 4 \cdot e^{-x} \cdot \dfrac{d}{dx}(-x)$
$f'(x) = -4 \qquad \cdot e^{-x}$
$f'(x) = \dfrac{-4}{e^x}$

5. $f(x) = e^x \cdot \sin(x)$
$f'(x) = e^x \cdot \cos(x) + \sin(x) \cdot e^x$
$f'(x) = e^x(\cos(x) + \sin(x))$

6. $f(x) = \ln(4x^2)$
$u = 4x^2 \qquad \dfrac{du}{dx} = 8x$
$f'(x) = \dfrac{1}{4x^2} \cdot (8x)$
$f'(x) = \dfrac{2}{x}$

7. $f(x) = \ln\sqrt{x}$
$u = \sqrt{x} \qquad \dfrac{du}{dx} = \dfrac{1}{2\sqrt{x}}$
$f'(x) = \dfrac{1}{\sqrt{x}} \cdot \dfrac{1}{2\sqrt{x}}$
$f'(x) = \dfrac{1}{2x}$

8. $f(x) = (\ln(5x))^5$
$f'(x) = 5(\ln(5x))^4 \dfrac{d}{dx}(\ln(5x))$
$f'(x) = 5(\ln(5x))^4 \cdot \dfrac{1}{5x} \cdot 5$
$f'(x) = \dfrac{5(\ln(5x))^4}{x}$

9. $f(x) = (1-e^{2x})^2$
$\dfrac{d}{dx}(1-e^{2x}) = -e^{2x}\dfrac{d}{dx}(2x) = -2e^{2x}$
$f'(x) = 2(1-e^{2x})^1(-2e^{2x})$
$f'(x) = -4e^{2x}(1-e^{2x})$

10. $f(x) = e^{-x}\sin(x)$
$f'(x) = e^{-x}\dfrac{d}{dx}\sin(x) + \sin(x)\dfrac{d}{dx}(e^{-x})$
$f'(x) = e^{-x} \cdot \cos(x) + \sin(x) \cdot e^{-x}(-1)$
$f'(x) = e^{-x} \cdot \cos(x) - \sin(x) \cdot e^{-x}$
$f'(x) = e^{-x}(\cos(x) - \sin(x))$

11. $f(x) = e^{2x}\ln(2x)$
$f'(x) = e^{2x}\dfrac{d}{dx}\ln(2x) + \ln(2x)\dfrac{d}{dx}(e^{2x})$
$f'(x) = e^{2x}\dfrac{2}{2x} + \ln(2x) \cdot 2e^{2x}$
$f'(x) = e^{2x}\left[\dfrac{1}{x} + 2\ln(2x)\right]$

12. $f(x) = \cos[\ln(3x)]$
$f'(x) = -\sin[\ln(3x)] \cdot \frac{d}{dx}\ln(3x)$
$f'(x) = -\sin[\ln(3x)] \cdot \frac{1}{3x} \cdot 3$
$f'(x) = \frac{-\sin[\ln(3x)]}{x}$

Lesson Practice 13B

1. $f(x) = \ln(2x) + e^{2x}$
$f'(x) = \frac{1}{2x} \cdot \frac{2}{1} + \frac{d}{dx}(e^{2x})$
$f'(x) = \frac{1}{x} + 2e^{2x}$

2. $f(x) = (e^{-2x} + 1)^4$
$f'(x) = 4(e^{-2x} + 1)^3(-2e^{-2x})$
$f'(x) = -8(e^{-2x} + 1)^3(e^{-2x})$

3. $f(x) = \sin(e^{3x})$
$f'(x) = \cos(e^{3x}) \frac{d}{dx}(e^{3x})$
$f'(x) = \cos(e^{3x})(3e^{3x})$
$f'(x) = 3e^{3x}\cos(e^{3x})$

4. $f(x) = e^{3x} - \cos(2x) + \csc(4x)$
$f'(x) = 3e^{3x} - (2)(-\sin(2x)) - 4\csc(4x)\cot(4x)$
$f'(x) = 3e^{3x} + 2\sin(2x) - 4\csc(4x)\cot(4x)$

5. $f(x) = xe^{\sin(x)}$
$f'(x) = x\frac{d}{dx}(e^{\sin(x)}) + e^{\sin(x)}\frac{d}{dx}(x)$
$f'(x) = xe^{\sin(x)}(\cos(x)) + e^{\sin(x)}(1)$
$f'(x) = e^{\sin(x)}(x\cos(x) + 1)$

6. $f(x) = \frac{e^x}{x} = e^x x^{-1}$
$f'(x) = e^x \frac{d}{dx}(x^{-1}) + x^{-1}\frac{d}{dx}(e^x)$
$f'(x) = e^x(-1)(x^{-2}) + x^{-1}(e^x)$
$f'(x) = \frac{-e^x}{x^2} + \frac{e^x}{x}$
$f'(x) = \frac{-e^x + xe^x}{x^2} = \frac{e^x(x-1)}{x^2}$

7. $f(x) = e^x \cos(2x)$
$f'(x) = e^x \frac{d}{dx}(\cos(2x)) + \cos(2x)\frac{d}{dx}(e^x)$
$f'(x) = e^x(-2\sin(2x)) + e^x\cos(2x)$
$f'(x) = e^x(-2\sin(2x) + \cos(2x))$

8. $f(x) = \sqrt{\ln\left(\frac{x}{2}\right)}$
$f'(x) = \frac{1}{2}\left[\ln\left(\frac{x}{2}\right)\right]^{-\frac{1}{2}} \frac{d}{dx}[\ln\left(\frac{x}{2}\right)]$
$f'(x) = \frac{1}{2\sqrt{\ln\left(\frac{x}{2}\right)}} \cdot \frac{1}{\frac{x}{2}} \cdot \frac{1}{2}$
$f'(x) = \frac{1}{2\sqrt{\ln\left(\frac{x}{2}\right)}} \cdot \frac{2}{x} \cdot \frac{1}{2}$
$f'(x) = \frac{1}{2x\sqrt{\ln\left(\frac{x}{2}\right)}}$

9. $f(x) = \sin[\ln(2x-1)]$
$f'(x) = \cos[\ln(2x-1)] \frac{d}{dx}\ln(2x-1)$
$f'(x) = \cos[\ln(2x-1)] \cdot \frac{1}{2x-1} \cdot \frac{2}{1}$
$f'(x) = \frac{2\cos[\ln(2x-1)]}{2x-1}$

10. $f(x) = \frac{\ln(x^2)}{e^{(2x)}} = \ln(x^2) \cdot e^{-2x}$
$f'(x) = \ln(x^2)\frac{d}{dx}(e^{-2x}) + e^{-2x}\frac{d}{dx}(\ln(x^2))$
$f'(x) = -2\ln(x^2)(e^{-2x}) + e^{-2x} \cdot \frac{1}{x^2} \cdot \frac{2x}{1}$
$f'(x) = -2e^{-2x}\left(\ln(x^2) - \frac{1}{x}\right)$

11. $f(x) = \ln[\tan(2x)]$

$f'(x) = \dfrac{1}{\tan(2x)} \dfrac{d}{dx}[\tan(2x)]$

$f'(x) = \cot(2x)(2)\sec^2(2x)$

$f'(x) = \dfrac{2\cos(2x)}{\sin(2x)} \cdot \dfrac{1}{\cos^2(2x)}$

$= \dfrac{2}{\sin(2x)\cos(2x)}$

12. $f(x) = \ln[\sin(e^{2x})]$

$f'(x) = \dfrac{1}{\sin(e^{2x})} \dfrac{d}{dx}\sin(e^{2x})$

$f'(x) = \dfrac{1}{\sin(e^{2x})} \cdot \cos(e^{2x}) \dfrac{d}{dx}(e^{2x})$

$f'(x) = \dfrac{\cos(e^{2x})}{\sin(e^{2x})} \cdot 2e^{2x} = 2(e^{2x})\cot(e^{2x})$

Lesson Practice 13C

1. $f(x) = \ln(3x) + \ln(2x^2)$

$f'(x) = \dfrac{1}{3x} \cdot \dfrac{3}{1} + \dfrac{1}{2x^2} \cdot \dfrac{4x}{1}$

$f'(x) = \dfrac{1}{x} + \dfrac{2}{x} = \dfrac{3}{x}$

2. $f(x) = e^{\sqrt{x}} - e^{4x}$

$f'(x) = e^{\sqrt{x}} \dfrac{d}{dx} x^{\left(\frac{1}{2}\right)} - e^{4x} \dfrac{d}{dx}(4x)$

$f'(x) = e^{\sqrt{x}}\left(\dfrac{1}{2}\right)\left(x^{-\frac{1}{2}}\right) - 4e^{4x}$

$f'(x) = \dfrac{e^{\sqrt{x}}}{2\sqrt{x}} - 4e^{4x}$

3. $f(x) = \dfrac{e^{2x}}{x} = e^{2x} \cdot x^{-1}$

$f'(x) = e^{2x} \dfrac{d}{dx}(x^{-1}) + x^{-1} \dfrac{d}{dx}(e^{2x})$

$f'(x) = e^{2x}(-1)(x^{-2}) + x^{-1}(2)(e^{2x})$

$f'(x) = \dfrac{-e^{2x}}{x^2} + \dfrac{2e^{2x}}{x}$

$f'(x) = \dfrac{-e^{2x}}{x^2} + \dfrac{2xe^2}{x^2}$

$f'(x) = \dfrac{e^{2x}(2x-1)}{x^2}$

4. $f(x) = e^{-x}\cos(2x)$

$f'(x) = e^{-x}\dfrac{d}{dx}(\cos(2x)) + \cos(2x)\dfrac{d}{dx}(e^{-x})$

$f'(x) = e^{-x}(-2)(\sin(2x)) + (\cos(2x))(-1)(e^{-x})$

$f'(x) = e^{-x}(-2\sin(2x) - \cos(2x))$

5. $f(x) = \tan^2(e^{2x}) = (\tan(e^{2x}))^2$

$f'(x) = 2(\tan(e^{2x}))^1 \dfrac{d}{dx}(\tan(e^{2x}))$

$f'(x) = 2\tan(e^{2x})(\sec^2(e^{2x}))(2)(e^{2x})$

$f'(x) = 4e^{2x}\tan(e^{2x})\sec^2(e^{2x})$

6. $f(x) = \ln(x^2 + e^x)$

$f'(x) = \dfrac{1}{x^2 + e^x} \cdot \dfrac{d}{dx}(x^2 + e^x)$

$f'(x) = \dfrac{1}{x^2 + e^x} \cdot \dfrac{2x + e^x}{1}$

$f'(x) = \dfrac{2x + e^x}{x^2 + e^x}$

7. $f(x) = [\ln(2 - 3x)]^3$

$f'(x) = 3[\ln(2-3x)]^2 \dfrac{d}{dx}\ln(2-3x)$

$f'(x) = 3[\ln(2-3x)]^2 \dfrac{1}{2-3x} \cdot \dfrac{-3}{1}$

$f'(x) = \dfrac{-9[\ln(2-3x)]^2}{2-3x}$

8. $f(x) = \dfrac{\ln(2x)}{\ln(x^2)} = \ln(2x) \cdot (\ln x^2)^{-1}$

$f'(x) = \ln(2x)\dfrac{d}{dx}(\ln(x^2))^{-1} + (\ln(x^2))^{-1}\dfrac{d}{dx}(\ln(2x))$

$f'(x) = \ln(2x)(-1)(\ln(x^2))^{-2}\left(\dfrac{2x}{x^2}\right) + (\ln(x^2))^{-1}\left(\dfrac{2}{2x}\right)$

$f'(x) = \dfrac{-2\ln(2x)}{x(\ln(x^2))^2} + \dfrac{1}{x\ln(x^2)}\dfrac{(\ln(x^2))}{(\ln(x^2))}$

$f'(x) = \dfrac{-2\ln(2x) + \ln(x^2)}{x(\ln(x^2))^2}$

9. $f(x) = e^x - \frac{x^2}{2} = e^x - \frac{1}{2}x^2$

   $f'(x) = e^x - x$

   $f''(x) = e^x - 1$

   $e^x - 1 = 0$ and $e^x = 1$

   $x = \ln(1) = 0$

   $0 = x$

   Enter into the original equation.

   $f(0) = e^0 - \frac{0^2}{2} = 1 - 0 = 1$

   When x is 0, $f(0)$ is 1,
   and the point is (0, 1).

10. $f(x) = xe^{2x}$

    $f'(x) = x \frac{d}{dx}(e^{2x}) + e^{2x} \frac{d}{dx}(x)$

    $f'(x) = x \cdot 2e^{2x} + e^{2x}(1) = 2xe^{2x} + e^{2x}$

    $f''(x) = [2x(2)(e^{2x}) + e^{2x}(2)] + (2)e^{2x}$

    $f''(x) = 4Xe^{2x} + 2e^{2x} + 2e^{2x}$

    $f''(x) = 4xe^{2x} + 4e^{2x} = 4e^{2x}(x+1)$

    Then $f''(x) = 0 = 4e^{2x}(x+1)$

    $4e^{2x}(x+1) = 0$

    $e^{2x} = 0 \qquad (x+1) = 0$

    $\ln(e^{2x}) = \ln(0) \qquad x = -1 \rightarrow f(-1) = (-1)e^{2(-1)}$

    (no solution) $\qquad f(-1) = -e^{-2}$

    When x is -1, $f(-1)$ is $-e^{-2}$ and the point
    is $\left(-1, -\frac{1}{e^2}\right)$.

Lesson Practice 13D

1. $f(x) = 2\ln(x^5)$

   $f'(x) = 2\left(\frac{1}{x^5}\right)\left(\frac{5x^4}{1}\right)$

   $f'(x) = \frac{10}{x}$

2. $f(x) = \tan(\ln(2x))$

   $f'(x) = \sec^2(\ln(x)) \frac{d}{dx}(\ln(2x))$

   $f'(x) = \sec^2(\ln(2x))\left(\frac{1}{2x}\right)(2)$

   $f'(x) = \frac{\sec^2(\ln(2x))}{x}$

3. $f(x) = 3\ln(x^2 - 2x)$

   $f'(x) = 3\left(\frac{1}{x^2 - 2x}\right)\frac{d}{dx}(x^2 - 2x)$

   $f'(x) = \left(\frac{3}{x^2 - 2x}\right)\left(\frac{2-2}{1}\right)$

   $f'(x) = \frac{6x - 6}{x^2 - 2x}$

4. $f(x) = \csc(3x) - 3\ln(3x) + e^{3x}$

   $f'(x) = -3\csc(3x)\cot(3x) - 3\left(\frac{1}{3x}\right)(3) + 3e^{3x}$

   $f'(x) = -3\csc(3x)\cot(3x) - \frac{3}{x} + 3e^{3x}$

   $f'(x) = -3\left(\csc(3x)\cot(3x) + \frac{1}{x} - e^{3x}\right)$

5. $f(x) = e^{\sqrt{x} - x}$

   $f'(x) = \left(e^{\sqrt{x} - x}\right)\frac{d}{dx}(\sqrt{x} - x)$

   $f'(x) = \left(e^{\sqrt{x} - x}\right)\left(\frac{1}{2}x^{-\frac{1}{2}} - 1\right)$

   $f'(x) = \left(e^{\sqrt{x} - x}\right)\left(\frac{1}{2\sqrt{x}} - 1\right)$

6. $f(x) = \frac{e^x - 1}{e^x + 1}$

   $f'(x) = \frac{(e^x + 1)(e^x) - (e^x - 1)(e^x)}{(e^x + 1)^2}$

   $f'(x) = \frac{e^x[(e^x + 1) - (e^x - 1)]}{(e^x + 1)^2}$

   $f'(x) = \frac{e^x(2)}{(e^x + 1)^2} = \frac{2e^x}{(e^x + 1)^2}$

7. $f(x) = e^{2x} \sec(2x)$

$f'(x) = e^{2x} 2\sec(2x)\tan(2x) + \sec(2x)(2)e^{2x}$

$f'(x) = 2e^{2x}(\sec(2x))(\tan(2x) + 1)$

8. $f(x) = \sin(2x)\csc(e^x)$

$f'(x) = \sin(2x)\frac{d}{dx}\csc(e^x) + \csc(e^x)\frac{d}{dx}(\sin(2x))$

$f'(x) = \sin(2x)(-\csc(e^x)\cot(e^x))(e^x) + (\csc(e^x))(2\cos(2x))$

$f'(x) = (\csc(e^x))[-e^x \sin(2x)\cot(e^x) + 2\cos(2x)]$

9. $f(x) = e^{3x} - x^2$

$f'(x) = 3e^{3x} - 2x$

$f''(x) = 9e^{3x} - 2$

$9e^{3x} - 2 = 0$

$9e^{3x} = 2$

$e^{3x} = \frac{2}{9}$

$\ln(e^{3x}) = \ln\left(\frac{2}{9}\right)$

$3x = \ln\left(\frac{2}{9}\right)$

$x = \frac{1}{3}\ln\left(\frac{2}{9}\right)$

Enter into the original equation.

$f\left(\frac{1}{3}\ln\left(\frac{2}{9}\right)\right) = e^{3\left(\frac{1}{3}\ln\left(\frac{2}{9}\right)\right)} - \left(\frac{1}{3}\ln\left(\frac{2}{9}\right)\right)^2$

$= e^{\left(\ln\left(\frac{2}{9}\right)\right)} - \frac{1}{9}\left(\ln\left(\frac{2}{9}\right)\right)^2$

When x is $\frac{1}{3}\left(\ln\left(\frac{2}{9}\right)\right)$,

$f\left(\frac{1}{3}\ln\left(\frac{2}{9}\right)\right)$ is $e^{\left(\ln\left(\frac{2}{9}\right)\right)} - \frac{1}{9}\left(\ln\left(\frac{2}{9}\right)\right)^2$

The point is $\left(\frac{1}{3}\ln\left(\frac{2}{9}\right), \frac{2}{9} - \frac{1}{9}\left(\ln\left(\frac{2}{9}\right)\right)^2\right)$

10. $f(x) = \ln(x^2+1)$

$f'(x) = \left(\dfrac{1}{x^2+1}\right)\left(\dfrac{2x}{1}\right) = \dfrac{2x}{x^2+1}$

$f''(x) = \dfrac{(x^2+1)(2) - 2x(2x)}{(x^2+1)^2}$

$f''(x) = \dfrac{2x^2+2-4x^2}{(x^2+1)^2}$

$f''(x) = \dfrac{2-2x^2}{(x^2+1)^2}$

$0 = \dfrac{2(1-x^2)}{(x^2+1)^2}$

$0 = 1-x^2$

$0 = (1-x)(1+x)$

$x = 1, -1$

Enter both values into the original equation.

$f(1) = \ln\left[(1)^2+1\right] = \ln(2)$

$f(-1) = \ln\left[(-1)^2+1\right] = \ln(2)$

The points are $(1, \ln(2))$ and $(-1, \ln(2))$.

Lesson Practice 14A

1. $2x + 3y = 4$

$2 + 3y' = 0$

$3y' = -2$

$y' = -\dfrac{2}{3}$

2. $3xy = 2x$

$3[x \cdot y' + y \cdot (1)] = 2$ product rule

$3xy' + 3y = 2$

$3xy' = 2 - 3y$

$y' = \dfrac{2-3y}{3x}$

3. $\ln(y) = x^2$

$\dfrac{1}{y}y' = 2x$

$y' = 2xy$

4. $\sin(y) = x^2 - 1$

$\cos(y) \cdot y' = 2x$

$y' = \dfrac{2x}{\cos(y)}$

5. $(y^2 - 2y)^3 = x$

$3(y^2-2y)^2 \cdot \dfrac{d}{dx}(y^2-2y) = 1$

$3(y^2-2y)^2 (2y-2)y' = 1$

$y' = \dfrac{1}{3(y^2-2y)^2(2y-2)}$

6. $15x = 15y + 5y^3 + 3y^5$

$15 = 15y' + 15y^2 y' + 15y^4 y'$

$1 = y' + y^2 y' + y^4 y'$

$1 = y'(1 + y^2 + y^4)$

$y' = \dfrac{1}{1+y^2+y^4}$

7. $x = \sqrt{y} + \sqrt[3]{y}$

$1 = \dfrac{1}{2}y^{-\frac{1}{2}}y' + \dfrac{1}{3}y^{-\frac{2}{3}}y'$

$1 = y'\left(\dfrac{1}{2\sqrt{y}} + \dfrac{1}{3y^{\frac{2}{3}}}\right)$

$y' = \dfrac{1}{\dfrac{1}{2\sqrt{y}} + \dfrac{1}{3y^{\frac{2}{3}}}}$

8.
$$x^3 + x^2y + y^3 = 3$$
$$3x^2 + x^2y' + y(2x) + 3y^2y' = 0$$
$$x^2y' + 3y^2y' = -3x^2 - 2xy$$
$$y'(x^2 + 3y^2) = -3x^2 - 2xy$$
$$y' = \frac{-3x^2 - 2xy}{x^2 + 3y^2}$$

9.
$$\sin(x) + \cos(y) = y$$
$$\cos(x) - (\sin(y))y' = y'$$
$$\cos(x) = y' + (\sin(x))y'$$
$$\cos(x) = y'(1 + \sin(y))$$
$$\frac{\cos(x)}{1 + \sin(y)} = y'$$
$$y' = \frac{\cos(x)}{1 + \sin(y)}$$

10.
$$e^x + e^y = 2y$$
$$e^x + e^y y' = 2y'$$
$$e^x = 2y' - e^y y'$$
$$e^x = y'(2 - e^y)$$
$$y' = \frac{e^x}{2 - e^y}$$

11.
$$-3 + xy = y$$
$$xy' + y = y'$$
$$xy' - y' = -y$$
$$y'(x - 1) = -Y$$
$$y' = \frac{-y}{x - 1}$$

At (2, 3):
$$y' = \frac{-(3)}{(2) - 1}$$
$$y' = \frac{-3}{1}$$
$$y' = -3$$

12.
$$xy^2 + \frac{1}{x} = 2y + x$$
$$x \cdot 2yy' + y^2 - x^{-2} = 2y' + 1$$
$$x \cdot 2yy' - 2y' = 1 + x^{-2} - y^2$$
$$y'(2xy - 2) = 1 + x^{-2} - y^2$$
$$y' = \frac{1 + x^{-2} - y^2}{2xy - 2}$$

At (1, 2):
$$y' = \frac{1 + (1)^{-2} - (2)^2}{2(1)(2) - 2}$$
$$y' = \frac{1 + 1 - 4}{4 - 2}$$
$$y' = \frac{-2}{2}$$
$$y' = -1$$

Lesson Practice 14B

1.
$$2y^3 - 1 = x^2 + 2$$
$$6y^2 y' = 2x$$
$$y' = \frac{2x}{6y^2} = \frac{x}{3y^2}$$

2.
$$\sqrt{y} - 2x = x^2$$
$$\frac{1}{2} y^{-\frac{1}{2}} y' - 2 = 2x$$
$$\frac{y'}{2\sqrt{y}} - 2 = 2x$$
$$\frac{y'}{2\sqrt{y}} = 2x + 2$$
$$y' = (2x + 2)(2\sqrt{y})$$

3.
$$2 \ln(4y) + 3 = e^{4x}$$
$$2 \frac{1}{4Y} \cdot 4 \cdot y' = 4e^{4x}$$
$$\frac{2y'}{y} = 4e^{4x}$$
$$y' = \frac{4e^{4x} \cdot y}{2} = 2e^{4x} \cdot y$$

4. $\tan(y) + \sec(y) = \sin(x)$
$(\sec^2(y))y' + \sec(y)(\tan(y))y' = \cos(x)$
$y'(\sec^2(y) + \sec(y)\tan(y)) = \cos(x)$
$y' = \dfrac{\cos(x)}{\sec^2(y) + \sec(y)\tan(y)}$

5. $y\cos(2y) = x^3$
$y \cdot \dfrac{d}{dx}(\cos(2y)) + \cos(2y) \cdot y' = 3x^2$
$-2y\sin(2y) \cdot y' + \cos(2y) \cdot y' = 3x^2$
$y' = \dfrac{3x^2}{\cos(2y) - 2y\sin(2y)}$

6. $x^2 + 2y^2 = 4$
$2x + 4yy' = 0$
$2x = -4yy'$
$y' = -\dfrac{x}{2y}$

7. $\ln(y) + \ln(x) = 2x$
$\dfrac{1}{y} \cdot y' + \dfrac{1}{x} = 2$
$\dfrac{y'}{y} + \dfrac{1}{x} = 2$
$\dfrac{y'}{y} = 2 - \dfrac{1}{x}$
$\dfrac{y'}{y} = \dfrac{2x}{x} - \dfrac{1}{x}$
$y' = y\left(\dfrac{2x-1}{x}\right)$
$= \dfrac{y(2x-1)}{x}$
$= \dfrac{2xy - y}{x}$

8. $e^{xy} + \tan(x) = \dfrac{1}{3}x^3$
$e^{xy} \cdot \dfrac{d}{dx}(xy) + \sec^2(x) = x^2$
$e^{xy} \cdot (xy' + y) = x^2 - \sec^2(x)$
$xy'e^{xy} + ye^{xy} = x^2 - \sec^2(x)$
$xy'e^{xy} = x^2 - \sec^2(x) - ye^{xy}$
$y' = \dfrac{x^2 - \sec^2(x) - ye^{xy}}{xe^{xy}}$

9. $(xy)^3 + 3X = Y$
$3(xy)^2 \cdot \dfrac{d}{dx}(xy) + 3 = y'$
$3(xy)^2(xy' + y) + 3 = y'$
$3x^3y^2y' - y' = -3 - 3X^2Y^3$
$y'(3x^3y^2 - 1) = -3 - 3X^2Y^3$
$y' = \dfrac{-3 - 3x^2y^3}{3x^3y^2 - 1}$
$= \dfrac{3 + 3x^2y^3}{1 - 3x^3y^2}$

10. $2y^2 - 3y^3 + y = x$
$4yy' - 9y^2y' + y' = 1$
$y'(4y - 9y^2 + 1) = 1$
$y' = \dfrac{1}{4y - 9y^2 + 1}$

11. $x^2 + xy + 2y^2 = 28$
$2x + xy' + y + 4yy' = 0$
$y'(x + 4y) = -2x - y$
$y' = \dfrac{-2x - y}{x + 4y}$

At $(2, 3)$:
$y' = \dfrac{-2(2) - (3)}{(2) + 4(3)}$
$y' = \dfrac{-4 - 3}{2 + 12}$
$y' = \dfrac{-7}{14}$
$y' = -\dfrac{1}{2}$

12.

$$x^3 - 3xy^2 + y^3 = 1$$
$$3x^2 - 3(x \cdot 2yy' + y^2) + 3y^2y' = 0$$
$$3x^2 - 6xyy' - 3y^2 + 3Y^2Y' = 0$$
$$x^2 - 2xyy' - y^2 + y^2y' = 0$$
$$y'(y^2 - 2xy) = y^2 - x^2$$
$$y' = \frac{y^2 - x^2}{y^2 - 2xy}$$

At $(2, -1)$:

$$y' = \frac{(-1)^2 - (2)^2}{(-1)^2 - 2(2)(-1)} = \frac{1-4}{1+4} = -\frac{3}{5}$$

Lesson Practice 14C

1.
$$\ln(x^2+1) + e^y = 4$$
$$\frac{1}{x^2+1}(2x) + e^y y' = 0$$
$$\frac{2x}{x^2+1} = -e^y \cdot y'$$
$$y' = \frac{-2x}{(e^y)(x^2+1)}$$

2.
$$\frac{\ln(y)}{y} + 20 = -x^2$$
$$\frac{y\left(\frac{d}{dx}\ln(y)\right) - \ln(y) \cdot y'}{y^2} = -2x$$
$$\frac{y\left(\frac{1}{y}\right)y' - y'\ln(y)}{y^2} = -2x$$
$$y'[1 - \ln(y)] = -2xy^2$$
$$y' = \frac{-2xy^2}{1 - \ln(y)}$$

3.
$$x^2 = 3y^3 + y$$
$$2x = 9y^2 y' + y'$$
$$2x = y'(9y^2 + 1)$$
$$y' = \frac{2x}{9y^2 + 1}$$

4.
$$\sqrt{xy} = 2$$
$$\frac{1}{2}(xy)^{-\frac{1}{2}} \cdot (xy' + y) = 0$$
$$xy' + y = 0$$
$$xy' = -y$$
$$y' = \frac{-y}{x}$$

5.
$$\sin(e^y) = \ln(x)$$
$$\cos(e^y) \cdot \frac{d}{dx}e^y = \frac{1}{x}$$
$$\cos(e^y) \cdot e^y y' = \frac{1}{x}$$
$$y' = \frac{1}{xe^y \cos(e^y)}$$

6. $\ln(y) - x^2 = -9$ at $(3, 1)$

$\dfrac{y'}{y} - 2x = 0$

$y' = 2xy$

$y'(3, 1) = 2(3)(1) = 6$

Slope of tangent line at $(3, 1)$ is 6.

$y = mx + b$
$(1) = (6)(3) + b$
$1 = 18 + b$
$-17 = b$

Equation of tangent to curve at $(3, 1)$ is $y = 6x - 17$.

Slope of normal line at $(3, 1)$ to curve is $-\dfrac{1}{6}$.

$(1) = \left(-\dfrac{1}{6}\right)(3) + b$

$1 = -\dfrac{3}{6} + b$

$1 + \dfrac{3}{6} = b$

$1\dfrac{1}{2} = b$

$\dfrac{3}{2} = b$

Equation of normal line is:

$y = -\dfrac{1}{6}x + \dfrac{3}{2}$

7. $\sin(y) + x^2 = 4$ at $(2, 0)$

$\cos(y)y' + 2x = 0$

$y' = \dfrac{-2x}{\cos(y)}$

$y'(2, 0) = \dfrac{-2(2)}{\cos(0)} = \dfrac{-4}{1} = -4$

Slope of tangent line at $(2, 0)$ is -4.

$(0) = (-4)(2) + b$
$0 = -8 + b$
$8 = b$

Equation of tangent line is $y = -4x + 8$.

Slope of normal line is $\dfrac{1}{4}$.

$(0) = \left(\dfrac{1}{4}\right)(2) + b$

$0 = \dfrac{1}{2} + b$

$-\dfrac{1}{2} = b$

Equation of the normal line to the curve at $(2, 0)$ is:

$y = \dfrac{1}{4}x - \dfrac{1}{2}$

$4y = x - 2$

$x - 4y = 2$

8. $y = \dfrac{1}{\sqrt{4-x}}$

domain: $(-\infty, 4)$

range: $(0, \infty)$

9. $y = \dfrac{1}{1+x^2}$

domain: $(-\infty, \infty)$

range: $(0, 1]$

10. $y = \ln(1-x)$

domain: $(-\infty, 1)$

range: $(-\infty, \infty)$

LESSON PRACTICE 14C - LESSON PRACTICE 14D

11. $y = x^2 \quad x = 2t - 5$

$$\frac{dy}{dt} = \frac{dy}{dx} \cdot \frac{dx}{dt}$$
$$= (2x)(2)$$
$$= 4x$$
$$= 4(2t - 5)$$
$$= 8t - 20$$

12. $y = x^4 \quad x = \ln(t)$

$$\frac{dy}{dt} = \frac{dy}{dx} \cdot \frac{dx}{dt}$$
$$= 4x^3 \cdot \frac{1}{t}$$
$$= \frac{4x^3}{t}$$
$$= \frac{4\ln^3 t}{t}$$

Lesson Practice 14D

1. $\ln(2x - e^x) = y$

$$\frac{dy}{dx} = \frac{1}{2x - e^x} \cdot \frac{d}{dx}(2x - e^x)$$
$$= \frac{2 - e^x}{2x - e^x}$$

2. $2y^2 - y^4 = x^2 - x$

$4yy' - 4y^3 y' = 2x - 1$

$y'(4y - 4y^3) = 2x - 1$

$y' = \frac{2x - 1}{4y - 4y^3}$

3. $\ln(x) + 2y^2 = 4$

$\frac{1}{x} + 4yy' = 0$

$4yy' = -\frac{1}{x}$

$y' = \frac{-1}{4xy}$

4. $y = x^2 - 3y^2 + 7$

$y' = 2x - 6yy'$

$y' + 6yy' = 2x$

$y'(1 + 6y) = 2x$

$y' = \frac{2x}{1 + 6y}$

5. $\tan(y) = \sec(y) + x$

$y' \sec^2(y) = \sec(y)\tan(y) y' + 1$

$y'(\sec^2(y) - \sec(y)\tan(y)) = 1$

$y' = \frac{1}{\sec(y)(\sec(y) - \tan(y))}$

6. $\ln(x) + y^2 = y + 2$

$\frac{1}{x} + 2yy' = y'$

$\frac{1}{x} = y'(1 - 2y)$

$\frac{1}{x(1 - 2y)} = y'$

Slope of tangent line at $(1, 2)$ is $\frac{1}{(1)(1 - 2(2))} = \frac{1}{-3} = -\frac{1}{3}$.

$y = mx + b$

$(2) = \left(-\frac{1}{3}\right)(1) + b$

$2 = -\frac{1}{3} + b$

$2 + \frac{1}{3} = b$

$b = \frac{7}{3}$

Equation of tangent line is $y = -\frac{1}{3}x + \frac{7}{3}$

Slope of normal line is 3

$y = mx + b$

$(2) = (3)(1) + b$

$2 = 3 + b$

$-1 = b$

Equation of normal line is
$y = 3x - 1$

7. $e^x + y^2 = 4$ at $(0, 2)$
    $e^x + 2yy' = 0$
    $e^x = -2yy'$
    $y' = \dfrac{e^x}{-2y}$

    Slope of tangent line at $(0, 2)$.
    is $\dfrac{e^{(0)}}{-2(2)} = \dfrac{1}{-4} = -\dfrac{1}{4}$

    $(2) = \left(-\dfrac{1}{4}\right)(0) + b$
    $2 = b$

    Equation of tangent line is
    $y = -\dfrac{1}{4}x + 2$.

    Slope of normal line is 4.
    $(2) = (4)(0) + b$
    $2 = b$

    Equation of normal line is
    $y = 4x + 2$.

8. $y = \sqrt{1-x^2}$
    domain $[-1, 1]$
    range $[0, 1]$

9. $y = \sin^2(x)$
    domain $(-\infty, \infty)$
    range $[0, 1]$

10. $y = -\tan(x)$
    domain $(-\infty, \infty)$ except odd multiples of $\dfrac{\pi}{2}$
    range $(-\infty, \infty)$

11. $y = e^x \quad x = \ln(t)$
    $\dfrac{dy}{dt} = \dfrac{dy}{dx} \cdot \dfrac{dx}{dt}$
    $= e^x \cdot \dfrac{1}{t}$
    $= e^{\ln(t)} \cdot \dfrac{1}{t} = (t)\left(\dfrac{1}{t}\right) = 1$

12. $y = \sin(x) \quad x = \ln(\sin(t))$
    $\dfrac{dy}{dt} = \dfrac{dy}{dx} \cdot \dfrac{dx}{dt}$
    $= \cos(x) \cdot \dfrac{1}{\sin(t)} \cdot \cos(t)$
    $= \cos(x) \cdot \cot(t)$
    $= \cos(\ln(\sin(t))) \cdot \cot(t)$

Lesson Practice 15A

1. $f(x) = x^3 - 3x^2 + 1$
    $f'(x) = 3x^2 - 6x$
    $f'(x) = (3x)(x-2)$
    $(3x)(x-2) = 0$
    Critical values: $x = 0, x = 2$

    When $x = -1$, $f'(-1)$ is positive.
    When $x = 1$, $f'(1)$ is negative.
    When $x = 3$, $f'(3)$ is positive.
    There is a local max at $(0, 1)$.
    There is a local min at $(2, -3)$.
    $f(x)$ is increasing on $(-\infty, 0)$ and $(2, \infty)$.
    $f(x)$ is decreasing on $(0, 2)$.

2. $f(x) = \dfrac{1}{3-x}$
    $f'(x) = \dfrac{(3-x)(0) - (1)(-1)}{(3-x)^2}$
    $f'(x) = \dfrac{1}{(3-x)^2}$
    Critical value: $x = 3$

    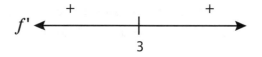

When x = 0, $f'(0)$ is positive.
When x = 4, $f'(4)$ is positive.
$f(x)$ is increasing everywhere.
There are no local maxima or minima.

3. $f(x) = x^2 - 4x + 2$
$f'(x) = 2x - 4$
$2x - 4 = 0$
$2x = 4$
$x = 2$

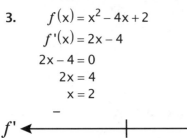

Critical value: x = 2
When x = 0, $f'(0)$ is negative.
When x = 3, $f'(3)$ is positive.
There is a local min at $(2, -2)$.
$f(x)$ is decreasing on $(-\infty, 2)$ and increasing on $(2, \infty)$.

4. $f(x) = 2x^2 e^x$
$f'(x) = 2[x^2 e^x + 2x(e^x)]$
$f'(x) = (2xe^x)(x + 2)$
$(2xe^x)(x + 2) = 0$
$(x)(x + 2) = 0$
$x = 0, -2$

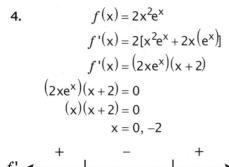

When x = -3, $f'(-3)$ is positive.
When x = -1, $f'(-1)$ is negative.
When x = 1, $f'(1)$ is positive.

There is a local max at $\left(-2, \dfrac{8}{e^2}\right)$.

There is a local min at $(0, 0)$.
$f(x)$ is increasing on $(-\infty, -2)$ and $(0, \infty)$.
$f(x)$ is decreasing on $(-2, 0)$.

5. $f(x) = x^4 - 2x^3$
$f'(x) = 4x^3 - 6x^2$
$f'(x) = (2x^2)(2x - 3)$
$(2x^2)(2x - 3) = 0$
$x = 0, x = \dfrac{3}{2}$ are critical points.

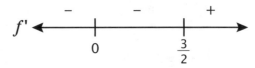

When x = -1, $f'(-1)$ is negative.
When x = 1, $f'(1)$ is negative.
When x = 2, $f'(2)$ is positive.

There is a local min at $\left(\dfrac{3}{2}, -\dfrac{27}{16}\right)$.

$f(x)$ is decreasing on $\left(-\infty, \dfrac{3}{2}\right)$.

$f(x)$ is increasing on $\left(\dfrac{3}{2}, \infty\right)$.

6. $f(x) = \dfrac{2}{2 + x^2}$

$f'(x) = \dfrac{(2 + x^2)(0) - (2)(2x)}{(2 + x^2)^2}$

$f'(x) = \dfrac{-4x}{(2 + x^2)^2}$

$\dfrac{-4x}{(2 + x^2)^2} = 0$

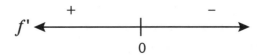

The only critical point is x = 0, because the denominator will always be positive.
When x = -1, $f'(-1)$ is positive.
When x = 1, $f'(1)$ is negative.
There is a local max at $(0, 1)$.
$f(x)$ is increasing on $(-\infty, 0)$.
$f(x)$ is decreasing on $(0, \infty)$.

7. $f(x) = -2x^3 - 3x^2 + 12x + 10$
$f'(x) = -6x^2 - 6x + 12$
$f'(x) = -6(x^2 + x - 2)$
$f'(x) = -6(x-1)(x+2)$
$0 = -6(x-1)(x+2)$
$x = 1, x = -2$ critical points

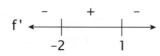

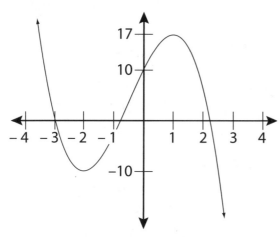

There is a local min at $(-2, -10)$.
There is a local max at $(1, 17)$.
$f(x)$ is increasing on $(-2, 1)$.
$f(x)$ is decreasing on $(-\infty, -2)$ and $(1, \infty)$.

X	Y
-3	1
-2	-10
-1	-3
0	10
1	17
2	6

Lesson Practice 15B

1. $f(x) = 2x^2 - x^4$
$f'(x) = 4x - 4x^3$
$f'(x) = (4x)(1 - x^2)$
$(4x)(1-x)(1+x) = 0$
Critical values: $x = 0, x = -1, x = 1$

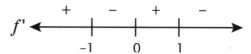

When $x = -2$, $f'(-2)$ is positive.
When $x = -\frac{1}{2}$, $f'\left(-\frac{1}{2}\right)$ is negative.
When $x = \frac{1}{2}$, $f'\left(\frac{1}{2}\right)$ is positive.
When $x = 2$, $f'(2)$ is negative.
There is a local max at $(-1, 1)$ and $(1, 1)$.
There is a local min at $(0, 0)$.

2. $f(x) = x^4 + 3$
$f'(x) = 4x^3$
$4x^3 = 0$
$x = 0$
Critical value: $x = 0$

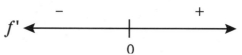

When $x = -1$, $f'(-1)$ is negative.
When $x = 1$, $f'(1)$ is positive.
$f(x)$ is decreasing on $(-\infty, 0)$.
$f(x)$ is increasing on $(0, \infty)$.
There is a local min at $(0, 3)$.

3. $f(x) = x^5 - 5x^4$
    $f'(x) = 5x^4 - 20x^3$
    $f'(x) = 5x^3(x-4)$
    $(5x^3)(x-4) = 0$
    $x - 4 = 0 \quad x = 0$
    $x = 4$
    Critical value at $x = 4$ and $x = 0$.

When $x = -1$, $f'(-1)$ is positive.
When $x = 1$, $f'(1)$ is negative.
When $x = 5$, $f'(5)$ is positive.
There is a local max at $(0, 0)$.
There is a local min at $(4, -256)$.
$f(x)$ is increasing on $(-\infty, 0)$ and $(4, \infty)$.
$f(x)$ is decreasing on $(0, 4)$.

4. $f(x) = 3x^4 - 4x^3 - 12x^2 + 2$
    $f'(x) = 12x^3 - 12x^2 - 24x$
    $f'(x) = 12x(x^2 - x - 2)$
    $(12x)(x^2 - x - 2) = 0$
    $(12x)(x-2)(x+1)$
    $x = 0 \quad x - 2 = 0 \quad x + 1 = 0$
    $\qquad \qquad x = 2 \qquad x = -1$

    Critical values at $x = 0$, $x = 2$, $x = -1$.

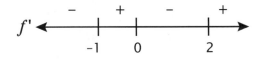

When $x = -2$, $f'(-2)$ is negative.
When $x = -\frac{1}{2}$, $f'\left(-\frac{1}{2}\right)$ is positive.
When $x = 1$, $f'(1)$ is negative.
When $x = 3$, $f'(3)$ is positive.
There is a local min at $(-1, -3)$ and $(2, -30)$.
There is a local max at $(0, 2)$.
$f(x)$ is increasing on $(-1, 0)$ and $(2, \infty)$.
$f(x)$ is decreasing on $(-\infty, -1)$ and $(0, 2)$.

5. $f(x) = \dfrac{x^2}{e^x}$
    $f'(x) = \dfrac{(e^x)(2x) - (x^2)(e^x)}{(e^x)^2}$
    $f'(x) = \dfrac{(e^x)(2x - x^2)}{(e^x)^2}$
    $f'(x) = \dfrac{2x - x^2}{e^x}$
    $f'(x) = \dfrac{(x)(2 - x)}{e^x}$
    $\dfrac{(x)(2-x)}{e^x} = 0$
    Critical values are $x = 0$, $x = 2$.

When $x = -1$, $f'(-1)$ is negative.
When $x = 1$, $f'(1)$ is positive.
When $x = 3$, $f'(3)$ is negative.
There is a local min at $(0, 0)$.
There is a local max at $\left(2, \dfrac{4}{e^2}\right)$.
$f(x)$ is decreasing at $(-\infty, 0)$ and $(2, \infty)$.
$f(x)$ is increasing at $(0, 2)$.

6. $f(x) = x^2 - x + 2$
    $f'(x) = 2x - 1$
    $2x - 1 = 0$
    $2x = 1$
    $x = \frac{1}{2}$

    Critical point: $x = \frac{1}{2}$
    When $x = 0$, $f'(0)$ is negative.
    When $x = 1$, $f'(1)$ is positive.

    $f(x)$ is increasing on $\left(\frac{1}{2}, \infty\right)$.

    $f(x)$ is decreasing on $\left(-\infty, \frac{1}{2}\right)$.

    There is a local min at $\left(\frac{1}{2}, 1\frac{3}{4}\right)$.

7. $f(x) = x^3 - 6x^2 + 9x$
    $f'(x) = 3x^2 - 12x + 9$
    $3x^2 - 12x + 9 = 0$
    $x^2 - 4x + 3 = 0$
    $(x-1)(x-3) = 0$
    $x = 1, x = 3$ critical points

In order to see what is happening with $f'(x)$, take values around the critical values 1 and 3. A table is included so that you can see how the +'s and −'s were determined. We can draw a sketch of this curve. It must rise until $x = 1$ and fall until $x = 3$ and then rise again.

Increasing on $(-\infty, 1)$ and $(3, \infty)$.
Decreasing on $(1, 3)$.

When $x = 0$, $f'(0)$ is positive.
When $x = 2$, $f'(2)$ is negative.
When $x = 4$, $f'(4)$ is positive.
$f(x)$ is increasing on $(-\infty, 1)$ and $(3, \infty)$.
$f(x)$ is decreasing on $(1, 3)$.
There is a max at $(1, 4)$.
There is a min at $(3, 0)$.

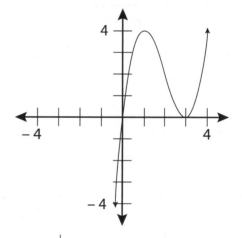

x	y
0	0
1	4
2	2
3	0
4	4

Lesson Practice 15C

1. $f(x) = 3x^4 - 4x^3 - 12x^2 + 1$
   $f'(x) = 12x^3 - 12x^2 - 24x$
   $f'(x) = (12x)(x^2 - x - 2)$
   $(12x)(x^2 - x - 2) = 0$
   $(12x)(x+1)(x-2) = 0$
   $x = 0, -1, 2$

   Critical points: $x = 0, -1, 2$

   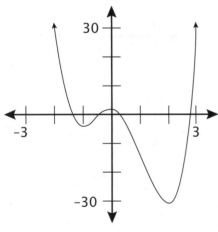

   $f'(x)$ is increasing on $(-1, 0)$ and $(2, \infty)$.
   $f'(x)$ is decreasing on $(-\infty, -1)$ and $(0, 2)$.

x	y
-2	33
-1	-4
0	1
1	-12
2	-31
3	28

   Local max at $(0, 1)$.
   Local min at $(-1, -4)$ and $(2, -31)$.

2. $f(x) = \frac{1}{3}x^3 - \frac{1}{2}x^2 - 2x + 2$
   $f'(x) = x^2 - x - 2$
   $f'(x) = (x+1)(x-2)$
   $(x+1)(x-2) = 0$
   $x = -1, 2$ critical points

   $f(x)$ is increasing on $(-\infty, -1)$ and $(2, \infty)$.
   $f(x)$ is decreasing on $(-1, 2)$.

   There is a local max at $\left(-1, \frac{19}{6}\right)$.
   There is a local min at $\left(2, -\frac{4}{3}\right)$.

   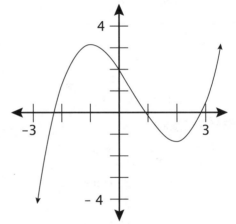

x	y
0	2
-2	$\frac{4}{3}$
-1	$\frac{19}{16}$
2	$-\frac{4}{3}$
3	$\frac{1}{2}$

LESSON PRACTICE 15C - LESSON PRACTICE 15D

3. $f(x) = 3x^2 + bx + 2$
$f'(x) = 6x + b$
$6x + b = 0$
$6x = -b$
$x = -\dfrac{b}{6}$

When $x = -1$:
$(-1) = -\dfrac{b}{6}$
$(-1)(-6) = b$
$b = 6$

4. $f(x) = ax^2 + 2x + 4$
$f'(x) = 2ax + 2$
$f'(x) = 2(ax + 1)$
$2(ax + 1) = 0$
$ax + 1 = 0$
$ax = -1$
$x = -\dfrac{1}{a}$

When $x = 3$:
$(3) = -\dfrac{1}{a}$
$3a = -1$
$a = -\dfrac{1}{3}$

Check: $f(x) = -\dfrac{1}{3}x^2 + 2x + 4$
$f'(x) = -\dfrac{2}{3}x + 2$
$f'(x) = -2\dfrac{1}{3}x - 1$

$-2\dfrac{1}{3}x - 1 = 0$
$\dfrac{1}{3}x - 1 = 0$
$\dfrac{1}{3}x = 1$
$x = 3$

3 is a max

5. $\displaystyle\sum_{i=-2}^{4} 3i = -6 - 3 + 0 + 3 + 6 + 9 + 12 = 21$

6. $\displaystyle\sum_{i=1}^{3} 2i^2 = 2 + 8 + 18 = 28$

7. $\displaystyle\sum_{i=1}^{2} (i^3 + 2) = 3 + 10 = 13$

8. $\displaystyle\sum_{i=-3}^{-1} (i^4 + 3) = 84 + 19 + 4 = 107$

Lesson Practice 15D

1. $f(x) = -2x^3 - 3x^2 + 12x + 10$
$f'(x) = -6x^2 - 6x + 12$
$f'(x) = -6(x^2 + x - 2)$
$f'(x) = -6(x - 1)(x + 2)$
$-6(x - 1)(x + 2) = 0$
$x = 1$, $x = -2$ critical points

$f(x)$ is increasing on $(-2, 1)$.
$f(x)$ is decreasing on $(-\infty, -2)$ and $(1, \infty)$.
There is a local max at $(1, 17)$.
There is a local min at $(-2, -10)$.

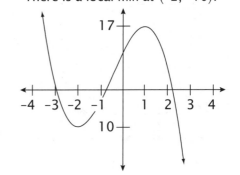

2. $f(x) = x^3 + 2x^2 - 15x + 20$
$f'(x) = 3x^2 + 4x - 15$
$f'(x) = (3x - 5)(x + 3)$
$(3x - 5)(x + 3) = 0$
$x = \frac{5}{3}$, $x = -3$ critical points

There is a local max at $(-3, 56)$.
There is a local min at $\left(\frac{5}{3}, \frac{140}{27}\right)$.
$f(x)$ is increasing on $(-\infty, -3)$
and $\left(\frac{5}{3}, \infty\right)$.
$f(x)$ is decreasing on $\left(-3, \frac{5}{3}\right)$.

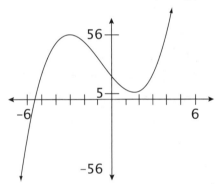

3. $f(x) = x^2 e^x$
$f'(x) = x^2 e^x + (e^x)(2x)$
$f'(x) = (e^x)(x^2 + 2x)$
$(e^x)(x^2 + 2x) = 0$
$x^2 + 2x = 0$
$(x)(x + 2) = 0$
$x = 0$, $x = -2$ critical points

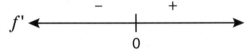

When $x = -10$, $f'(-10)$ is positive.
When $x = -1$, $f'(-1)$ is negative.
When $x = 1$, $f'(1)$ is positive.
$f(x)$ is increasing on $(-\infty, -2)$ and $(0, \infty)$.
$f(x)$ is decreasing on $(-2, 0)$.
There is a local min at $(0, 0)$.
There is a local max at $\left(-2, \frac{4}{e^2}\right)$.

4. $f(x) = \ln(x^2 + 2)$
$f'(x) = \frac{2x}{x^2 + 2}$
$\frac{2x}{x^2 + 2} = 0$

f' — | +
     0

$x = 0$ critical point
When $x = -1$, $f'(-1)$ is negative.
When $x = 1$, $f'(1)$ is positive.
There is a local min at $(0, \ln 2)$.
$f(x)$ is decreasing on $(\infty, 0)$.
$f(x)$ is increasing on $(0, \infty)$.

5. $f(x) = 2x^2 + bx + 2$
$f'(x) = 4x + b$
$4x = -b$
$x = -\frac{b}{4}$
$(2) = -\frac{b}{4}$
$b = -8$

6. $f(x) = 4x^2 + ax + 3$
    $f'(x) = 8x + a$

    $8x + a = 0$
    $8x = -a$
    $x = -\dfrac{a}{8}$

    When $x = -1$:
    $-1 = -\dfrac{a}{8}$
    $-8 = -a$
    $a = 8$

7. $\displaystyle\sum_{i=1}^{3} (i^3 + 3) = 4 + 11 + 30 = 45$

8. $\displaystyle\sum_{i=-2}^{0} \dfrac{1}{i+3} = \dfrac{1}{1} + \dfrac{1}{2} + \dfrac{1}{3} = 1\dfrac{5}{6}$

9. $\displaystyle\sum_{i=1}^{4} e^i = e^1 + e^2 + e^3 + e^4$

10. $\displaystyle\sum_{i=-5}^{-4} (i^3 + 2i^2) = [(-5)^3 + 2(25)] + [(-4)^3 + 2(16)]$
    $= -125 + 50 - 64 + 32$
    $= -107$

Lesson Practice 16A

1. $f(x) = x^2 - 5x + 6$
    $f'(x) = 2x - 5$;

    critical point: $x = 2\dfrac{1}{2}$

    $f''(x) = 2$;
    no inflection pts.; concave up everywhere

x	$f(x)$	$f'(x)$	$f''(x)$	Conclusions
2	0	–	+	
$2\dfrac{1}{2}$	$-\dfrac{1}{4}$	0	+	minimum-absolute
3	0	+	+	

    This is a parabola. ("smile")

    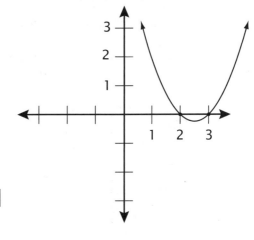

2. $f(x) = x^{\frac{5}{3}} - 1$

$f'(x) = \frac{5}{3} x^{\frac{2}{3}};$

critical point: $x = 0$

$f''(x) = \frac{5}{3} \cdot \frac{2}{3} x^{-\frac{1}{3}} = \frac{10}{9\sqrt[3]{x}};$

possible inflection pt $x = 0$

x	$f(x)$	$f'(x)$	$f''(x)$	Conclusions
−1	−2	+	−	concave down
0	−1	0	undefined	inflection point
1	0	+	+	concave up

$(-\infty, 0)$ concave down; $(0, \infty)$ concave up

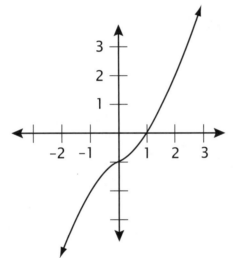

There are no maximum or minimum values.

3. $f(x) = x^3 - 3x$

$f'(x) = 3x^2 - 3$
$= 3(x^2 - 1)$
$= 3(x + 1)(x - 1);$ critical pts ± 1

$f''(x) = 6x$

possible inflection pt, $x = 0$

x	$f(x)$	$f'(x)$	$f''(x)$	Conclusions
−2	−2	+	−	
−1	2	0	−	max−local, concave down
$-\frac{1}{2}$	$1\frac{3}{8}$	−	−	
0	0	−	0	inflection point
$\frac{1}{2}$	$-1\frac{3}{8}$	−	+	
1	−2	0	+	min−local, concave up
2	2	+	+	

$(-\infty, 0)$ concave down; $(0, \infty)$ concave up

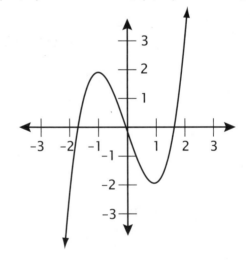

4. $f(x) = \frac{1}{4}x^4$

$f'(x) = x^3$;   0 is a critical point

$f''(x) = 3x^2$;   0 is a possible inflection pt

x	f(x)	f'(x)	f''(x)	Conclusions
-1	1/4	-	+	
0	0	0	0	min– changes from dec to inc } absolute concave up everywhere
1	1/4	+	+	

There is no inflection point.
There is an absolute minimum at $(0, 0)$.

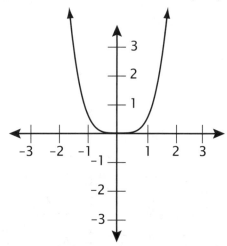

x	f(x)	f'(x)	f''(x)	Conclusions
0	-1	+	+	} concave up
$\frac{\pi}{4}$	0	+	0	inflection pt
$\frac{\pi}{2}$	1	+	-	
$\frac{3\pi}{4}$	$\sqrt{2}$	0	-	max absolute } concave down
π	1	-	-	
$\frac{5\pi}{4}$	0	-	0	inflection point
$\frac{3\pi}{2}$	-1	-	+	
$\frac{7\pi}{4}$	$-\sqrt{2}$	0	+	min absolute } concave up
2π	-1	+	+	

concave down $\left(\frac{\pi}{4}, \frac{5\pi}{4}\right)$

concave up $\left[\left(0, \frac{\pi}{4}\right) \text{ and } \left(\frac{5\pi}{4}, 2\pi\right)\right]$

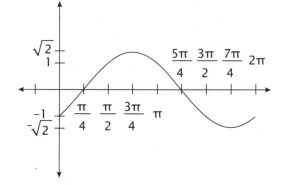

5. $f(x) = \sin(x) - \cos(x);\ [0, 2\pi]$

$f'(x) = \cos(x) + \sin(x) = 0$,

$\cos(x) = -\sin(x)$,

$x = \frac{3\pi}{4}, \frac{7\pi}{4}$; critical points

$f''(x) = -\sin(x) + \cos(x) = 0$,

$\sin(x) = \cos(x)$,

$x = \frac{\pi}{4}, \frac{5\pi}{4}$; possible inflection points

6. $f(x) = \sqrt{1-x^2}$

$f'(x) = \frac{1}{2}(1-x^2)^{-\frac{1}{2}}(-2x)$

$= -x(1-x^2)^{-\frac{1}{2}}$

critical pts $x = 0, \pm 1$

$= \frac{-x}{\sqrt{1-x^2}} = (-x)(1-x^2)^{-\frac{1}{2}}$

$f''(x) = (-x) \cdot \left[-\frac{1}{2}(1-x^2)^{-\frac{3}{2}}(-2x)\right] + (1-x^2)^{-\frac{1}{2}}(-1)$

$= -x^2(1-x^2)^{-\frac{3}{2}} - (1-x^2)^{-\frac{1}{2}}$

$= (1-x^2)^{-\frac{3}{2}}\left[-x^2 - (1-x^2)\right]$

$= (1-x^2)^{-\frac{3}{2}}(-1)$

$= \frac{-1}{(1-x^2)^{\frac{3}{2}}}$

possible inflection pts $x = \pm 1$

x	$f(x)$	$f'(x)$	$f''(x)$	Conclusions
-1	0	und.	und.	min– absolute
$-\frac{1}{2}$	.87	+	–	
0	1	0	-1	max– absolute
$\frac{1}{2}$	.87	–	–	
1	0	und.	und.	min– absolute

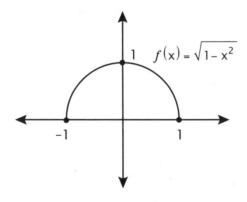

This function is constrained. It has to be positive under the square root sign. The interval is [-1, 1]. The function is concave down on this interval. There are two locations where the function has an absolute minimum: $(-1, 0)$ and $(1, 0)$.
This graph is a semi-circle.

7.

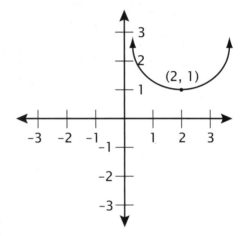

8.

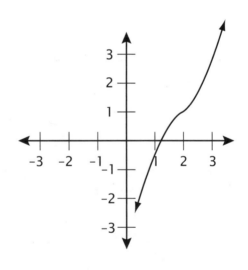

Lesson Practice 16B

1. $f(x) = x^3 - 9x$

$f'(x) = 3x^2 - 9$
$= 3(x^2 - 3);$
$x = \pm\sqrt{3}$; critical points

$f''(x) = 6x;$
$x = 0$; possible inflection points

x	$f(x)$	$f'(x)$	$f''(x)$	Conclusions
−3	0	+	−	
−√3	+6√3	0	−	max–local } concave down
−1	8	−	−	
0	0	−	0	inflection pt
+1	−8	−	+	
√3	−6√3	0	+	min–local } concave up
3	0	+	+	

concave down $(-\infty, 0)$
concave up $(0, +\infty)$

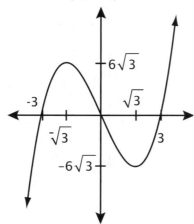

2. $f(x) = x^2 - 6x + 3$

$f'(x) = 2x - 6;$ $x = 3$, critical point
$f''(x) = 2;$ no inflection points

x	$f(x)$	$f'(x)$	$f''(x)$	Conclusions
0	3	−	+	
3	−6	0	+	min–absolute } concave up everywhere
4	−5	+	+	

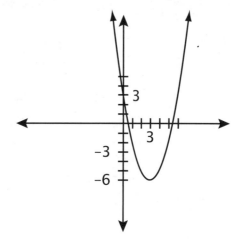

3.
$$f(x) = x^4 - 6x^2$$
$$f'(x) = 4x^3 - 12x$$
$$= 4x(x^2 - 3)$$
$x = 0, \pm\sqrt{3}$; critical points
$$f''(x) = 12x^2 - 12$$
$$= 12(x^2 - 1)$$
$x = \pm 1$ possible inflection pts

x	$f(x)$	$f'(x)$	$f''(x)$	Conclusions
-2	-8	$-$	$+$	
$-\sqrt{3}$	-9	0	$+$	min– absolute
-1	-5	$+$	0	inflection point
0	0	0	$-$	max– local
1	-5	$-$	0	inflection point
$\sqrt{3}$	-9	0	$+$	min– absolute
2	-8	$+$	$+$	

concave down $(-1, 1)$
concave up $(-\infty, -1)$ and $(1, +\infty)$

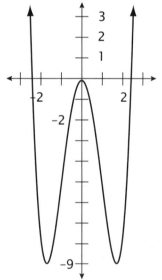

4.
$$f(x) = 12x - x^3$$
$$f'(x) = 12 - 3x^2$$
$$= 3(4 - x^2)$$
critical points $x = \pm 2$
$$f''(x) = -6x$$
possible inflection pt $x = 0$

x	$f(x)$	$f'(x)$	$f''(x)$	Conclusions
-3	-9	$-$	$+$	
-2	-16	0	$+$	min-local
0	0	$+$	0	inflection pt
2	16	0	$-$	max-local
3	9	$-$	$-$	

concave up $(-\infty, 0)$, concave down $(0, +\infty)$

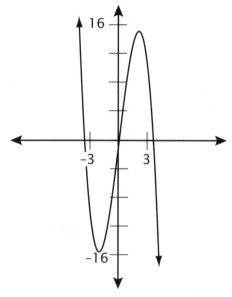

5. $f(x) = \dfrac{\ln(x)}{x}$

Hint: Find vertical and horizontal asymptotes.
When $x = 0$, we have a vertical asymptote.

$$\lim_{x \to \infty} \frac{\ln(x)}{x} = 0$$

There is a horizontal asymptote at $y = 0$.

$$f'(x) = \frac{x\left(\frac{1}{x}\right) - [\ln(x)](1)}{x^2} = \frac{1 - \ln(x)}{x^2}$$

$\dfrac{1 - \ln(x)}{x^2} = 0$ when:

$$1 - \ln(x) = 0$$
$$1 = \ln(x)$$
$$e^1 = x$$

critical points $x = 0, e$

$$f''(x) = \frac{x^2\left(-\frac{1}{x}\right) - (1 - \ln(x))(2x)}{x^4}$$
$$= \frac{-x - 2x + 2x\ln(x)}{x^4}$$
$$= \frac{-3x + 2x\ln(x)}{x^4}$$
$$= \frac{-3 + 2\ln(x)}{x^3} \text{ or } \frac{2\ln(x) - 3}{x^3}$$

$f''(x) = 0$ when:

$$2\ln(x) - 3 = 0$$
$$2\ln(x) = 3$$
$$\ln(x) = \frac{3}{2}$$
$$x = e^{\frac{3}{2}} = \left(\sqrt{e}\right)^3$$

possible inflection points $-\left(x = 0, e^{\frac{3}{2}}\right)$

x	$f(x)$	$f'(x)$	$f''(x)$	Conclusions
0	DNE	DNE	DNE	vertical asymptote
1	0	+	−	
e	$\frac{1}{e}$	0	−	max– absolute
$e^{\frac{3}{2}}$	$\frac{3}{2}\left(e^{-\frac{3}{2}}\right)$	−	0	point of inflection
e^2	$2(e^{-2})$	−	+	

concave down $\left(0, e^{\frac{3}{2}}\right)$

concave up $\left(e^{\frac{3}{2}}, \infty\right)$

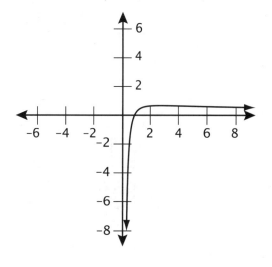

LESSON PRACTICE 16B - LESSON PRACTICE 16B

6. $f(x) = x^4 - 4x$
$f'(x) = 4x^3 - 4$
$\quad = 4(x^3 - 1) = (x-1)(x^2 + x + 1)$
critical point $x = 1$
$f''(x) = 12x^2$
Possible inflection point pt $(x = 0)$

x	$f(x)$	$f'(x)$	$f''(x)$	Conclusions
−1	5	−	+	
0	0	−	0	
1	−3	0	+	min-absolute at x = 1
2	8	+	+	

No inflection point: because it is concave up everywhere

7.
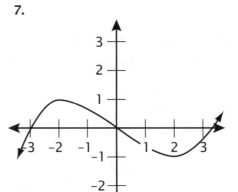

8.
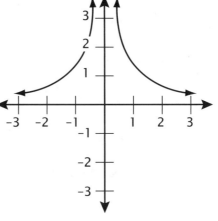

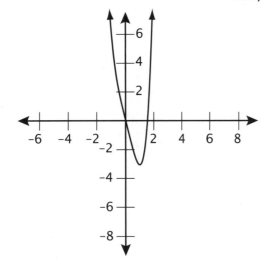

Lesson Practice 16C

1.
$f(x) = x^3 - 6x^2 + 9x$
$f'(x) = 3x^2 - 12x + 9$
$\quad = 3(x^2 - 4x + 3)$
$\quad = 3(x-1)(x-3)$
$\quad$ critical points $x = 1, 3$
$f''(x) = 6x - 12$
$\quad = 6(x-2)$
$\quad$ possible inflection pt $x = 2$

x	$f(x)$	$f'(x)$	$f''(x)$	Conclusions
0	0	+	−	
1	4	0	−	max-local
2	2	−	0	inflection pt
3	0	0	+	min-local
4	4	+	+	

$(-\infty, 2)$ concave down
$(2, \infty)$ concave up

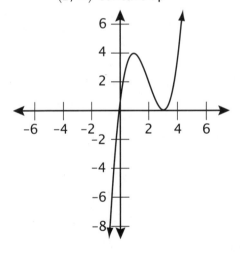

2.
$f(x) = 10 + 12x - 3x^2 - 2x^3$
$f'(x) = 12 - 6x - 6x^2$
$\quad = 6(2 - x - x^2)$
$\quad = 6(2 + x)(1 - x)$
$\quad$ critical points $x = 1, -2$
$f''(x) = -6 - 12x$
$\quad = -6(1 + 2x)$
$\quad$ possible inflection pt $x = -\frac{1}{2}$

x	$f(x)$	$f'(x)$	$f''(x)$	Conclusions
−3	1	−	+	
−2	−10	0	+	min-local
$-\frac{1}{2}$	$3\frac{1}{2}$	+	0	inflection pt.
1	17	0	−	max-local
2	6	−	−	

concave up $\left(-\infty, -\frac{1}{2}\right)$
concave down $\left(-\frac{1}{2}, \infty\right)$

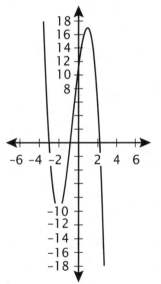

3. $f(x) = \dfrac{x^2 + x + 4}{x + 1}$

   $(x = -1)$ vertical asymptote

   $f'(x) = \dfrac{(x+1)(2x+1) - (x^2 + x + 4)}{(x+1)^2}$

   $= \dfrac{2x^2 + x + 2x + 1 - x^2 - x - 4}{(x+1)^2}$

   $= \dfrac{x^2 + 2x - 3}{(x+1)^2}$

   $= \dfrac{(x-1)(x+3)}{(x+1)^2}$

   critical pts $x = -1, 1, -3$

$f''(x) = \dfrac{(x+1)^2(2x+2) - (x^2 + 2x - 3)2(x+1)(1)}{(x+1)^4}$

$= \dfrac{(x^2 + 2x + 1)(2x+2) - (x^2 + 2x - 3)(2x+2)}{(x+1)^4}$

$= \dfrac{2x^3 + 4x^2 + 2x + 2x^2 + 4x + 2 - 2x^3 - 2x^2 - 4x^2 - 4x + 6x + 6}{(x+1)^4}$

$= \dfrac{8x + 8}{(x+1)^4} = \dfrac{8(x+1)}{(x+1)^4} = \dfrac{8}{(x+1)^3}$

$x = -1$ possible inflection point

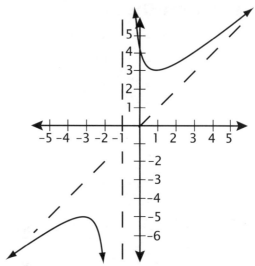

Oblique asymptote at y = x.

x	$f(x)$	$f'(x)$	$f''(x)$	Conclusions
-4	$\dfrac{-16}{3}$	+	−	
-3	−5	0	−	max
-2	−6	−	−	$(-1, \infty)$ concave up
-1	und	und	und	vertical asymptote
0	4	−	+	$(-\infty, -1)$ concave down
1	3	0	+	min
2	$\dfrac{10}{3}$	+	+	

Since a point does not exist when $x = -1$, there is no true inflection point.
The graph is concave down for all $x < -1$ and concave up for all $x > -1$.

4. $f(x) = 3x - 2x^2 - \frac{4}{3}x^3$

$f'(x) = 3 - 4x - 4x^2$
$= -(4x^2 + 4x - 3)$
$= -(2x - 1)(2x + 3)$

critical points $x = \frac{1}{2}, -\frac{3}{2}$

$f''(x) = -4 - 8x$
$= -4(1 + 2x)$

possible inflection pt $x = -\frac{1}{2}$

x	f(x)	f'(x)	f''(x)	Conclusions
-2	$-3\frac{1}{3}$	$-$	$+$	
$-\frac{3}{2}$	$-4\frac{1}{2}$	0	$+$	min-local
$-\frac{1}{2}$	$-1\frac{5}{6}$	$+$	0	inflection pt
0	0	$+$	$-$	
$\frac{1}{2}$	$\frac{5}{6}$	0	$-$	max-local
1	$-\frac{1}{3}$	$-$	$-$	

$\left(-\infty, -\frac{1}{2}\right)$ concave up

$\left(-\frac{1}{2}, \infty\right)$ concave down

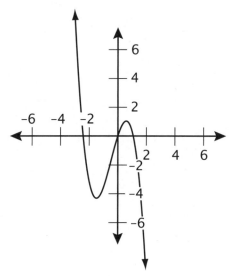

5. $f(x) = \frac{2x^2 - 5}{x^2 + 4x + 4} = \frac{2x^2 - 5}{(x+2)(x+2)}$

vertical asymptote at $x = -2$

$\lim_{x \to \infty} \frac{2x^2 - 5}{x^2 + 4x + 5} = 2$

horizontal asymptote at $y = 2$

There are no oblique asymptotes.

6. $f(x) = \frac{2x^2}{x+1}$

There is a vertical asymptote at $x = -1$.

$$\begin{array}{r} 2x -2 \\ x+1 \overline{\smash{)}2x^2 + 0x +0} \\ \underline{2x^2 + 2x} \\ -2x \\ \underline{-2x -2} \\ 2R \end{array}$$

oblique asymptote at $y = 2x - 2$

7. $f(x) = \frac{x^2 + x - 2}{x + 2}$

$f(x) = \frac{(x-1)\cancel{(x+2)}}{\cancel{(x+2)}}$

There is no vertical asymptote, but there is a hole at $x = -2$.

There are no horizontal or oblique asymptotes.

8. $f(x) = \dfrac{2x^2}{x+1}$

$f'(x) = \dfrac{(x+1)(4x) - 2x^2(1)}{(x+1)^2}$

$= \dfrac{4x^2 + 4x - 2x^2}{(x+1)^2}$

$= \dfrac{2x^2 + 4x}{(x+1)^2} = \dfrac{2x(x+2)}{(x+1)^2}$

critical pts $0, -2, -1$

$f''(x) = \dfrac{(x+1)^2(4x+4) - (2x^2+4x)\,2(x+1)(1)}{(x+1)^4}$

$= \dfrac{4(x+1)[(x+1)^2 - x(x+2)]}{(x+1)^4}$

$= \dfrac{4(x+1)[x^2+2x+1-x^2-2x]}{(x+1)^4}$

$= \dfrac{4(1)}{(x+1)^3} = \dfrac{4}{(x+1)^3}$

possible inflection pt $x = -1$

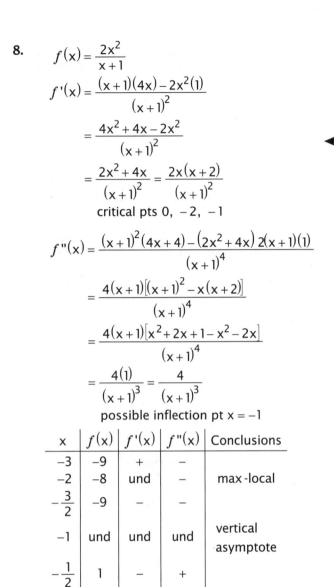

x	$f(x)$	$f'(x)$	$f''(x)$	Conclusions
-3	-9	$+$	$-$	
-2	-8	und	$-$	max-local
$-\dfrac{3}{2}$	-9	$-$	$-$	
-1	und	und	und	vertical asymptote
$-\dfrac{1}{2}$	1	$-$	$+$	
0	0	0	$+$	min-local
1	1	$+$	$+$	

$y = 2x - 2$ is an oblique asymptote

Lesson Practice 16D

1. $f(x) = x^3 + 3x^2 - 2$

$f'(x) = 3x^2 + 6x$
$= 3x(x+2)$

critical pts $x = 0, -2$

$f''(x) = 6x + 6$
$= 6(x+1)$

possible inflection pt $x = -1$

x	$f(x)$	$f'(x)$	$f''(x)$	Conclusions
-3	-2	+	-	
-2	2	0	-	max-local
-1	0	-	0	inflection pt
0	-2	0	+	min-local
1	2	+	+	

$(-\infty, -1)$ concave down
$(-1, \infty)$ concave up

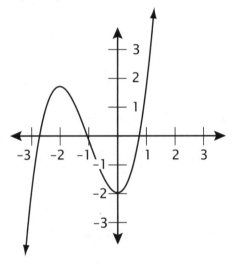

2. $f(x) = x^4 - 4x^2 + 4$

$f'(x) = 4x^3 - 8x$
$= 4x(x^2 - 2)$

critical pts $x = 0, \pm\sqrt{2}$

$f''(x) = 12x^2 - 8$
$= 4(3x^2 - 2)$

possible inflection pts $x = \pm\sqrt{\frac{2}{3}}$

x	$f(x)$	$f'(x)$	$f''(x)$	Conclusions
-2	4	-	+	
$-\sqrt{2}$	0	0	+	min-absolute (global)
$-\sqrt{\frac{2}{3}}$	$1\frac{7}{9}$	+	0	inflection pt
0	4	0	-	max-local
$\sqrt{\frac{2}{3}}$	$1\frac{7}{9}$	-	0	inflection pt
$\sqrt{2}$	0	0	+	min-absolute (global)
2	4	+	+	

$\left(-\infty, -\sqrt{\frac{2}{3}}\right)$ concave up
$\left(-\sqrt{\frac{2}{3}}, \sqrt{\frac{2}{3}}\right)$ concave down

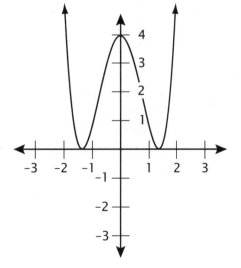

3.

$f(x) = 5x - x^5$

$f'(x) = 5 - 5x^4$

$\quad = 5(1 - x^4)$

$\quad = 5(1 - x^2)(1 + x^2)$

critical pts $x = \pm 1$

$\quad = 5(1-x)(1+x)(1+x^2)$

$f''(x) = -20x^3$

possible inflection pt $x = 0$

x	$f(x)$	$f'(x)$	$f''(x)$	Conclusions
-2	$+22$	$-$	$+$	
-1	-4	0	$+$	min-local
0	0	$+$	0	inflection pt
1	4	0	$-$	max-local
2	-22	$-$	$-$	

$(-\infty, 0)$ concave up $(0, \infty)$ concave down

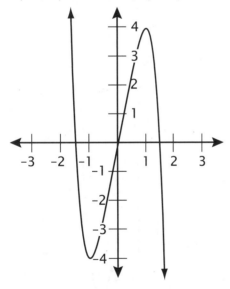

4.

$f(x) = \dfrac{6x}{x^2 + 3}$

no vertical asymptotes

$\lim\limits_{x \to -\infty} \dfrac{6x}{x^2 + 3} = 0$ and $\lim\limits_{x \to +\infty} \dfrac{6x}{x^2 + 3} = 0$

horizontal asymptote at $y = 0$

Note: These limits can be obtained easily with L'Hôpital's Rule from Chapter 17. It can be seen by substituting larger x values. The denominator will increase much more quickly than the numerator. Thus the value of the limit will approach zero.

$f'(x) = \dfrac{(x^2 + 3)(6) - 6x(2x)}{(x^2 + 3)^2}$

$\quad = \dfrac{6x^2 + 18 - 12x^2}{(x^2 + 3)^2}$

$\quad = \dfrac{18 - 6x^2}{(x^2 + 3)^2} = \dfrac{6(3 - x^2)}{(x^2 + 3)^2}$

critical pts $x = \pm \sqrt{3}$

$f''(x) = \dfrac{(x^2 + 3)^2(-12x) - (18 - 6x^2)2(x^2 + 3)(2x)}{(x^2 + 3)^4}$

$\quad = \dfrac{(x^2 + 3)(-4x)[3(x^2 + 3) + (18 - 6x^2)]}{(x^2 + 3)^4}$

$\quad = \dfrac{-4x(x^2 + 3)[3x^2 + 9 + 18 - 6x^2]}{(x^2 + 3)^4}$

$\quad = \dfrac{-4x(x^2 + 3)(27 - 3x^2)}{(x^2 + 3)^4}$

$\quad = \dfrac{-4x(27 - 3x^2)}{(x^2 + 3)^3} = \dfrac{-12x(9 - x^2)}{(x^2 + 3)^3}$

possible inflection points at $x = 0, \pm 3$

-4	$-\frac{24}{19}$	$-$	$-$	
-3	$-1\frac{1}{2}$	$-$	0	inflection pt
$-\sqrt{3}$	$-\sqrt{3}$	0	$+$	min-absolute
0	0	$+$	0	inflection pt
$\sqrt{3}$	$\sqrt{3}$	0	$-$	max-absolute
3	$1\frac{1}{2}$	$-$	0	inflection pt
4	$\frac{24}{19}$	$-$	$+$	

$(-\infty, -3)$ concave down
$(-3, 0)$ concave up
$(0, 3)$ concave down
$(3, \infty)$ concave up

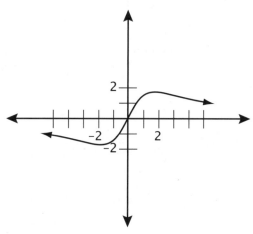

5. $f(x) = \dfrac{x^2-4}{x^2-1} = \dfrac{(x+2)(x-2)}{(x-1)(x+1)}$
Vertical asymptotes at $x = \pm 1$

$\displaystyle\lim_{x \to \infty} \dfrac{x^2-4}{x^2-1} = 1$
Horizontal asymptote at $y = 1$

6. $f(x) = \dfrac{x^2}{x-2}$
Vertical asymptote at $x = 2$

No horizontal asymptotes.

$$\begin{array}{r} x+2\\ x-2\overline{)x^2+0x+0}\\ \underline{x^2-2x}\\ 2x+0\\ \underline{+2x-4}\\ 4\text{ R}\end{array}$$

Oblique asymptote at $f(x) = x+2$

7. $f(x) = \dfrac{x^2+3x+2}{x+2}$
$= \dfrac{(x+1)(x+2)}{(x+2)} = (x+1)$

There are no asympotes.
There is a hole at $x = -2$.

8. $f(x) = \dfrac{x^2}{x-2}$
$f'(x) = \dfrac{(x-2)(2x) - x^2(1)}{(x-2)^2}$
$= \dfrac{2x^2 - 4x - x^2}{(x-2)^2}$
$= \dfrac{x^2 - 4x}{(x-2)^2}$
$= \dfrac{x(x-4)}{(x-2)^2}$ critical pts $x = 4, 2, 0$

$$f''(x) = \frac{(x-2)^2(2x-4) - (x^2-4x)(x-2)(2)}{(x-2)^4}$$

$$= \frac{2(x-2)[(x-2)^2 - x(x-4)]}{(x-2)^4}$$

$$= \frac{2(x-2)[x^2 - 4x + 4 - x^2 + 4x]}{(x-2)^4}$$

$$= \frac{2(x-2)4}{(x-2)^4}$$

$$= \frac{8}{(x-2)^3}$$

possible inflection pt x = 2

x	$f(x)$	$f'(x)$	$f''(x)$	Conclusions
-1	$-\frac{1}{3}$	+	-	
0	0	0	-	max– local
1	-1	-	-	
2	und	und	und	vertical asymptote
3	9	-	+	
4	8	0	+	min– local
5	$\frac{25}{3}$	+	+	

y = x + 2 is an oblique asymptote

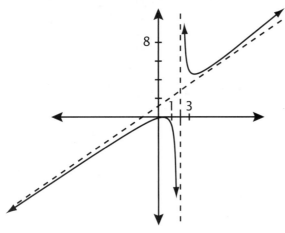

Lesson Practice 17A

1. $f(x) = \sqrt{x}$ [1, 4]
   $f(1) = 1; f(4) = 2$
   Slope between pts (1, 1) and (4, 2) is:
   $$m = \frac{2-1}{4-1} = \frac{1}{3}$$

   $$f'(x) = \frac{1}{2}x^{-\frac{1}{2}} = \frac{1}{2\sqrt{x}}$$

   $$\left(\frac{1}{2\sqrt{x}} = \frac{1}{3}\right)$$

   $$3 = 2\sqrt{x}$$

   $$\frac{3}{2} = \sqrt{x}$$

   $$\frac{9}{4} = x$$

2. $f(x) = x^3 + 3x$ [-1, 1]
   $f(-1) = -1 - 3 = -4$
   $f(1) = 1 + 3 = 4$
   Slope between (-1, -4) and (1, 4) is:
   $$\frac{4-(-4)}{1-(-1)} = \frac{8}{2} = 4.$$

   $$f'(x) = 3x^2 + 3$$
   $$3x^2 + 3 = 4$$
   $$3x^2 = 1$$

   $$x^2 = \frac{1}{3} \quad x = \pm\sqrt{\frac{1}{3}} = \pm\frac{\sqrt{3}}{3}$$

   There are two solutions for c.
   They are both found in [-1, 1].

LESSON PRACTICE 17A - LESSON PRACTICE 17B

3. $f(x) = \dfrac{1}{2x}$ [1, 3]

$f(1) = \dfrac{1}{2}$

$f(3) = \dfrac{1}{6}$ $\left(1, \dfrac{1}{2}\right)\left(3, \dfrac{1}{6}\right)$

$m = \dfrac{\frac{1}{6} - \frac{1}{2}}{3 - 1} = \dfrac{-\frac{2}{6}}{2} = \dfrac{-1}{6}$

$f(x) = (2x)^{-1}$

$f'(x) = -(2x)^{-2}(2) = \dfrac{-2}{(2x)^2} = \dfrac{-2}{4x^2} = \dfrac{-1}{2x^2}$

$\dfrac{-1}{2x^2} = \dfrac{-1}{6}$

so $6 = 2x^2$, $x^2 = 3$ $x = \pm\sqrt{3}$

Since $-\sqrt{3}$ is not in the interval, there is only one solution, $c = \sqrt{3}$.

4. $f(x) = x^2 - 4$ $[-3, -1]$

$f(-3) = 5$

$f(-1) = -3$

$(-3, 5)(-1, -3)$

$m = \dfrac{5 - (-3)}{-3 - (-1)} = \dfrac{8}{-2} = -4$

$f'(x) = 2x$

$2x = -4$ when $x = -2$ so $c = -2$

5. According to the MVT there is an instantaneous rate of change where:

$\dfrac{22°F - 2°F}{5 \text{ hours}} = \dfrac{20}{5} \; \dfrac{°F}{\text{hour}}$

                    $= 4°F$ per hour

6. $\dfrac{200 \text{ miles}}{2\frac{3}{4} \text{ hours}} = 72.7 \text{ miles/hour}$

Since the speed limit was 55 mph, the police could cite the MVT to say that at least once the driver was going 17+ miles per hour over the limit. Yes, he was ticketed.

7. $\lim\limits_{x \to 3} \dfrac{x-3}{x^2-9}$ Has form $\dfrac{0}{0}$ Use LR

$\lim\limits_{x \to 3} \dfrac{1}{2x} = \dfrac{1}{2(3)} = \dfrac{1}{6}$

8. $\lim\limits_{x \to 0} \dfrac{\sin(4x)}{x}$ Has form $\dfrac{0}{0}$ Use LR

$\lim\limits_{x \to 0} \dfrac{4\cos(4x)}{1} = 4$

9. $\lim\limits_{\theta \to 0} \dfrac{2\cos(\theta) - 2}{\theta^2}$ Has form $\dfrac{0}{0}$ Use LR

$= \lim\limits_{\theta \to 0} \dfrac{-2\sin(\theta)}{2\theta}$ Has form $\dfrac{0}{0}$ Use LR

$= \lim\limits_{\theta \to 0} \dfrac{-2\cos(\theta)}{2} = \dfrac{-2}{2} = -1$

10. $\lim\limits_{\alpha \to 0} \dfrac{2(-\sin(\alpha) + \alpha)}{\alpha^2}$ Has form $\dfrac{0}{0}$ Use LR

$= \lim\limits_{\alpha \to 0} \dfrac{2(-\cos(\alpha) + 1)}{2\alpha}$ Has form $\dfrac{0}{0}$ Use LR

$= \lim\limits_{\alpha \to 0} \dfrac{2(+\sin(\alpha))}{2} = 0$

11. $\lim\limits_{x \to \infty} \dfrac{2x}{e^x}$ Has form $\dfrac{\infty}{\infty}$ Use LR

$= \lim\limits_{x \to \infty} \dfrac{2}{e^x} = 0$

12. $\lim\limits_{x \to 0} \dfrac{\sin^2(x)}{1 - \cos(x)}$ Has form $\dfrac{0}{0}$ Use LR

$= \lim\limits_{x \to 0} \dfrac{2(\sin(x))(\cos(x))}{\sin(x)} = \lim\limits_{x \to 0} 2\cos(x) = 2$

Lesson Practice 17B

1. $f(x) = e^x$ $[0, 1]$

$f(0) = 1$

$f(1) = e$

$(0, 1)(1, e)$ $m = \dfrac{e - 1}{1 - 0} = e - 1$

$f'(x) = e^x$

$e^x = e - 1$

$x = \ln(e - 1) \approx .54$

.54 lies within the interval $[0, 1]$ so $x = \ln(e - 1)$.

CALCULUS

LESSON PRACTICE 17B - LESSON PRACTICE 17B

2. $f(x) = x^{\frac{2}{3}}$ $[-1, 1]$
$f(-1) = 1$
$f(1) = 1$
$(-1, 1)(1, 1)$ $m = \dfrac{1-1}{1-(-1)} = 0$
$f'(x) = \dfrac{2}{3}x^{-\frac{1}{3}} = \dfrac{2}{3\sqrt[3]{x}}$
$\dfrac{2}{3\sqrt[3]{x}} = 0$ nowhere.
There is no point c.

The function $f(x) = x^{\frac{2}{3}}$ is not continuous. $f'(0)$ does not exist so the MVT does not apply. The conditions of the MVT should be checked first.

3. $f(x) = 6x^2 - x^3$ $[-1, 1]$
$f(-1) = 6 + 1 = 7$ $(-1, 7)(1, 5)$

$m = \dfrac{5-7}{1-(-1)} = \dfrac{-2}{2} = -1$
$f(1) = 6 - 1 = 5$
$f'(x) = 12x - 3x^2$
$12x - 3x^2 = -1$
$3x^2 - 12x - 1 = 0$
$x = \dfrac{12 \pm \sqrt{144 - 4(3)(-1)}}{2(3)}$
$= \dfrac{12 \pm \sqrt{156}}{6}$
$= \dfrac{12 \pm 2\sqrt{39}}{6}$
$= \dfrac{6 \pm \sqrt{39}}{3}$
$x = \dfrac{6 \pm \sqrt{39}}{3} = \dfrac{6 \pm 6.24}{3}$
Only one of these points is found in $[-1, 1]$.
$c = \dfrac{6 - \sqrt{39}}{3}$

4. $f(x) = x\ln(x)$ $[1, 4]$
$f(1) = 0$ $(1, 0)(4, 4\ln(4))$
$f(4) = 4\ln(4)$
$m = \dfrac{4\ln(4) - 0}{4 - 1} = \dfrac{4}{3}\ln(4)$
$f'(x) = x \cdot \dfrac{1}{x} + \ln(x) = 1 + \ln(x)$
$1 + \ln(x) = \dfrac{4}{3}\ln(4)$
$\ln(x) = \dfrac{4}{3}\ln(4) - 1$
$x = e^{\frac{4}{3}\ln(4) - 1} \approx 2.34$
c is approximately 2.34

5. The MVT tells us that Joe was running at a pace of $\dfrac{26.2}{2.5} = 10.48$ m/hr. at one or more times during the race. We know that Joe began the race at 0 mph. In order to reach 10.48 m/hr, he would have to run at 10 m/hr before and after that time.

6. No, there is no derivative at the point which makes a corner between a and b.

7. $\lim\limits_{x \to 2} \dfrac{x^3 - 8}{x^2 - 2x}$ Has form $\dfrac{0}{0}$ Use LR
$= \lim\limits_{x \to 2} \dfrac{3x^2}{2x - 2} = \dfrac{12}{2} = 6$

8. $\lim\limits_{\theta \to 0} \dfrac{\tan(\theta)}{\tan(2\theta)}$ Has form $\dfrac{0}{0}$ Use LR
$= \lim\limits_{\theta \to 0} \dfrac{\sec^2(\theta)}{2\sec^2(2\theta)} = \dfrac{1}{2}$

9. $\lim\limits_{\theta \to \frac{\pi}{2}} \dfrac{\cos^2(\theta)}{1 - \sin(\theta)}$ Has form $\dfrac{0}{0}$ Use LR
$= \lim\limits_{\theta \to \frac{\pi}{2}} \dfrac{2\cos(\theta)(-\sin(\theta))}{-\cos(\theta)}$
$= \lim\limits_{\theta \to \frac{\pi}{2}} 2\sin(\theta) = 2$

10. $\lim\limits_{x\to\infty}\dfrac{\cos(x)}{x}$ does **not** have the form $\dfrac{\infty}{\infty}$ or $\dfrac{0}{0}$. You can **not** use LR. $\cos(x)$ bounces back and forth from -1 and 1 and all values between. The denominator grows without bound, so the answer is 0.

11. $\lim\limits_{\theta\to 0}\dfrac{\tan(2\theta)}{3\tan(\theta)}$ Has form $\dfrac{0}{0}$ Use LR

    $=\lim\limits_{\theta\to 0}\dfrac{2\sec^2(2\theta)}{3\sec^2(\theta)}=\dfrac{2}{3}$

12. $\lim\limits_{x\to -1}\dfrac{x+1}{\sqrt{x^2-1}}$ Has form $\dfrac{0}{0}$ Use LR

    $=\lim\limits_{x\to -1}\dfrac{1}{\frac{1}{2}(x^2-1)^{-\frac{1}{2}}(2x)}$

    $=\lim\limits_{x\to -1}\dfrac{\sqrt{x^2-1}}{x}=0$

Lesson Practice 17C

1. a. $f(x)=\sqrt{x-1}$ [1, 5]

   The graph is both continuous and differentiable on [1, 5] as can be seen by the drawing.

   $f(x)=(x-1)^{1/2}$

   $f'(x)=\dfrac{1}{2}(x-1)^{-\frac{1}{2}}=\dfrac{1}{2\sqrt{x-1}}$

   The derivative exists for all points on [1, 5] except when $x=1$.
   The theorem still applies. The derivative must exist on the open interval $(1, 5)$.

b.

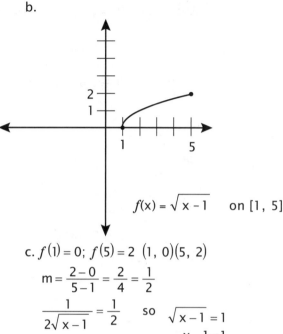

$f(x)=\sqrt{x-1}$ on [1, 5]

c. $f(1)=0;\ f(5)=2\ (1,0)(5,2)$

   $m=\dfrac{2-0}{5-1}=\dfrac{2}{4}=\dfrac{1}{2}$

   $\dfrac{1}{2\sqrt{x-1}}=\dfrac{1}{2}$ so $\sqrt{x-1}=1$

   $x-1=1$

   $x=2$ so $c=2$

2. $g(x)=f(x)\sin(x)$

   $g(x)=0$ when $\sin(x)=0$ so $g(x)=0$ when $x=0,\ \pi$.

   $g(x)=0$ when $x=\dfrac{\pi}{4}$ from given definition.

   Therefore the requirements for Rolle's theme are met only in c. Both endpoints have g values of 0. Since $f(x)$ and $\sin(x)$ are differentiable, c is true.
   a and b might be true, but c must be true.

3. $\lim\limits_{\theta\to 0}\dfrac{\sin(\theta)}{\theta}$ Has form $\dfrac{0}{0}$ Use LR

   $=\lim\limits_{\theta\to 0}\dfrac{\cos(\theta)}{1}=1$

4. $\lim\limits_{x\to 0}\dfrac{\sin(ax)}{x}$ Has form $\dfrac{0}{0}$ Use LR

   $=\lim\limits_{x\to 0}\dfrac{a\cos(ax)}{1}=a$

LESSON PRACTICE 17C - LESSON PRACTICE 17D

5. $\lim\limits_{\theta \to 0} \dfrac{\sec(\theta) - \cos(\theta)}{\theta^2}$ Has form $\dfrac{0}{0}$ Use LR

$\lim\limits_{\theta \to 0} \dfrac{\sec(\theta)\tan(\theta) + \sin(\theta)}{2\theta}$ Has form $\dfrac{0}{0}$ Use LR

$\lim\limits_{\theta \to 0} \dfrac{\left(\sec(\theta)\sec^2(\theta) + \tan(\theta)\sec(\theta)\tan(\theta)\right) + \cos(\theta)}{2}$

$= \dfrac{2}{2} = 1$

6. $\lim\limits_{\theta \to 0} \dfrac{\tan(\theta) - \sin(\theta)}{\theta^2}$ Has form $\dfrac{0}{0}$ Use LR

$= \lim\limits_{\theta \to 0} \dfrac{\sec^2(\theta) - \cos(\theta)}{2\theta}$

Has form $\dfrac{0}{0}$ Use LR

$= \lim\limits_{\theta \to 0} \dfrac{2\sec(\theta)\sec(\theta)\tan(\theta) + \sin(\theta)}{2}$

$= 0$

7. $\lim\limits_{x \to \infty} \dfrac{3x^2}{e^{2x}}$ Has form $\dfrac{\infty}{\infty}$ Use LR

$= \lim\limits_{x \to \infty} \dfrac{6x}{2e^{2x}}$ Has form $\dfrac{\infty}{\infty}$ Use LR

$= \lim\limits_{x \to \infty} \dfrac{6}{4e^{2x}} = 0$

8. $\lim\limits_{x \to \infty} \dfrac{\sin x}{e^x}$ Does **not** have the form $\dfrac{\infty}{\infty}$.
Cannot use LR

$= \dfrac{\text{small number}}{\infty} = 0$

9. Even because all the powers of x are even.

10. $f(-x) = \dfrac{4}{1+x}$

$-f(x) = -\dfrac{4}{1-x}$

This function is neither odd nor even.

11. $f(-x) = -\dfrac{3}{x^3}$

$-f(x) = -\dfrac{-3}{x^3}$ so this function is odd.

12. $f(x) = 3x^6 - 2x^4 + x^2$
All the powers of x are even so it is an even function.

Lesson Practice 17D

1. $f(x) = x^3 + 2x + 1$ is differentiable for all x
$f'(x) = 3x^2 + 2$ which is always positive.
The values of $f(x)$ are always increasing.
Looking at several values of the function such as $f(-2) = -11, f(-1) = -2, f(0) = 1,$ and $f(2) = 13$, we know that between $x = -1$ and $x = 0$ there is one root to the polynomial. Since f' is always positive we know that f has at most one zero. If f had 2 or more zeros, by Rolle's Thm, f' would have a zero between them. Therefore, f must have exactly one zero. (or one real root)

2. $x - \dfrac{2}{x} = 0$ [1, 3]

This function is differentiable and continuous on [1, 3].

$f(1) = 1 - 2 = -1$

$f(3) = 3 - \dfrac{2}{3} = 2\dfrac{1}{3}$

We know that the function will have to cross the x-axis at least one time.

$f'(x) = 1 + 2x^{-2} = 1 + \dfrac{2}{x^2}$

When is $1 + \dfrac{2}{x^2} = 0$?

$x^2 + 2 = 0$

$x^2 = -2$ Never.

Because there is no time that f' is zero, there must be at most one real root. Finally, there would be exactly one real solution.

3. $\lim_{\theta \to 0} \dfrac{1-\cos(\theta)}{\theta \sin(\theta)}$ Has form $\dfrac{0}{0}$ Use LR

$= \lim_{\theta \to 0} \dfrac{\sin(\theta)}{\theta \cos(\theta) + \sin(\theta)}$ Has form $\dfrac{0}{0}$ Use LR

$= \lim_{\theta \to 0} \dfrac{+\cos(\theta)}{-\theta \sin(\theta) + \cos(\theta) + \cos(\theta)} = \dfrac{1}{2}$

4. $\lim_{\theta \to 0} \dfrac{\tan(3\theta)}{\theta}$ Has form $\dfrac{0}{0}$ Use LR

$\lim_{\theta \to 0} \dfrac{3\sec^2(3\theta)}{1} = 3$

5. $\lim_{x \to \infty}\left(1 + \cos\left(\dfrac{1}{x}\right)\right)$

Does **not** have form $\dfrac{0}{0}$ or $\dfrac{\infty}{\infty}$.

Can not use LR

$\dfrac{1}{x}$ approaches 0 when x approaches infinity, so $\cos \dfrac{1}{x}$ approaches 1.

The answer is 2.

6. $\lim_{x \to 2} \dfrac{x^3 - 8}{x - 2}$ Has form $\dfrac{0}{0}$ Use LR

$= \lim_{x \to 2} \dfrac{3x^2}{1} = 12$

7. $\lim_{x \to 1} \dfrac{1 - \sqrt{x}}{1 - x}$ Has form $\dfrac{0}{0}$ Use LR

$= \lim_{x \to 1} \dfrac{-\dfrac{1}{2}x^{-\tfrac{1}{2}}}{-1}$

$= \lim_{x \to 1} \dfrac{1}{2\sqrt{x}} = \dfrac{1}{2}$

8. $\lim_{x \to 0} \dfrac{x}{\tan(3x)}$ Has form $\dfrac{0}{0}$ Use LR

$= \lim_{x \to 0} \dfrac{1}{3\sec^2(3x)} = \dfrac{1}{3}$

9. $f(-x) = -x - 2$
$-f(x) = -x + 2$
This function is neither odd nor even.

10. $f(x) = \dfrac{3}{x^2}$

This function is even because it has even exponents.

11. $f(x) = 2x^2 - x$
$f(-x) = 2x^2 + x$
$-f(x) = -2x^2 + x$
This function is not odd or even.

12. $f(x) = \dfrac{2x^2 + 1}{x}$

$f(-x) = \dfrac{2x^2 + 1}{-x}$ This function is odd.

$-f(x) = \dfrac{2x^2 + 1}{-x}$

Lesson Practice 18A

1. $d = 16.1t^2$
   a. $v(t) = d'(t) = 32.2t$
   b. $a(t) = v'(t) = 32.2$
   c. $v(5) = 32.2(5) = 161$ ft/sec

2. a. $v(t) = 32t - 64$
   b. $v\left(\dfrac{1}{2}\right) = 32\left(\dfrac{1}{2}\right) - 64 = 16 - 64 = -48$

   The object is falling at 48 ft/sec.

   c. $a(t) = 32$

   The acceleration is constant at 32 ft/sec².

   d. This object will stop moving when the distance is zero.

   $d(t) = 16t^2 - 64t + 64 = 0$
   $t^2 - 4t + 4 = 0$
   $(t - 2)^2 = 0$
   $t = 2$

   The object is moving for 2 seconds.

3. $d = d_0 + v_0 t - 4.9t^2$

   $v(t) = d'(t) = v_0 - 9.8t$

   $= -9.8t$

   The penny will strike the ground when $d = 0$.

   $d = 427 - 4.9t^2 = 0$

   $427 = 4.9t^2$

   $t = 9.34$ seconds

   The penny will strike the ground in 9.34 seconds.

   $v(9.34) = -9.8(9.34) = -91.53$ m/sec

   The penny struck the ground at a speed of 91.53 m/sec.

4. a. $s(t) = 40 + 30t - 5t^2$

      The rock will hit its peak when $v(t) = 0$.

      $v(t) = 30 - 10t = 0$

      $t = 3$

   b. The maximum height will occur when $t = 3$.

      $s(3) = 40 + 30(3) - 5(3)^2$

      $= 40 + 90 - 45 = 85$ meters

   c. The rock will hit the ground when $s(t) = 0$.

      $s(t) = -5t^2 + 30t + 40 = 0$

      $t^2 - 6t - 8 = 0$

      $t = \dfrac{6 \pm \sqrt{36 - 4(1)(-8)}}{2(1)}$

      $= \dfrac{6 \pm \sqrt{68}}{2}$

      $t = 7.1, -1.1$

      The rock will hit the ground in 7.1 seconds.

   d. $v(7.1) = 30 - 10(7.1) = -41$ m/sec

   e. $a(t) = -10$

Lesson Practice 18B

1. a. $x(t) = -16t^2 + v_0 t + x_0$

      $x(t) = -16t^2 + 48t + 100$

      $v(t) = x'(t) = -32t + 48$

      At the maximum height, the velocity will be zero.

      $v(t) = -32t + 48 = 0$

      $32t = 48$

      $t = \dfrac{3}{2}$

      When $t = \dfrac{3}{2}$ seconds, the ball will be at its peak.

      Calculating $x\left(\dfrac{3}{2}\right)$ will tell us how high the ball will go.

      $x\left(\dfrac{3}{2}\right) = -16\left(\dfrac{3}{2}\right)^2 + 48\left(\dfrac{3}{2}\right) + 100$

      $= -36 + 72 + 100$

      $= 136$ ft

   b. The ball will hit the ground when $x(t) = 0$.

      $0 = -16t^2 + 48t + 100$

      $0 = -4t^2 + 12t + 25$

      Using the quadratic formula:

      $t = \dfrac{-12 \pm \sqrt{144 - 4(-4)(25)}}{2(-4)}$

      $= \dfrac{-12 \pm \sqrt{544}}{-8}$

      $= \dfrac{-12 \pm 4\sqrt{34}}{-8}$

      $= \dfrac{-3 \pm \sqrt{34}}{-2}$

      $t = \dfrac{3 \pm \sqrt{34}}{2} \approx 4.42$ seconds

      The negative answer is meaningless. The ball will hit the ground 4.42 seconds after being thrown.

2. We are asked to calculate v_0.

x_0 is 0 because Seth is on the ground.
Therefore our distance equation is

$$x(t) = -16t^2 + v_0 t + 0$$

When the ball reaches the top of the building:

$$x(t) = -16t^2 + v_0 t$$
$$100 = -16t^2 + v_0 t$$
$$v(t) = x'(t) = -32t + v_0$$
$$0 = -32t + v_0$$
$$v_0 = 32t$$
$$t = \frac{v_0}{32}$$

Plugging this into the distance equation, we get:

$$100 = -16\left(\frac{v_0}{32}\right)^2 + v_0\left(\frac{v_0}{32}\right)$$
$$= \left(\frac{-v_0}{64}\right)^2 + \left(\frac{v_0}{32}\right)^2$$
$$= \left(\frac{v_0}{64}\right)^2$$
$$6400 = (v_0)^2$$
$$80 = v_0$$

Therefore, the initial velocity of the throw would be $\frac{80 \text{ ft}}{\text{sec}}$.

3. a. $d_{police} = d_0 + v_0 t + \frac{1}{2}at^2$

$$d_{police} = \frac{1}{2}6t^2$$
$$d_{police} = 3t^2$$
$$d_{Ima} = \text{rate} \times \text{time}$$
$$= 40t$$

The distances will be equal when the policeman catches Ima.

$$40t = 3t^2$$
$$3t^2 - 40t = 0$$
$$t(3t - 40) = 0$$
$$3t = 40$$
$$t = \frac{40}{3} \approx 13\frac{1}{3} \text{ seconds}$$

The policeman will catch Ima in $13\frac{1}{3}$ seconds.

b. $d\left(\frac{40}{3}\right) = 40\left(\frac{40}{3}\right) = 533\frac{1}{3}$ meters

It will take $533\frac{1}{3}$ meters for the policeman to catch Ima.

c. $v_{police} = 6t$

$$v\left(\frac{40}{3}\right) = 6\left(\frac{40}{3}\right) = 80 \text{ m/sec}$$

The velocity of the policeman when he reached Ima was 80 m/sec.

4. a. $d(t) = -4.9t^2 + 49t + 15$
$v(t) = -9.8t + 49$

When the velocity is zero, the object will be at its peak.

$$-9.8t + 49 = 0$$
$$t = 5 \text{ seconds}$$

$$d(5) = -4.9(5)^2 + 49(5) + 15$$
$$= 137.5 \text{ meters}$$

b. The object will be at its peak in 5 seconds.

c. The object will hit the ground when d = 0.

$$d(t) = -4.9t^2 + 49t + 15 = 0$$

$$t = \frac{-49 \pm \sqrt{(49)^2 - 4(-4.9)(15)}}{2(-4.9)}$$

$$= \frac{-49 \pm 51.9}{-9.8} \approx 10.3 \text{ seconds}$$

The object will hit the ground in 10.3 seconds.

d. $v(10.3) = -9.8(10.3) + 49 = -51.9$
The velocity upon impact would be -51.9 m/sec.

Lesson Practice 18C

1. a. $x(t) = 6t + t^3$
   $v(t) = x'(t) = 6 + 3t^2$
   b. $v(1) = 6 + 3(1)^2 = 9$ ft/sec
   c. $v_0(t) = v(0) = 6 + 3(0)^2 = 6$ ft/sec
   d. $a(t) = 6t$
   e. $a(2) = 6(2) = 12$ ft/sec^2

2. a. $x(t) = -16t^2 + v_0 t + x_0$
   $x_0 = 16$ft $v_0 = 0$
   $x(t) = -16t^2 + 16$
   When the coconut hits the ground, the distance = 0.
   $x(t) = -16t^2 + 16 = 0$
   $t^2 = 1$
   It will take 1 sec for the coconut to hit the ground.

   b. The question asks for $v(t)$ when $x(t) = 0$.
   $x(t) = -16t^2 + 16$
   $v(t) = -32t$
   $v(1) = -32$
   The coconut will be traveling at 32 ft/sec when it hits the ground.

3. $s(t) = -8.5t^2 + 66t$
   a. $v(t) = -17t + 66$
   $a(t) = -17$
   b. Car will come to a stop when $v(t) = 0$
   $-17t + 66 = 0$
   $17t = 66$
   $t = 3.9$ seconds
   The truck will stop in 3.9 seconds.

c. $s(3.9) = -8.5(3.9)^2 + 66(3.9)$
   $= -129.3 + 257.4$
   $= 128.1$ ft
   The truck will go 128.1 ft before stopping once the brake is applied.

4. a. $d(t) = d_0 + v_0 t + \frac{1}{2}at^2$
   $= 512 - 64t - \frac{32}{2}t^2$
   $= -16t^2 - 64t + 512$
   The ball will hit the ground when the distance is zero.
   $-16t^2 - 64t + 512 = 0$
   $-16(t^2 + 4t - 32) = 0$
   $-16(t + 8)(t - 4) = 0$
   $t = -8, t = 4$
   Therefore the ball will hit the ground 4 seconds after the ball was thrown.

   b. $v(t) = -32t - 64$
   $v(4) = -32(4) - 64 = -192$ ft/sec
   The ball will hit the ground with a velocity of -192 ft/sec.

Lesson Practice 18D

1. a. At $t = 0$, the rock is at $s(0) = 320$. (It is 320 ft high). The rock will hit the ground after it has traveled 320 feet.
   $0 = -16t^2 + 16t + 320$
   $0 = -t^2 + t + 20$
   $0 = t^2 - t - 20$
   $0 = (t - 5)(t + 4)$
   $t = 5$ seconds
   When $t = 5$ seconds, the rock will hit the ground.

   b. $v(t) = s'(t) = -32t + 16$
   $v(5) = -32(5) + 16$
   $= -160 + 16 = -144$

LESSON PRACTICE 18D - LESSON PRACTICE 19A

The velocity of the rock on impact is -144 ft/sec.

c. $a(t) = v'(t) = -32$ ft/sec^2
The acceleration of the rock is a constant -32 ft/sec^2.

2. a. $s = 48t - 16t^2$
$v(t) = 48 - 32t$
$v(1) = 48 - 32 = 16$
The object is moving upward at a velocity of 16 ft/sec after 1 second.

b. Maximum height is achieved when $v = 0$.
$0 = 48 - 32t$
$0 = 16(3 - 2t)$
$t = \frac{3}{2}$

The maximum height is achieved in $\frac{3}{2}$ sec.

$s\left(\frac{3}{2}\right) = 48\left(\frac{3}{2}\right) - 16\left(\frac{3}{2}\right)^2$
$= 72 - 36 = 36$ ft

The maximum height is 36 ft.

c. When the distance is zero, the object is starting or stopping.
$48t - 16t^2 = 0$
$16t(3 - t) = 0$
$t = 0, 3$
The object returns to its original position after 3 seconds.

d. $v(3) = 48 - 32(3) = 48 - 96 = -48$
The object has a velocity of 48 ft/sec just before it strikes the ground.

3. a. $s(t) = 12t^2 + 120t + 3$
$v(t) = \frac{d}{dt}(s) = 24t + 120$
$a(t) = \frac{d}{dt}(v) = 24$

Therefore the acceleration is constant at 24 meters/minute. The velocity is $24t + 120$.

b. $a(3) = 24$ meters/minute
$v(t) = 24(3) + 120$
$= 192$ meters/minute

c. $192 \frac{\text{meters}}{\text{minute}} \cdot \frac{60 \text{ minutes}}{1 \text{ hour}} \cdot \frac{1 \text{ km}}{1,000 \text{ meter}}$
$= 11\frac{1}{2}$ km/hour

4. a. The missile began its journey from 35 meters below ground.

b. $v(t) = 2t^2$
$a(t) = 4t$

c. $s(6) = \frac{2}{3}(6)^3 - 35 = 109$ m
The missile will be 109 meters above the ground after 6 seconds.

d. $v(6) = 2(6)^2 = 72$ m/sec

Lesson Practice 19A

1. Solution $C(x) = 18,000 + 5.2x$
$R(x) = 7.6x$
The break-even point will be where the cost = revenue.
$18,000 + 5.2x = 7.6x$
$x = 7500$
When $x = 7500$, $C(x) = R(x) = \$57,000$.

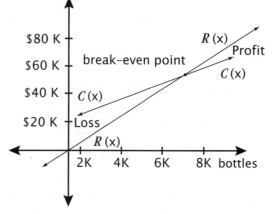

2. a. $C(x) = \frac{1}{8}x^2 + 4x + 200$
$p(x) = 49 - x$
$R(x) = x \cdot p(x) = 49x - x^2$

$P(x) = R(x) - C(x)$
$= 49x - x^2 - \frac{1}{8}x^2 - 4x - 200$

$P(x) = -\frac{9}{8}x^2 + 45x - 200$

$P'(x) = 45 - \frac{9}{4}x = 0$

$\frac{4}{9} \cdot 45 = \frac{9}{4}x \cdot \frac{4}{9}$

$20 = x$

$P''(x) = -\frac{9}{4}$

$P''(20) = $ maximum

It would be advisable to make 20 sweaters to maximize profit. The maximum profit would be:

$P(20) = 45(20) - \frac{9}{8}(20)^2 - 200$
$= 900 - 450 - 200$
$= \$250$

3. a. $R(x) = x(200 - .005x)$
$= 200x - .005x^2$

b. $P(x) = R(x) - C(x)$
$= 200x - .005x^2 - [75,000 + 100x - .03x^2 + .000004x^3]$
$P(x) = -75,000 + 100x + .025x^2 - .000004x^3$

c. $C'(x) = 100 - .06x + .000012x^2$

d. $R'(x) = 200 - 0.01x$

e. $P'(x) = 100 + .05x - .000012x^2$

f. You are asked to find $C'(2,500)$, $R'(2,500)$ and $P'(2,500)$.

$C'(2,500) = 100 - .06(2,500) + .000012(2,500)^2$
$= 100 - 150 + 75 = 25$

$R'(2500) = 200 - .01(2,500) = 175$

$P'(2,500) = 100 + .05(2,500) - .000012(2,500)^2$
$= 150$

When selling the 2,501st item, the cost to the company will be \$25. The revenue generated is \$175 of which approximately \$150 is profit.

Lesson Practice 19B

1. $5x = 375 - 3p$
$5x - 375 = -3p$
$p = \frac{5x - 375}{-3} = 125 - \frac{5}{3}x$

$R(x) = p \cdot x = 125x - \frac{5}{3}x^2$

$C(x) = 500 + 15x + \frac{1}{5}x^2$

$P(x) = R(x) - C(x)$
$= 125x - \frac{5}{3}x^2 - 500 - 15x - \frac{1}{5}x^2$

$P(x) = 110x - \frac{28}{15}x^2 - 500$

To find the maximum profit, find $P'(x)$.

$P'(x) = 110 - \frac{56}{15}x$

$110 = \frac{56}{15}x$

$x = 29.46$

Production should be approximately 30 instruments per week.

2. a. $C'(x) = 250 + .004x$

b. $R'(x) = 1250 - .01x$

c. $C'(10,000) = 250 + .004(10,000)$
$= 290$

$R'(10,000) = 1250 - .01(10,000)$
$= 1150$

d. $P(x) = R(x) - C(x)$
$= 1250x - .005x^2 - (5,000,000 + 250x + .002x^2)$
$= -.007x^2 + 1000x - 5,000,000$

e. $P'(x) = -.014x + 1000 = 0$
$1000 = .014x$
$x = 71,428.57$

Check: $P''(x) = -.014$
$P''(71,429)$ is a max.
71,429 units should be produced to maximize profits.

3. $R(x) = 2000x - 60x^2$
$C(x) = 4000 + 500x$

The peak will occur when $R'(x) = 0$.
$R'(x) = 2000 - 120x = 0$
$120x = 2000$
$x = 16\frac{2}{3}$

$R\left(16\frac{2}{3}\right) \approx \$16,667$

Break-even will occur when cost = revenue.
$4000 + 500x = 2000x - 60x^2$
$60x^2 - 1500x + 4000 = 0$
$3x^2 - 75x + 200 = 0$
$x = \dfrac{75 \pm \sqrt{75^2 - 4(3)(200)}}{2(3)}$
$= \dfrac{75 \pm \sqrt{3225}}{6} \approx 3.04 \text{ and } 21.96$

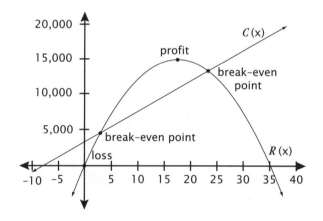

Lesson Practice 19C

1. $P(x) = -8x^2 + 3200x - 80,000$
$P'(x) = -16x + 3200$
$x = 200$
$P''(x) = -16$

$P''(200)$ is a maximum
Renting 200 apartments results in a maximum profit.

$P(200) = -8(200)^2 + 3200(200) - 80,000$
$= -320,000 + 640,000 - 80,000 = 240,000$

If they rent 200 apartments, they will realize $240,000 in profit. Notice that there are 250 apartments and the maximum profit is realized when 50 apartments are left unrented. In order to make even more profit, the owner will need to reduce the maintenance costs. It is hoped that the maximum profit would be obtained when all the apartments are rented.

LESSON PRACTICE 19C - LESSON PRACTICE 19D

2. $C(x) = 250,000 + .08x + \dfrac{200,000,000}{x}$

$C'(x) = .08 - \dfrac{200,000,000}{x^2} = 0$

$.08 = \dfrac{200,000,000}{x^2}$

$.08x^2 = 200,000,000$

$x^2 = \dfrac{200,000,000}{.08}$

$x = 50,000$

$x = 50,000$ is a critical point.

$C''(x) = \dfrac{400,000,000}{x^3}$

$C''(50,000)$ is positive so $x = 50,000$ is a minimum.

$C(50,000)$

$= 250,000 + .08(50,000) + \dfrac{200,000,000}{(50,000)}$

$= 250,000 + 4000 + 4000$

$= 258,000$

Make 50,000 thingamabobs and your costs will be at a minimum of $258,000.

3. a. $R(x) = -\dfrac{x^2}{2} + 500x$

$R'(x) = -\dfrac{1}{2}(2x) + 500$

$= -x + 500$

b. $R'(x) = 0 \quad -x + 500 = 0$

$x = 500$

500 units should be produced to maximize profit revenue.

c. $R'(100) =$ is the revenue from selling the 101st item.

$R'(100) = -100 + 500 = 400$

$400 is the revenue from the sale of the 101st item.

Lesson Practice 19D

1. a. $C(301) - C(300) =$

$500 + 350(301) - .09(301)^2 -$
$[500 + 350(300) - .09(300)^2]$
$= 97,695.91 - 97,400 = \$295.91$

b. $C'(x) = 350 - .18x$

$C'(200) = 350 - .18(200)$
$= 314$

c. $C'(300) = 350 - .18(300)$
$= 296$

d. The approximate cost of the 201st item is $314.

The approximate cost of the 301st item is $296.

This tells us that the costs are improving with an increase in production.

2. a. $R(x) = p(x) \cdot x$

$= \left(20 - \dfrac{x}{10}\right)x$

$= 20x - \dfrac{x^2}{10}$

b. $P(x) = R(x) - C(x)$

$= \left(20x - \dfrac{x^2}{10}\right) - (50 + 3x)$

$= 17x - \dfrac{x^2}{10} - 50$

c. $P'(x) = 17 - \dfrac{x}{5}$

d. $17 = \dfrac{x}{5}$

$x = 85$

e. $p(85) = 20 - \dfrac{85}{10}$

$= \$11.50$ per skate rental.

Lesson Practice 20A

1. $x = $ 1st number
   $y = $ 2nd number

   Constraint $x \cdot y = 30$ $y = \dfrac{30}{x}$

   Optimization Min $S = x + y$ $x, y > 0$

   Min $S = x + \dfrac{30}{x} = x + 30x^{-1}$

   $S' = 1 - 30x^{-2}$

   $S' = 1 - \dfrac{30}{x^2} = 0$

   $1 = \dfrac{30}{x^2}$

   $x^2 = 30$

   $x = \pm\sqrt{30}$; choose $x = \sqrt{30}$

   $\sqrt{30}$ is a minimum because values of the first derivative are negative to the left of $\sqrt{30}$ and positive to the right of $\sqrt{30}$; $\sqrt{30} \cdot y = 30$ so $y = \sqrt{30}$.
   $x = \sqrt{30}$ and $y = \sqrt{30}$ are the numbers we are seeking.
   Their sum is $2\sqrt{30}$.

2. $x = $ width
   $y = $ height

   [Rectangle with $A = xy$, width x, height y]

   Perimeter $= 2x + 2y$

   $2y = P - 2x$; $y = \dfrac{P - 2x}{2}$

   Max Area $= xy$

   $A = x\left(\dfrac{P - 2x}{2}\right)$

   $A = \dfrac{1}{2}xP - x^2$

   $A' = \dfrac{1}{2}P - 2x = 0$

   $2x = \dfrac{1}{2}P$

   $x = \dfrac{P}{4}$

   The width needs to be 1/4 of the perimeter.

   $y = \dfrac{P - 2\left(\dfrac{P}{4}\right)}{2}$

   $= \dfrac{P - \dfrac{P}{2}}{2}$

   $= \dfrac{P}{2} - \dfrac{P}{4} = \dfrac{P}{4}$

   The height needs to be 1/4 of the perimeter.

   $A'' = -2$

   $A''\left(\dfrac{P}{4}\right)$ is a maximum

   Therefore $x = y$ and the rectangle with the largest area is a square.

3. Volume $= 10 = x^2 L$; $L = \dfrac{10}{x^2}$

   Min Surface Area $= 2x^2 + 3xL$

   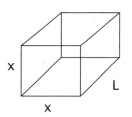

   S.A. $= 2x^2 + 3x\left(\dfrac{10}{x^2}\right)$

   $= 2x^2 + \dfrac{30}{x}$

   S.A.$' = 4x - \dfrac{30}{x^2}$

   $= \dfrac{4x^3 - 30}{x^2}$

   $4x^3 - 30 = 0$

   $4x^3 = 30$

   $x = \sqrt[3]{\dfrac{30}{4}} \approx 1.96$ ft

$S.A." = 4 + \dfrac{60}{x^3}$

$S.A."\left(\sqrt[3]{\dfrac{30}{4}}\right)$ is positive, so $\sqrt[3]{\dfrac{30}{4}}$ is a minimum

$L = \dfrac{10}{\left(\sqrt[3]{\dfrac{30}{4}}\right)^2} \approx 2.61$ ft

dimensions ≈ 1.96 ft $\times 2.61$ ft $\times 1.96$ ft

4. $80 = x + x + L \quad 0 < x < 40$
$\qquad\qquad\qquad\qquad 0 < L < 80$
$ = 2x + L \,;\, L = 80 - 2x$

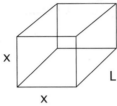

Max $V = x \cdot x \cdot L = x^2 L$
$V = x^2(80 - 2x)$
$ = 80x^2 - 2x^3$
$V' = 160x - 6x^2$
$2x(80 - 3x) = 0$
$\qquad x = 0, \dfrac{80}{3}$ critical points

$x = \dfrac{80}{3}$ is only critical point in our domain.
$V" = -12x + 160$
$V"\left(\dfrac{80}{3}\right)$ is negative, so $x = \dfrac{80}{3}$ is a maximum.

$L = 80 - 2\left(\dfrac{80}{3}\right) = 80 - \dfrac{160}{3} = \dfrac{80}{3}$

Dimensions $\dfrac{80}{3}$ in $\times \dfrac{80}{3}$ in $\times \dfrac{80}{3}$ in

5. Total Area $= (x + 12)(y + 6)$
Constraint: $800 = xy$
$y = \dfrac{800}{x}$

Min Total Area
$= (x + 12)\left(\dfrac{800}{x} + 6\right)$
$= 800 + 6x + \dfrac{9,600}{x} + 72$
$= 872 + 6x + \dfrac{9,600}{x}$

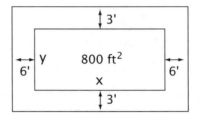

$T.A.' = 6 + 9,600(-x^{-2})$
$ = 6 - \dfrac{9,600}{x^2} = 0$

$6 = \dfrac{9,600}{x^2} ; \quad 6x^2 = 9,600$
$\qquad\qquad\qquad x^2 = 1,600$
$\qquad\qquad\qquad x = 40$

$y = \dfrac{800}{40} = 20$

$T.A." = \dfrac{19,200}{x^3}$

$T.A."(40)$ is positive, so $x = 40$ is a minimum.
The dimensions of the flower bed are 40 ft by 20 ft.

6. Find the point on the graph $y = \frac{1}{x}$ which is closest to the origin in the first quadrant.
domain $(0, \infty)$

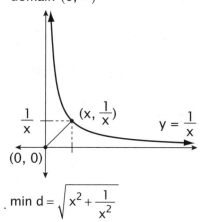

min $d = \sqrt{x^2 + \frac{1}{x^2}}$

Minimizing the quantity under the square root is the same as minimizing the square root of the quantity.

Minimize $D = x^2 + \frac{1}{x^2}$

$D' = 2x - 2x^{-3}$
$ = 2x^{-3}(x^4 - 1)$
$ = \frac{2(x^2 + 1)(x - 1)(x + 1)}{x^3}$

Critical points are $x = 0, 1, -1$.
Only $x = 1$ is in our domain.
$D'' = 2 + 6x^{-4}$
$D''(1) =$ is positive, so $x = 1$ is a minimum.

Closest point is $\left(1, \frac{1}{1}\right) = (1, 1)$.

Lesson Practice 20B

1. $x = $ 1st number $\quad y = $ 2nd number
    Constraint $\quad x + y = 10 \quad y = 10 - x$
    Optimization $\quad$ Max $P = xy$
    $P = x(10 - x) = 10x - x^2 \quad\quad 0 \le x \le 10$
    $P' = 10 - 2x \quad\quad\quad\quad\quad\quad\quad 0 \le y \le 10$
    $x = 5$

    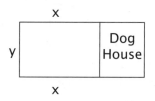

    $x = 5$ is a maximum.
    The two numbers are $x = 5$ and $y = 5$;
    25 is the maximum product.

2. $20 = 2x + y$; $y = 20 - 2x \quad 0 < x < 10$
    Max $A = xy \quad\quad\quad\quad\quad\quad 0 < y < 20$

    ```
         x
    ┌─────────┬────┐
    │         │Dog │
  y │         │House│
    │         │    │
    └─────────┴────┘
         x
    ```

    $A = x(20 - 2x)$
    $ = 20x - 2x^2$
    $A' = 20 - 4x$
    $x = 5 \quad$ critical point
    $A'' = -4$

    $A''(5)$ is a maximum.
    $y = 20 - 2(5) = 10$
    Dimensions 5 ft × 10 ft
    Area $= (5 \text{ ft})(10 \text{ ft}) = 50 \text{ ft}^2$

3. $V = 108 = x^2 y$
    Min S.A. $= x^2 + 4xy$

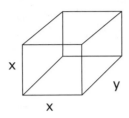

$y = \dfrac{108}{x^2}$ So Min S.A. $= x^2 + 4x\left(\dfrac{108}{x^2}\right)$

Min S.A. $= x^2 + \dfrac{432}{x}$

S.A.' $= 2x - 432x^{-2}$

$2x^3 - 432 = 0$
$2x^3 = 432$
$x^3 = 216$
$x = 6$ cm

S.A.'' $= 2 + 864x^{-3}$
S.A.''(6) is positive,
so $x = 6$ cm is a minimum.
$y = \dfrac{108}{36} = 3$ cm
The dimensions of the box would be
3 cm × 6 cm × 6 cm.

4. $158 = x + x + h$

    Max V $= x^2 h$

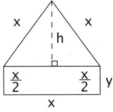

$158 = 2x + h$
$h = 158 - 2x$

Max V $= x^2(158 - 2x)$
$= 158x^2 - 2x^3$

$V' = 316x - 6x^2$
$= 2x(158 - 3x) = 0$
$x = 0,\ x = \dfrac{158}{3} = 52\dfrac{2}{3}$
$h = 158 - 2\left(\dfrac{158}{3}\right) = 52\dfrac{2}{3}$

$V'' = 316 - 12x$
$V''\left(\dfrac{158}{3}\right) = $ negative,
so $\dfrac{158}{3}$ is a maximum.
Dimensions =
$52\dfrac{2}{3}$ cm × $52\dfrac{2}{3}$ cm × $52\dfrac{2}{3}$ cm

5. Perimeter $= 3x + 2y = 15;\ y = \dfrac{15 - 3x}{2}$

Max Area $= xy + \dfrac{1}{2}(x)\left(\dfrac{\sqrt{3}}{2}x\right)$
$= \dfrac{x(15 - 3x)}{2} + \dfrac{\sqrt{3}(x^2)}{4}$
$= \dfrac{15}{2}x - \dfrac{3x^2}{2} + \dfrac{\sqrt{3}(x^2)}{4}$

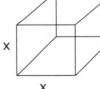

$A' = \dfrac{15}{2} - 3x + \dfrac{\sqrt{3}}{2}x$

$\dfrac{15}{2} - 3x + \dfrac{\sqrt{3}}{2}x = 0$

$\dfrac{15}{2} = 3x - \dfrac{\sqrt{3}}{2}x$

$\dfrac{15}{2} = \dfrac{6x - \sqrt{3}x}{2}$

$15 = 6x - \sqrt{3}x$
$15 = x(6 - \sqrt{3})$

$h^2 + \left(\dfrac{1}{2}x\right)^2 = x^2$

$h^2 + \dfrac{1}{4}x^2 = x^2$

$h^2 = \dfrac{3}{4}x^2$

$h = \dfrac{\sqrt{3}}{2}x$

LESSON PRACTICE 20B - LESSON PRACTICE 20C

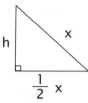

$x = \dfrac{15}{6-\sqrt{3}} \approx 3.51$ ft;

$y = \dfrac{15-3(3.51)}{2} \approx 2.24$ ft

So $x = 3.51$ is the max.

Dimensions are 3.51 ft $\times 2.24$ ft for the rectangle and 3.51 ft for each side of the triangle.

6. Area of rectangle

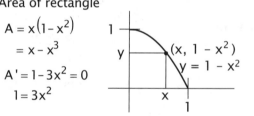

$A = x(1-x^2)$
$= x - x^3$
$A' = 1 - 3x^2 = 0$
$1 = 3x^2$

$x^2 = \dfrac{1}{3}$;

$x = \sqrt{\dfrac{1}{3}} = \dfrac{\sqrt{3}}{3}$ (critical point)

$y = 1 - \left(\dfrac{\sqrt{3}}{3}\right)^2 = 1 - \dfrac{1}{3} = \dfrac{2}{3}$

$A'' = -6x$ so $A''\left(\dfrac{\sqrt{3}}{3}\right)$ is a maximum

$A\left(\dfrac{\sqrt{3}}{3}\right) = \dfrac{\sqrt{3}}{3}\left(1 - \dfrac{3}{9}\right)$
$= \dfrac{\sqrt{3}}{3} \cdot \dfrac{2}{3} = \dfrac{2\sqrt{3}}{9}$

Dimensions are $\dfrac{\sqrt{3}}{3} \times \dfrac{2}{3}$.

Lesson Practice 20C

1. $(0, 0) \quad (x, 4x+6)$

$d = \sqrt{(x-0)^2 + (4x+6-0)^2}$
$= \sqrt{x^2 + 16x^2 + 48x + 36}$
$= \sqrt{17x^2 + 48x + 36}$

Min $17x^2 + 48x + 36$

$d' = 34x + 48 = 0$

$34x = -48$

$x = -\dfrac{48}{34} = -\dfrac{24}{17}$

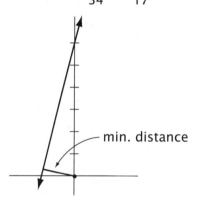

$d'' = 34$

$d''\left(-\dfrac{24}{17}\right)$ is a positive number so this is a minimum.

$y = 4\left(-\dfrac{24}{17}\right) + 6$
$= -\dfrac{96}{17} + 6$
$= \dfrac{-96 + 102}{17} = \dfrac{6}{17}$

The point which is closest to $y = 4x + 6$ and the origin is $\left(-\dfrac{24}{17}, \dfrac{6}{17}\right)$.

2. $108 = 4x + L$; $L = 108 - 4x$
Note: Girth = length (largest measurement) + (2 × width) + (2 × height) for UPS and USPS

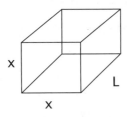

Max $V = x^2 L$ $0 < x < 27$
$= x^2(108 - 4x)$ $0 < L < 108$
$= 108x^2 - 4x^3$

$V' = 216x - 12x^2$
$= 12x(18 - x) = 0$
$x = 0, 18$ are critical points
$x = 18$ is only critical point in our domain

$V'' = 216 - 24x$
$V''(18)$ is negative so $x = 18$ is a maximum
$L = 108 - 4(18) = 36$
Dimensions; $18\text{ in} \times 18\text{ in} \times 36\text{ in}$

Maximum volume $= (18\text{ in})^2(36\text{ in})$
$= 11{,}664$ cu in

3. $x^2 + y^2 = 9$
$y^2 = 9 - x^2$
$y = \sqrt{9 - x^2}$

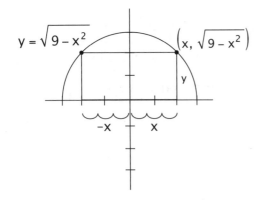

Area = length · width
$= 2xy$ $0 < x < 3$
$= 2x\sqrt{9 - x^2}$ $0 < y < 3$

$A' = 2x\left[\frac{1}{2}(9 - x^2)^{-\frac{1}{2}}(-2x)\right] + (9 - x^2)^{\frac{1}{2}}(2)$

$= -2x^2(9 - x^2)^{-\frac{1}{2}} + 2(9 - x^2)^{\frac{1}{2}}$

$= 2(9 - x^2)^{-\frac{1}{2}}\left[-x^2 + (9 - x^2)\right]$

$= 2(9 - x^2)^{-\frac{1}{2}}(-2x^2 + 9)$

$= \frac{2(9 - 2x^2)}{\sqrt{9 - x^2}}$

$= 0$ when:

$9 - 2x^2 = 0$
$2x^2 = 9$
$x^2 = \frac{9}{2}$
$x = \frac{3}{\sqrt{2}} = \frac{3\sqrt{2}}{2}$

$x = \frac{3\sqrt{2}}{2}$ is the only critical point in the domain.
$x \approx 2.12$

```
                              undefined
      +                          f'    −
   ───┼──────────┼────────┼──────┼───
      0        3√2       2.5     3
                ───
                 2
```

There is a maximum at $x = \frac{3\sqrt{2}}{2}$

$y = \sqrt{9 - \left(\frac{3\sqrt{2}}{2}\right)^2}$

$= \sqrt{9 - \frac{9}{2}}$

$= \frac{3}{\sqrt{2}} = \frac{3\sqrt{2}}{2}$

The dimensions are $2x \cdot y = 3\sqrt{2}$ by $\dfrac{3\sqrt{2}}{2}$

The area $= 2\left(\dfrac{3\sqrt{2}}{2}\right)\left(\dfrac{3\sqrt{2}}{2}\right)$

$= \dfrac{9 \cdot 2}{2} = 9$ sq units

4. The most economical can would use the least amount of material, so we'd want to minimize the surface area for the given volume.

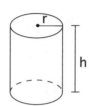

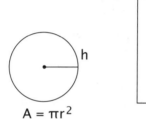

$A = \pi r^2 \qquad 2\pi r$

Min S.A. $= 2\pi r^2 + 2\pi rh$

Volume $= \pi r^2 h$; $h = \dfrac{v}{\pi r^2}$

Min S.A. $= 2\pi r^2 + 2\pi r \left(\dfrac{v}{\pi r^2}\right)$

$= 2\pi r^2 + \dfrac{2v}{r}$

SA' $= 4\pi r + 2v(-r^{-2}) = 0$

$4\pi r = \dfrac{2v}{r^2}$

$4\pi r^3 = 2v$

$r^3 = \dfrac{2v}{4\pi} = \dfrac{v}{2\pi}$;

$r = \sqrt[3]{\dfrac{v}{2\pi}}$ this is a critical point

SA'' $= 4\pi + 4vr^{-3}$

SA'' $\left(\sqrt[3]{\dfrac{v}{2\pi}}\right)$ is positive, so there is a minimum there

$h = \dfrac{v}{\pi r^2} = \dfrac{v}{\pi \left(\sqrt[3]{\dfrac{v}{2\pi}}\right)^2}$

$= \dfrac{v(2\pi)^{\frac{2}{3}}}{\pi v^{\frac{2}{3}}} = \dfrac{v^{\frac{1}{3}} 2^{\frac{2}{3}}}{\pi^{\frac{1}{3}}}$

Notice, $2r = h$

$2 \cdot \dfrac{v^{\frac{1}{3}}}{2^{\frac{1}{3}} \pi^{\frac{1}{3}}} = \dfrac{2^{\frac{2}{3}} v^{\frac{1}{3}}}{\pi^{\frac{1}{3}}}$

Since $2r =$ diameter, the most economical can has its diameter equal to its height.

Lesson Practice 20D

1. $x =$ width
   $y =$ height

$x + 2y = 1,000 \qquad$ domain $0 < x < 500$

$y = \dfrac{(1,000 - x)}{2} \qquad 0 < y < 1,000$

Max area $= xy$

$= x\left(\dfrac{1,000 - x}{2}\right)$

$= 500x - \dfrac{x^2}{2}$

A' $= 500 - x$

$x = 500$ is a critical point

LESSON PRACTICE 20D - LESSON PRACTICE 20D

$A'' = -1$
$A''(500)$ is negative
so $x = 500$ is a maximum

$y = \dfrac{1{,}000 - 500}{2}$

$= \dfrac{500}{2} = 250$

$A = xy = 500 \cdot 250 = 125{,}000 \text{ ft}^2$

2. $A = xy = 216 \text{ ft}^2$
$P = 2x + 3y$

Min $P = 2x + 3y$

$= 2x + 3\left(\dfrac{216}{x}\right)$

$= 2x + \dfrac{648}{x}$

$P' = 2 - 648x^{-2} = 0$

$2 = \dfrac{648}{x^2}$

$x^2 = 324$

$x = 18$

so $y = \dfrac{216}{18} = 12$

$P'' = 1296x^{-3}$

$P''(18)$ is a positive number,
so it is a minimum.

$P = 2(18) + 3(12)$
$= 36 + 36 = 72$

Dimensions are 12 ft × 18 ft.
You will need 72 ft of fence.

3. Volume: $125 = x^2 y$

$y = \dfrac{125}{x^2}$

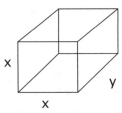

$C = 4x^2 + 2(4xy)$

$C = 4x^2 + 8x\left(\dfrac{125}{x^2}\right)$

$= 4x^2 + \dfrac{1{,}000}{x}$

$C' = 8x - \dfrac{1{,}000}{x^2} = 0$

$\dfrac{1{,}000}{x^2} = 8x$

$1{,}000 = 8x^3$

$x^3 = 125 \quad \text{so} \quad x = 5$

$C'' = 8 - 1{,}000(-2x^{-3})$

$= 8 + \dfrac{2{,}000}{x^3}$

$C''(5)$ is positive, so $x = 5$ is a minimum

$y = \dfrac{125}{25} = 5$

Dimensions: $x = y = 5$

Cost $= C(5) = 4(25) + \dfrac{1{,}000}{5} = 300$

A 5 yd × 5 yd × 5 yd tank would cost
$300 and would be the least expensive.

4. Volume $= 100\pi = \pi r^2 h$

$$h = \frac{100\pi}{\pi r^2} = \frac{100}{r^2}$$

Min $A = 2\pi r^2 + 2\pi rh$

$$= 2\pi r^2 + 2\pi r\left(\frac{100}{r^2}\right)$$

$$= 2\pi r^2 + \frac{200\pi}{r}$$

$A' = 4\pi r - 200\pi r^{-2} = 0$

$$4\pi r = \frac{200\pi}{r^2}$$

$4\pi r^3 = 200\pi$

$$r^3 = \frac{200}{4} = 50$$

$r = \sqrt[3]{50}$

$$h = \frac{100}{\left(\sqrt[3]{50}\right)^2} = \frac{100}{50^{\frac{2}{3}}}$$

$$A'' = 4\pi + \frac{400\pi}{r^3}$$

$A''\left(\sqrt[3]{50}\right)$ is positive so $\sqrt[3]{50}$ is a minimum.

$r = \sqrt[3]{50}$ and $h = \dfrac{100}{50^{\frac{2}{3}}}$

are the dimensions of the minimum can.

Lesson Practice 21A

1. A. Determine R

$$\left[\frac{1}{R} = \frac{1}{40} + \frac{1}{50} + \frac{1}{100}\right]200R$$

$200 = 5R + 4R + 2R$

$200 = 11R$

$R = 18.18$

B. Differentiate implicitly

$$\frac{-1}{R^2}R' = \frac{-1}{(R_1)^2}R_1' - \frac{1}{(R_2)^2}R_2' - \frac{1}{(R_3)^2}R_3'$$

C. Solving for R'

$$R' = R^2\left[\frac{R_1'}{(R_1)^2} + \frac{R_2'}{(R_2)^2} + \frac{R_3'}{(R_3)^2}\right]$$

D. Plugging in all the known values

$$R' = \left[\left(\frac{-.1}{1,600} - \frac{.2}{2,500} + \frac{.6}{10,000}\right)(330.51)\right]$$

$R' \approx -.03$

Therefore R decreases at a rate of .03 ohm/min.

2. $A = \pi r^2$

The rate of change of the area with respect to time is

$$\frac{dA}{dt} = 2\pi r\frac{dr}{dt}$$

In this problem,

$\dfrac{dr}{dt} = 10\,\dfrac{in}{min}$ and $r = 4$ inches.

$$\frac{dA}{dt} = 2\pi(4\text{ in})\left(10\,\frac{in}{min}\right)$$

$$= 80\pi\,\frac{in^2}{min}$$

3. $\dfrac{d\theta}{dt} = .01$ degrees/sec

Find $\dfrac{dh}{dt}$.

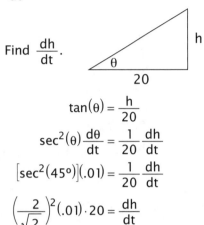

$$\tan(\theta) = \dfrac{h}{20}$$

$$\sec^2(\theta)\dfrac{d\theta}{dt} = \dfrac{1}{20}\dfrac{dh}{dt}$$

$$[\sec^2(45°)](.01) = \dfrac{1}{20}\dfrac{dh}{dt}$$

$$\left(\dfrac{2}{\sqrt{2}}\right)^2 (.01) \cdot 20 = \dfrac{dh}{dt}$$

$$\dfrac{dh}{dt} = .4$$

The balloon was rising at .4 ft/sec.

Lesson Practice 21B

1. A. Determine R.

$$\left[\dfrac{1}{R} = \dfrac{1}{50} + \dfrac{1}{75}\right] 150R$$

$$150 = 3R + 2R$$

$$150 = 5R$$

$$R = 30$$

B. Differentiate implicitly.

$$\dfrac{-1}{R^2}R' = \dfrac{-1}{(R_1)^2}R_1' - \dfrac{1}{(R_2)^2}R_2'$$

C. Solving for R'.

$$R' = R^2\left[\dfrac{R_1'}{(R_1)^2} + \dfrac{R_2'}{(R_2)^2}\right]$$

D. Plug in all the numbers:

$$R' = 30^2\left[\dfrac{-.3}{50^2} + \dfrac{.5}{75^2}\right]$$

$$R' = 900\left[\dfrac{-.3}{2500} + \dfrac{.5}{5625}\right]$$

$$R' \approx -.03$$

Therefore R decreases at a rate of .03 ohm/min.

2. $A = L \cdot w$;
Using the product rule implicitly.
In this problem,

$$\dfrac{dA}{dt} = L\dfrac{dw}{dt} + w \cdot \dfrac{dL}{dt}$$

$$\dfrac{dw}{dt} = 3\dfrac{in}{min}; \dfrac{dL}{dt} = -2\dfrac{in}{min}$$

$$w = 10 \text{ in}; \; L = 8 \text{ in}$$

$$\dfrac{dA}{dt} = (8 \text{ in})\left(3\dfrac{in}{min}\right) + (10 \text{ in})\left(-2\dfrac{in}{min}\right)$$

$$= 24\dfrac{in^2}{min} - 20\dfrac{in^2}{min} = 4\dfrac{in^2}{min}$$

The area of the rectangle was increasing at a rate of 4 in²/min.

3. $\dfrac{dh}{dt} = -30\dfrac{ft}{sec}$

Find $\dfrac{d\theta}{dt}$

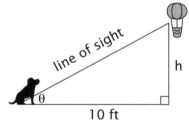

Using trig and the two sides we have labeled.

$$\tan(\theta) = \dfrac{h}{10}$$

Taking the derivative implicitly with respect to t.

$$\sec^2(\theta)\frac{d\theta}{dt} = \frac{1}{10}\frac{dh}{dt}$$

When the balloon hits the ground, $\theta = 0°$.

Plugging in $\frac{dh}{dt} = -30\frac{ft}{sec}$ as well yields

$$\sec^2(0)\cdot\frac{d\theta}{dt} = \frac{1}{10}(-30)$$

$$\frac{d\theta}{dt} = -3$$

The rate of the angle is decreasing at 3 radians per second.

Lesson Practice 21C

1. Determine R.

$$\left[\frac{1}{R} = \frac{1}{30} + \frac{1}{60}\right]60R$$

$$60 = 2R + R$$

$$R = 20$$

Differentiate implicitly.

$$\frac{-1}{R^2}R' = \frac{-1}{(R_1)^2}R_1' - \frac{1}{(R_2)^2}R_2'$$

Solve for R'.

$$R' = R^2\left[\frac{R_1'}{(R_1)^2} + \frac{R_2'}{(R_2)^2}\right]$$

Plugging in all the known values:

$$R' = 20^2\left[\frac{.2}{30^2} - \frac{.1}{60^2}\right]$$

$$R' = 400\left[\frac{.2}{900} - \frac{.1}{3600}\right]$$

$$R' = 400\left[\frac{.8 - .1}{3600}\right]$$

$$R' = .08\frac{ohm}{min}$$

Therefore R is increasing at a rate of .08 ohm/min.

2. Volume of a sphere $V = \frac{4}{3}\pi r^3$

Implicitly differentiating both sides we get:

$$\frac{dV}{dt} = 4\pi r^2\frac{dr}{dt}$$

Going back to the problem,

we know that: $\frac{dV}{dt} = 2\frac{cm^3}{min}$, $r = 5$ cm.

We are seeking $\frac{dr}{dt}$.

Substituting we get:

$$2\frac{cm^3}{min} = 4\pi(5\ cm)^2\cdot\frac{dr}{dt}$$

$$\frac{dr}{dt} = \frac{2\frac{cm^3}{min}}{100\pi\ cm^2} \approx .0064\frac{cm}{min}$$

The radius of the balloon is increasing at a rate of .0064 cm/min.

3. Find $\frac{dx}{dt}$

$$\frac{dy}{dt} = -3 \text{ ft/sec}$$

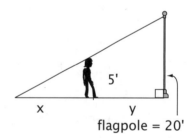

flagpole = 20'

By similar triangles

$$\frac{x+y}{20} = \frac{x}{5}$$

LESSON PRACTICE 21C - LESSON PRACTICE 21D

Implicitly differentiating with respect to time:

$$\frac{1}{20}\left(\frac{dx}{dt} + \frac{dy}{dt}\right) = \frac{1}{5}\frac{dx}{dt}$$

$$20\left[\frac{1}{20}\left(\frac{dx}{dt} - 3\right) = \frac{1}{5}\frac{dx}{dt}\right]$$

$$\frac{dx}{dt} - 3 = 4\frac{dx}{dt}$$

$$-3 = 3\frac{dx}{dt}$$

$$\frac{dx}{dt} = -1 \frac{ft}{sec}$$

Her shadow is shortening at a rate of 1 ft/sec.

Lesson Practice 21D

1. Implicitly differentiating with respect to time

$$\frac{-1}{R^2}R' = \frac{-1}{(R_1)^2}R_1' - \frac{1}{(R_2)^2}R_2'$$

$$R' = R^2\left[\frac{R_1'}{(R_1)^2} + \frac{R_2'}{(R_2)^2}\right]$$

$R_1' = .5\frac{ohm}{min}$, $R_2' = -.8\frac{ohm}{min}$

$R_1 = 80$ ohm $R_2 = 100$ ohm

We need only to compute R and then we can find out $\frac{dR}{dt}$

$$\frac{1}{R} = \frac{1}{R_1} + \frac{1}{R_2}$$

so $\frac{1}{R} = \frac{1}{80} + \frac{1}{100} = \frac{5+4}{400} = \frac{9}{400}$

$R \approx 44.44$

$$\frac{dR}{dt} = (44.44)^2\left[\frac{.5\frac{ohm}{min}}{(80\ ohm)^2} + \frac{\left(-.8\frac{ohm}{min}\right)}{(100\ ohm)^2}\right]$$

$$\approx -.0037\frac{ohm}{min}$$

So R is decreasing at a rate of .0037 ohm/min.

2. $\frac{dr}{dt} = 2\frac{cm}{min}$

$A = \pi r^2$

$\frac{dA}{dt} = 2\pi r\frac{dr}{dt}$

$\frac{dA}{dt} = 2\pi(15)(2)$

$\approx 188.5\frac{cm^2}{min}$

The area is expanding at a rate of 188.5 cm²/min.

3. x = distance from ball to pitcher and y = distance from ball to 1st base.

We want $\frac{dy}{dt}$ when the ball is at home plate.

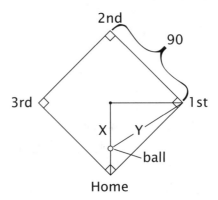

$\frac{dx}{dt} = 80\frac{m}{hr}$

$= 80\frac{miles}{hour} \cdot \frac{1\ hr}{60\ min} \cdot \frac{1\ min}{60\ sec} \cdot \frac{5{,}280\ ft}{mile}$

$= 117.3\frac{ft}{sec}$

We formed an isosceles right triangle. Thus, the distance from the pitcher's mound to any base is $\frac{90}{\sqrt{2}}$.

The relationship between x and y is:

$$x^2 + \left(\frac{90}{\sqrt{2}}\right)^2 = y^2$$

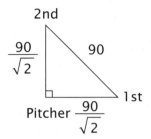

Differentiate implicitly with respect to time.

$$2x\frac{dx}{dt} = 2y\frac{dy}{dt}$$

$$\frac{dy}{dt} = \frac{2x\frac{dx}{dt}}{2y}$$

$$= \frac{x\frac{dx}{dt}}{y}$$

$$\frac{dy}{dt} = \frac{\left(\frac{90}{\sqrt{2}}\text{ ft}\right)\left(117.3\frac{\text{ft}}{\text{sec}}\right)}{(90\text{ ft})}$$

Find $\frac{dy}{dt}$ when the ball is at home.

$$\frac{dy}{dt} = \frac{117.3}{\sqrt{2}} \approx 82.9\frac{\text{ft}}{\text{sec}}$$

When the ball is traveling at 80 m/hr and reaches home base, the distance between the ball and first base is increasing at approximately 82.9 ft/sec.

Lesson Practice 22A

1. $\int dt = t + C$

   Check: $\frac{d}{dt}(t + C) = 1$

2. $\int x^7 dx = \frac{1}{7+1}x^{7+1} + C$

   $= \frac{1}{8}x^8 + C$

   Check: $\frac{d}{dx}\left(\frac{1}{8}x^8 + C\right)$

   $= \frac{1}{8} \cdot 8x^7$

   $= x^7$

3. $\int(v^4 - 2)dv = \int v^4 dv - \int 2 dv$

   $= \frac{1}{4+1}v^{4+1} - 2\int dv$

   $= \frac{1}{5}v^5 - 2v + C$

   Check: $\frac{d}{dv}\left(\frac{1}{5}v^5 - 2v + C\right)$

   $\frac{1}{5}5v^4 - 2 = v^4 - 2$

Note: Even though both derivatives would legitimately have their own constant, we can combine them into one final C as long as we are not asked to evaluate them further as in examples 3 and 4 in the instruction manual.

4. $\int(z^5 + 2z - 1)dz = \int z^5 dz + \int 2z dz - \int dz$

   $= \frac{1}{6}z^6 + 2\int z dz - \int dz$

   $= \frac{1}{6}z^6 + 2\frac{z^2}{2} - z + C$

   $= \frac{1}{6}z^6 + z^2 - z + C$

LESSON PRACTICE 22A - LESSON PRACTICE 22A

5. $\int (3r - \sqrt{r})\,dr = \int 3r\,dr - \int \sqrt{r}\,dr$

   $= 3\int r\,dr - \int r^{\frac{1}{2}}\,dr$

   $= 3\frac{r^2}{2} - \frac{r^{\frac{3}{2}}}{\frac{3}{2}} + C$

   $= \frac{3}{2}r^2 - \frac{2}{3}r^{\frac{3}{2}} + C$

6. $\int x^2(4 - x^3)\,dx = \int 4x^2\,dx - \int x^5\,dx$

   $= 4\int x^2\,dx - \frac{1}{6}x^6 + C$

   $= 4\frac{x^3}{3} - \frac{1}{6}x^6 + C$

   $= \frac{4}{3}x^3 - \frac{1}{6}x^6 + C$

7. $\int x^{\frac{1}{3}}\left(x^{\frac{1}{2}} + 1\right)dx = \int \left(x^{\frac{1}{3}+\frac{1}{2}}\right)dx + \int x^{\frac{1}{3}}\,dx$

   $= \int x^{\frac{5}{6}}\,dx + \frac{x^{\frac{4}{3}}}{\frac{4}{3}} + C$

   $= \frac{x^{\frac{11}{6}}}{\frac{11}{6}} + \frac{3}{4}x^{\frac{4}{3}} + C$

   $= \frac{6}{11}x^{\frac{11}{6}} + \frac{3}{4}x^{\frac{4}{3}} + C$

8. $\int \frac{dw}{w^2} = \int w^{-2}\,dw$

   $= \frac{w^{-1}}{-1} + C$

   $= -\frac{1}{w} + C$

9. $y = \int (2 - 4x)\,dx$

   $= \int 2\,dx - \int 4x\,dx$

   $= 2\int dx - 4\int x\,dx$

   $= 2x - 4\frac{x^2}{2} + C$

   $y = 2x - 2x^2 + C$

   $1 = 2(3) - 2(3)^2 + C$

   $1 = 6 - 18 + C$

   $13 = C$

   Curve is: $y = 2x - 2x^2 + 13$

10. $y = \int 6x\,dx$

    $= 6\int x\,dx$

    $= 6\frac{x^2}{2} + C_1$

    $= 3x^2 + C_1$

    $y = \int (3x^2 + C_1)\,dx$

    $= \int 3x^2\,dx + \int C_1\,dx$

    $= 3\int x^2\,dx + C_1\int dx$

    $= 3\frac{x^3}{3} + C_1 x + C_2$

    $= x^3 + C_1 x + C_2$

    $y = x^3 + C_1 x + C_2$; using points $(1, 2)$ and $(2, 3)$

    $2 = 1 + C_1 + C_2$ $3 = 8 + 2C_1 + C_2$
    $C_1 + C_2 = 1$ $2C_1 + C_2 = -5$
    Using elimination $2C_1 + C_2 = -5$
        $-C_1 - C_2 = -1$
        $C_1 = -6$

    $-6 + C_2 = 1$ So $y = x^3 - 6x + 7$ is the answer.
    $C_2 = 7$

Lesson Practice 22B

1. $\int 3\,dy = 3\int dy = 3y + C$

    Check: $\dfrac{d}{dy}(3y + C) = 3$

2. $\int p^8\,dp = \dfrac{1}{9}p^9 + C$

    Check:
    $\dfrac{d}{dp}\left(\dfrac{1}{9}p^9 + C\right) = \dfrac{1}{9}\cdot 9p^8 = p^8$

3. $\int\left(\dfrac{1}{2}x^2 - 4x\right)dx = \int\dfrac{1}{2}x^2\,dx - \int 4x\,dx$
    $= \dfrac{1}{2}\int x^2\,dx - 4\int x\,dx$
    $= \dfrac{1}{2}\dfrac{x^3}{3} - 4\dfrac{x^2}{2} + C$
    $= \dfrac{1}{6}x^3 - 2x^2 + C$

    Check:
    $\dfrac{d}{dx}\left(\dfrac{1}{6}x^3 - 2x^2 + C\right) = \dfrac{1}{6}\cdot 3x^2 - 2\cdot 2x$
    $= \dfrac{1}{2}x^2 - 4x$

4. $\int\left(-2w^3 - \dfrac{1}{4}w^2 + 4\right)dw$
    $= \int -2w^3\,dw - \int\dfrac{1}{4}w^2\,dw + \int 4\,dw$
    $= -2\int w^3\,dw - \dfrac{1}{4}\int w^2\,dw + 4\int dw$
    $= -2\dfrac{w^4}{4} - \dfrac{1}{4}\dfrac{w^3}{3} + 4w + C$
    $= -\dfrac{1}{2}w^4 - \dfrac{1}{12}w^3 + 4w + C$

5. $\int\left(2\sqrt{r} - 5\right)dr = \int 2\sqrt{r}\,dr - \int 5\,dr$
    $= 2\int r^{\frac{1}{2}}\,dr - 5\int dr$
    $= 2\dfrac{r^{\frac{3}{2}}}{\frac{3}{2}} - 5r + C$
    $= \dfrac{4}{3}r^{\frac{3}{2}} - 5r + C$

6. $\int y^3(2y^4 + y)\,dy = \int 2y^7\,dy + \int y^4\,dy$
    $= 2\int y^7\,dy + \int y^4\,dy$
    $= 2\dfrac{y^8}{8} + \dfrac{y^5}{5} + C$
    $= \dfrac{1}{4}y^8 + \dfrac{1}{5}y^5 + C$

7. $\int\dfrac{ds}{3s^3} = \dfrac{1}{3}\int s^{-3}\,ds$
    $= \dfrac{1}{3}\dfrac{s^{-2}}{-2} + C = -\dfrac{1}{6s^2} + C$

8. $\int r^{-\frac{1}{3}}\left(r^3 + 2r^{\frac{1}{3}} - 3\right)dr$
    $= \int r^{\frac{8}{3}}\,dr + \int 2r^0\,dr - \int 3r^{-\frac{1}{3}}\,dr$
    $= \dfrac{r^{\frac{11}{3}}}{\frac{11}{3}} + 2\int dr - 3\int r^{-\frac{1}{3}}\,dr$
    $= \dfrac{3}{11}r^{\frac{11}{3}} + 2r - 3\dfrac{r^{\frac{2}{3}}}{\frac{2}{3}} + C$
    $= \dfrac{3}{11}r^{\frac{11}{3}} + 2r - \dfrac{9}{2}r^{\frac{2}{3}} + C$

LESSON PRACTICE 22B - LESSON PRACTICE 22C

9. $y'' = 6x - x^2$

$$y' = \int(6x - x^2)dx$$
$$= \int 6x\,dx - \int x^2 dx$$
$$= 6\int x\,dx - \int x^2 dx$$
$$= 6\frac{x^2}{2} - \frac{x^3}{3} + C_1$$
$$= 3x^2 - \frac{1}{3}x^3 + C_1$$

$$y' = 3x^2 - \frac{1}{3}x^3 + C_1$$
$$-2 = 3 - \frac{1}{3} + C_1$$
$$-4\frac{2}{3} = C_1$$

$$y = \int\left(3x^2 - \frac{1}{3}x^3 - 4\frac{2}{3}\right)dx$$
$$= 3\int x^2 dx - \frac{1}{3}\int x^3 dx - 4\frac{2}{3}\int dx$$
$$= 3\frac{x^3}{3} - \frac{1}{3}\frac{x^4}{4} - 4\frac{2}{3}x + C_2$$

$$y = x^3 - \frac{1}{12}x^4 - \frac{14}{3}x + C_2$$
$$1 = 1 - \frac{1}{12} - \frac{14}{3} + C_2$$
$$\frac{19}{4} = C_2$$

$$y = x^3 - \frac{1}{12}x^4 - \frac{14}{3}x + \frac{19}{4}$$

10. $y'' = -12 + x$

$$y' = \int(-12 + x)dx$$
$$= -12\int dx + \int x\,dx$$
$$= -12x + \frac{x^2}{2} + C_1$$
$$60 = -12x + \frac{1}{2}x^2 + C_1$$
$$C_1 = 60 \text{ when } x = 0$$

$$y' = -12x + \frac{1}{2}x^2 + 60$$

$$y = \int\left(-12x + \frac{1}{2}x^2 + 60\right)dx$$
$$= -12\int x\,dx + \frac{1}{2}\int x^2 dx + \int 60\,dx$$

$$y = -12\frac{x^2}{2} + \frac{1}{2}\frac{x^3}{3} + 60x + C_2$$
$$C_2 = 0 \text{ when } x = y = 0$$

$$y = -6x^2 + \frac{1}{6}x^3 + 60x$$

Lesson Practice 22C

1. $\int(x^4 + x - 3)dx = \int x^4 dx + \int x\,dx - 3\int dx$
$$= \frac{x^5}{5} + \frac{x^2}{2} - 3x + C$$

Check:
$$\frac{d}{dx}\left(\frac{x^5}{5} + \frac{x^2}{2} - 3x + C\right) = \frac{1}{5}\cdot 5x^4 + \frac{1}{2}\cdot 2x - 3$$
$$= x^4 + x - 3$$

2. $\int \left(5x - \frac{4}{x^3} + \frac{3}{x^2}\right) dx$

$= 5\int x\, dx - 4\int x^{-3} dx + 3\int x^{-2} dx$

$= 5\frac{x^2}{2} - 4\left(\frac{x^{-2}}{-2}\right) + 3\left(\frac{x^{-1}}{-1}\right) + C$

$= \frac{5}{2}x^2 + 2x^{-2} - 3x^{-1} + C$

Check:

$\frac{d}{dx}\left(\frac{5}{2}x^2 + 2x^{-2} - 3x^{-1} + C\right)$

$= \frac{5}{2} \cdot 2x + 2(-2x^{-3}) - 3(-x^{-2})$

$= 5x - 4x^{-3} + 3x^{-2}$

3. $\int x\sqrt{x}\, dx = \int x^{\frac{3}{2}} dx$

$= \frac{x^{\frac{5}{2}}}{\frac{5}{2}} + C$

$= \frac{2}{5} x^{\frac{5}{2}} + C$

Check: $\frac{d}{dx}\left(\frac{2}{5}x^{\frac{5}{2}} + C\right) = \frac{2}{5} \cdot \frac{5}{2} x^{\frac{3}{2}}$

$= x^{\frac{3}{2}}$

$= x\sqrt{x}$

4. $\int \frac{1+x}{\sqrt{x}} dx = \int \frac{1}{\sqrt{x}} dx + \int \frac{x}{\sqrt{x}} dx$

$= \int x^{-\frac{1}{2}} dx + \int x^{\frac{1}{2}} dx$

$= \frac{x^{\frac{1}{2}}}{\frac{1}{2}} + \frac{x^{\frac{3}{2}}}{\frac{3}{2}} + C$

$= 2\sqrt{x} + \frac{2}{3} x^{\frac{3}{2}} + C$

Check:

$\frac{d}{dx}\left(2\sqrt{x} + \frac{2}{3} x^{\frac{3}{2}} + C\right)$

$= 2 \cdot \frac{1}{2} x^{-\frac{1}{2}} + \frac{2}{3} \cdot \frac{3}{2} x^{\frac{1}{2}}$

$= \frac{1}{\sqrt{x}} + \sqrt{x}$

$= \frac{1+x}{\sqrt{x}}$

5. $y' = 3\sqrt{x}$

$y = \int 3\sqrt{x} = 3\frac{x^{\frac{3}{2}}}{\frac{3}{2}} + C$

$y = 2x^{\frac{3}{2}} + C$ at point $(4, 9)$ we have

$9 = 2 \cdot 4^{\frac{3}{2}} + C$

$9 = 2 \cdot 8 + C$

$9 = 16 + C$ So $y = 2x^{\frac{3}{2}} - 7$

$-7 = C$

6. $\int 4\, dx = y' = 4x + C_1$

when $y' = 0$ and $x = 2$ we have

$0 = 4(2) + C_1$

$C_1 = -8$

LESSON PRACTICE 22C - LESSON PRACTICE 22D

$y' = 4x - 8$

$y = \int (4x - 8)\,dx$

$= 4\int x\,dx - 8\int dx$

$= 4\dfrac{x^2}{2} - 8x + C_2$

$y = 2x^2 - 8x + C_2$

When $x = 2$ and $y = 1$

$1 = 2 \cdot 4 - 16 + C_2$

$C_2 = 9$

$y = 2x^2 - 8x + 9$

Lesson Practice 22D

1. $\int (7 - x^6)\,dx = 7\int dx - \int x^6\,dx$

   $= 7x - \dfrac{x^7}{7} + C$

   Check:

   $\dfrac{d}{dx}\left(7x - \dfrac{x^7}{7} + C\right) = 7 - \dfrac{1}{7}(7x^6)$

   $= 7 - x^6$

2. $\int \left(\dfrac{1}{x^3} - \dfrac{2}{x^2} + 3\right)dx$

   $= \int x^{-3}\,dx - 2\int x^{-2}\,dx + 3\int dx$

   $= \dfrac{x^{-2}}{-2} - 2\dfrac{x^{-1}}{-1} + 3x + C$

   $= -\dfrac{1}{2}x^{-2} + 2x^{-1} + 3x + C$

   Check:

   $\dfrac{d}{dx}\left(-\dfrac{1}{2}x^{-2} + 2x^{-1} + 3x + C\right)$

   $= -\dfrac{1}{2}(-2x^{-3}) + 2(-x^{-2}) + 3$

   $= x^{-3} - \dfrac{2}{x^2} + 3$

3. $\int \dfrac{2}{\sqrt{x}}\,dx = 2\int x^{-\tfrac{1}{2}}\,dx$

   $= 2\dfrac{x^{\tfrac{1}{2}}}{\tfrac{1}{2}} + C$

   $= 4\sqrt{x} + C$

   Check:

   $\dfrac{d}{dx}(4\sqrt{x} + C) = 4 \cdot \dfrac{1}{2}x^{-\tfrac{1}{2}}$

   $= 2x^{-\tfrac{1}{2}}$

   $= \dfrac{2}{\sqrt{x}}$

4. $\int \dfrac{2 - \sqrt{x}}{x^2}\,dx = 2\int x^{-2}\,dx - \int \dfrac{\sqrt{x}}{x^2}\,dx$

   $= 2\dfrac{x^{-1}}{-1} - \int x^{-\tfrac{3}{2}}\,dx + C$

   $= -2x^{-1} - \left(\dfrac{x^{-\tfrac{1}{2}}}{-\tfrac{1}{2}}\right) + C$

   $= -2x^{-1} + 2x^{-\tfrac{1}{2}} + C$

   Check:

   $\dfrac{d}{dx}\left(-2x^{-1} + 2x^{-\tfrac{1}{2}} + C\right)$

   $= -2(-x^{-2}) + 2 \cdot \left(-\dfrac{1}{2}x^{-\tfrac{3}{2}}\right)$

   $= 2x^{-2} - x^{-\tfrac{3}{2}}$

   $= \dfrac{2 - \sqrt{x}}{x^2}$

5. $y'' = 6x - 6$

$\int (6x - 6) dx = \int 6x \, dx - \int 6 \, dx$

$= 6 \frac{x^2}{2} - 6x + C_1$

$y' = 3x^2 - 6x + C_1$

Critical point would mean that $y' = 0$ and $x = 1$.

$3 - 6 + C_1 = 0$

$C_1 = 3$

$y' = 3x^2 - 6x + 3$

$y = \int (3x^2 - 6x + 3) dx$

$= 3 \int x^2 dx - 6 \int x \, dx + 3 \int dx$

$= 3 \frac{x^3}{3} - 6 \frac{x^2}{2} + 3x + C_2$

$y = x^3 - 3x^2 + 3x + C_2$

$(1, 0)$ must satisfy the function $y = f(x)$

$0 = 1 - 3 + 3 + C_2$

$-1 = C_2$

$y = x^3 - 3x^2 + 3x - 1$

Lesson Practice 23A

1. Let $u = 2 + x$; $\frac{du}{dx} = 1$; $du = dx$

$\int (2 + x)^4 dx = \int u^4 du$

$= \frac{u^5}{5} + C$

$= \frac{1}{5}(2 + x)^5 + C$

Check:

$\frac{d}{dx} \left(\frac{1}{5}(2 + x)^5 \right) = \frac{1}{5} \cdot 5(2 + x)^4$

$= (2 + x)^4$

2. Let $u = t + 1$; $\frac{du}{dt} = 1$; $du = dt$.

$\int \frac{dt}{\sqrt{t + 1}} = \int \frac{du}{\sqrt{u}}$

$= \int u^{-\frac{1}{2}} du$

$= \frac{u^{\frac{1}{2}}}{\frac{1}{2}} + C$

$= 2\sqrt{u} + C$

$= 2\sqrt{t + 1} + C$

Check:

$\frac{d}{dt} \left(2\sqrt{t + 1} \right) = 2 \cdot \frac{1}{2}(t + 1)^{-\frac{1}{2}}$

$= \frac{1}{\sqrt{t + 1}}$

LESSON PRACTICE 23A

3. Let $u = 3 + x^2$; $\frac{du}{dx} = 2x$; $du = 2x\,dx$

$$\int (3+x^2)^{10} 2x\,dx = \int u^{10}\,du$$
$$= \frac{u^{11}}{11} + C$$
$$= \frac{1}{11}(3+x^2)^{11} + C$$

Check:
$$\frac{d}{dx}\left(\frac{1}{11}(3+x^2)^{11}\right) = \frac{1}{11} \cdot 11(3+x^2)^{10} \cdot 2x$$
$$= 2x(3+x^2)^{10}$$

4. Let $u = 4x$; $\frac{du}{dx} = 4$; $du = 4\,dx$

$$\int \csc^2(4x) \cdot 4\,dx = \int \csc^2(u)\,du$$
$$= -\cot(u) + C$$
$$= -\cot(4x) + C$$

Check:
$$\frac{d}{dx}(-\cot(4x)) = -(-\csc^2(4x) \cdot 4)$$
$$= 4\csc^2(4x)$$

5. Let $u = \sin(3\theta)$; $\frac{du}{d\theta} = \cos(3\theta) \cdot 3$;
$du = \cos(3\theta) \cdot 3(d\theta)$

$$\int \sin^3(3\theta)\cos(3\theta)\,d\theta = \frac{1}{3}\int \sin^3(3\theta)\cos(3\theta)\,d\theta \cdot 3$$
$$= \frac{1}{3}\int u^3\,du$$
$$= \frac{1}{3}\frac{u^4}{4} + C$$
$$= \frac{1}{12}u^4 + C$$
$$= \frac{1}{12}(\sin(3\theta))^4 + C$$

Check:
$$\frac{d}{d\theta}\left(\frac{1}{12}\sin(3\theta)\right)^4 = \frac{1}{12} \cdot 4(\sin(3\theta))^3 \cdot \cos(3\theta) \cdot 3$$
$$= (\sin(3\theta))^3 \cos(3\theta)$$

6. Let $u = a - y$; $\frac{du}{dy} = -1$; $du = -dy$

$$\int \frac{dy}{(a-y)^2} = -\int \frac{-dy}{(a-y)^2}$$
$$= -\int \frac{du}{u^2}$$
$$= -\int u^{-2}\,du$$
$$= -\frac{u^{-1}}{-1} + C$$
$$= \frac{1}{a-y} + C$$

Check:
$$\frac{d}{dy}\left(\frac{1}{a-y}\right) = \frac{d}{dy}(a-y)^{-1}$$
$$= -(a-y)^{-2}(-1)$$
$$= \frac{1}{(a-y)^2}$$

7. Let $u = 4x$; $\dfrac{du}{dx} = 4$; $du = 4dx$

$$\int e^{4x}\,dx = \dfrac{1}{4}\int e^{4x}\,dx \cdot 4$$
$$= \dfrac{1}{4}\int e^{u}\,du$$
$$= \dfrac{1}{4}e^{u} + C$$
$$= \dfrac{1}{4}e^{4x} + C$$

Check:
$$\dfrac{d}{dx}\left(\dfrac{1}{4}e^{4x}\right) = \dfrac{1}{4}e^{4x} \cdot 4$$
$$= e^{4x}$$

8. Let $u = 1 + e^x$; $\dfrac{du}{dx} = e^x$; $du = e^x\,dx$

$$\int e^x(1+e^x)^8\,dx = \int u^8\,du$$
$$= \dfrac{u^9}{9} + C$$
$$= \dfrac{1}{9}(1+e^x)^9 + C$$

Check:
$$\dfrac{d}{dx}\left[\left(\dfrac{1}{9}\cdot 1+e^x\right)^9\right] = \dfrac{1}{9}\cdot 9(1+e^x)^8 \cdot e^x$$
$$= e^x(1+e^x)^8$$

9. Let $u = \ln(x)$; $\dfrac{du}{dx} = \dfrac{1}{x}$; $du = \dfrac{dx}{x}$

$$\int \dfrac{(\ln(x))^2}{x}\,dx = \int u^2\,du$$
$$= \dfrac{u^3}{3} + C$$
$$= \dfrac{1}{3}(\ln(x))^3 + C$$

Check:
$$\dfrac{d}{dx}\left(\dfrac{1}{3}(\ln(x))^3\right) = \dfrac{1}{3}\cdot 3(\ln(x))^2 \cdot \dfrac{1}{x}$$
$$= \dfrac{\ln^2(x)}{x}$$

10. Let $u = 2t + 1$; $\dfrac{du}{dt} = 2$; $du = 2dt$

$$\int \dfrac{3\,dt}{\sqrt{2t+1}} = \dfrac{1}{2}\cdot 3\int \dfrac{dt}{\sqrt{2t+1}} \cdot 2$$
$$= \dfrac{3}{2}\int \dfrac{du}{\sqrt{u}}$$
$$= \dfrac{3}{2}\int u^{-\frac{1}{2}}\,du$$
$$= \dfrac{3}{2}\dfrac{u^{\frac{1}{2}}}{\frac{1}{2}} + C$$
$$= 3\sqrt{u} + C$$
$$= 3\sqrt{2t+1} + C$$

Check:
$$\dfrac{d}{dt}\left(3\sqrt{2t+1}\right) = 3\cdot \dfrac{1}{2}(2t+1)^{-\frac{1}{2}} \cdot 2$$
$$= \dfrac{3}{\sqrt{2t+1}}$$

Lesson Practice 23B

1. Let $u = x-3$; $\frac{du}{dx} = 1$; $du = dx$.

$$\int (x-3)^{21} dx = \int u^{21} du$$
$$= \frac{u^{22}}{22} + C$$
$$= \frac{1}{22}(x-3)^{22} + C$$

Check:

$$\frac{d}{dx}\left(\frac{1}{22}(x-3)^{22}\right) = \frac{1}{22} \cdot 22(x-3)^{21}$$
$$= (x-3)^{21}$$

2. Let $u = 2x^2 + 1$; $\frac{du}{dx} = 4x$; $du = 4x\,dx$

$$\frac{1}{4}\int x\sqrt{2x^2+1}\,dx \cdot 4 = \frac{1}{4}\int \sqrt{u}\,du$$
$$= \frac{1}{4}\int u^{\frac{1}{2}} du$$
$$= \frac{1}{4} \frac{u^{\frac{3}{2}}}{\frac{3}{2}} + C$$
$$= \frac{1}{6}(2x^2+1)^{\frac{3}{2}} + C$$

Check:

$$\frac{d}{dx}\left(\frac{1}{6}(2x^2+1)^{\frac{3}{2}}\right) = \frac{1}{6} \cdot \frac{3}{2}(2x^2+1)^{\frac{1}{2}} \cdot 4x$$
$$= x\sqrt{2x^2+1}$$

3. Let $u = 1-x^3$; $\frac{du}{dx} = -3x^2$; $du = -3x^2 dx$

$$-\frac{1}{3}\int \frac{x^2}{\sqrt{1-x^3}} dx(-3) = -\frac{1}{3}\int \frac{du}{\sqrt{u}}$$
$$= -\frac{1}{3}\int u^{-\frac{1}{2}} du$$
$$= -\frac{1}{3} \frac{u^{\frac{1}{2}}}{\frac{1}{2}} + C$$
$$= -\frac{2}{3}\sqrt{u} + C$$
$$= -\frac{2}{3}\sqrt{1-x^3} + C$$

Check:

$$\frac{d}{dx}\left(-\frac{2}{3}\sqrt{1-x^3}\right) = -\frac{2}{3} \cdot \frac{1}{2}(1-x^3)^{-\frac{1}{2}}(-3x^2)$$
$$= \frac{x^2}{\sqrt{1-x^3}}$$

4. Let $u = 1+x^3$; $\frac{du}{dx} = 3x^2$; $du = 3x^2 dx$

$$\int (1+x^3)^3 \cdot 3x^2 dx = \int u^3 du$$
$$= \frac{u^4}{4} + C$$
$$= \frac{1}{4}(1+x^3)^4 + C$$

Check:

$$\frac{d}{dx}\left(\frac{1}{4}(1+x^3)^4 + C\right) = \frac{1}{4} \cdot 4(1+x^3)^3 (3x^2)$$
$$= 3x^2(1+x^3)^3$$

5. Let $u = 2x - 3$; $\dfrac{du}{dx} = 2$; $du = 2dx$

$$\dfrac{1}{2}\int \cos(2x-3)dx \cdot 2 = \dfrac{1}{2}\int \cos(u)du$$
$$= \dfrac{1}{2}\sin(u) + C$$
$$= \dfrac{1}{2}\sin(2x-3) + C$$

Check:
$$\dfrac{d}{dx}\left(\dfrac{1}{2}\sin(2x-3)\right) = \dfrac{1}{2}\cos(2x-3)\cdot 2$$
$$= \cos(2x-3)$$

6. Let $u = \sqrt{\theta} = \theta^{\frac{1}{2}}$;

$\dfrac{du}{d\theta} = \dfrac{1}{2}\theta^{-\frac{1}{2}}$; $du = \dfrac{1}{2}\theta^{-\frac{1}{2}}d\theta$

$$2\int \dfrac{\cos(\sqrt{\theta})d\theta \cdot \dfrac{1}{2}}{\sqrt{\theta}} = 2\int (\cos(u))du$$
$$= 2\sin(u) + C$$
$$= 2\sin(\sqrt{\theta}) + C$$

Check:
$$\dfrac{d}{d\theta}\left(2\sin(\sqrt{\theta})\right) = 2\cos(\sqrt{\theta})\cdot \dfrac{1}{2}\theta^{-\frac{1}{2}}$$
$$= \dfrac{\cos(\sqrt{\theta})}{\sqrt{\theta}}$$

7. Let $u = \tan(\theta)$; $\dfrac{du}{d\theta} = \sec^2(\theta)$;

$du = \sec^2(\theta)d\theta$

$$\int \sqrt{\tan(\theta)}\,\sec^2(\theta)d\theta = \int u^{\frac{1}{2}}du$$
$$= \dfrac{u^{\frac{3}{2}}}{\frac{3}{2}} + C$$
$$= \dfrac{2}{3}u^{\frac{3}{2}} + C$$
$$= \dfrac{2}{3}(\tan(\theta))^{\frac{3}{2}} + C$$

Check:
$$\dfrac{d}{d\theta}\left(\dfrac{2}{3}(\tan(\theta))^{\frac{3}{2}}\right) = \dfrac{2}{3}\cdot\dfrac{3}{2}(\tan(\theta))^{\frac{1}{2}}\sec^2(\theta)$$
$$= \sqrt{\tan(\theta)}\,\sec^2(\theta)$$

8. Let $u = 2 - e^{2x}$; $\dfrac{du}{dx} = -e^{2x}\cdot 2$;

$du = -2e^{2x}dx$

$$-\dfrac{1}{2}\int e^{2x}(2-e^{2x})dx(-2) = -\dfrac{1}{2}\int u(du)$$
$$= -\dfrac{1}{2}\dfrac{u^2}{2} + C$$
$$= -\dfrac{1}{4}u^2 + C$$
$$= -\dfrac{1}{4}(2-e^{2x})^2 + C$$

You could also multiply through, which yields $2e^{2x} - e^{4x}$, and integrate each term.

Check:
$$\dfrac{d}{du}\left(-\dfrac{1}{4}(2-e^{2x})^2\right) = -\dfrac{1}{4}\cdot 2(2-e^{2x})(-e^{2x}\cdot 2)$$
$$= e^{2x}\cdot(2-e^{2x})$$

LESSON PRACTICE 23B - LESSON PRACTICE 23C

9. Let $u = 1-x$; $\frac{du}{dx} = -1$; $du = -dx$

$$-\int \frac{dx}{1-x}(-1) = -\int \frac{du}{u}$$
$$= -\ln(u) + C$$
$$= -\ln(1-x) + C$$

Check:
$$\frac{d}{dx}(-\ln(1-x)) = -\frac{1}{1-x}(-1)$$
$$= \frac{1}{1-x}$$

10. Let $u = x^2 + 2x - 1$; $\frac{du}{dx} = 2x+2$; $du = (2x+2)dx$

$$\frac{1}{2}\int \frac{(x+1)dx}{\sqrt[3]{x^2+2x-1}} \cdot 2 = \frac{1}{2}\int \frac{du}{u^{\frac{1}{3}}}$$
$$= \frac{1}{2}\int u^{-\frac{1}{3}} du$$
$$= \frac{1}{2} \frac{u^{\frac{2}{3}}}{\frac{2}{3}} + C$$
$$= \frac{3}{4} u^{\frac{2}{3}} + C$$
$$= \frac{3}{4}(x^2+2x-1)^{\frac{2}{3}} + C$$

Check:
$$\frac{d}{dx}\left(\frac{3}{4}(x^2+2x-1)^{\frac{2}{3}}\right) =$$
$$= \frac{3}{4} \cdot \frac{2}{3}(x^2+2x-1)^{-\frac{1}{3}}(2x+2)$$
$$= \frac{x+1}{\sqrt[3]{x^2+2-1}}$$

Lesson Practice 23C

1. Let $u = t+4$; $\frac{du}{dt} = 1$; $du = dt$

$$\int \frac{dt}{\sqrt{t+4}} = \int \frac{du}{\sqrt{u}}$$
$$= \int u^{-\frac{1}{2}} du$$
$$= \frac{u^{\frac{1}{2}}}{\frac{1}{2}} + C$$
$$= 2\sqrt{u} + C$$
$$= 2\sqrt{t+4} + C$$

Check:
$$\frac{d}{dt}(2\sqrt{t+4}) = 2 \cdot \frac{1}{2}(t+4)^{-\frac{1}{2}}$$
$$= \frac{1}{\sqrt{t+4}}$$

2. Let $u = 2-3y$; $\frac{du}{dy} = -3$; $du = -3dy$

$$-\frac{1}{3}\int (2-3y)^5 dy(-3) = -\frac{1}{3}\int u^5 (du)$$
$$= -\frac{1}{3} \frac{u^6}{6} + C$$
$$= -\frac{1}{18} u^6 + C$$
$$= -\frac{1}{18}(2-3y)^6 + C$$

Check:
$$\frac{d}{dy}\left(-\frac{1}{18}(2-3y)^6\right) = -\frac{1}{18} \cdot 6(2-3y)^5(-3)$$
$$= (2-3y)^5$$

3. If you let $u = x^2 + 1$, then $du = 2x\,dx$, which is not present. You must multiply the numerator and divide by the denominator before integrating.

$$\int \frac{(x^2+1)^2}{x^2}\,dx = \int \frac{x^4+2x^2+1}{x^2}\,dx$$
$$= \int x^2\,dx + \int 2\,dx + \int x^{-2}\,dx$$
$$= \frac{x^3}{3} + 2x + \frac{x^{-1}}{-1} + C$$
$$= \frac{1}{3}x^3 + 2x - \frac{1}{x} + C$$

Check:
$$\frac{d}{dx}\left(\frac{1}{3}x^3 + 2x - \frac{1}{x}\right) = \frac{1}{3}(3x^2) + 2 - (-x^{-2})$$
$$= x^2 + 2 + x^{-2}$$
$$= \frac{x^4+2x^2+1}{x^2}$$
$$= \frac{(x^2+1)^2}{x^2}$$

4. Let $u = 1 + \tan(\theta)$; $\frac{du}{d\theta} = \sec^2(\theta)$; $du = \sec^2(\theta)\,d\theta$

$$\int \frac{\sec^2(\theta)\,d\theta}{(1+\tan(\theta))^4} = \int \frac{du}{u^4}$$
$$= \int u^{-4}\,du$$
$$= \frac{u^{-3}}{-3} + C$$
$$= -\frac{1}{3}(1+\tan(\theta))^{-3} + C$$

Check: $\frac{d}{d\theta}\left(-\frac{1}{3}(1+\tan(\theta))^{-3}\right)$
$$= -\frac{1}{3}\left(-3(1+\tan(\theta))^{-4}\right)\sec^2(\theta)$$
$$= \frac{\sec^2(\theta)}{(1+\tan(\theta))^4}$$

5. Let $u = 1 + \ln(x)$; $du = \frac{1}{x}\,dx$

$$\int \frac{\sqrt{1+\ln(x)}}{x}\,dx = \int \sqrt{u}\,du$$
$$= \int u^{\frac{1}{2}}\,du$$
$$= \frac{u^{\frac{3}{2}}}{\frac{3}{2}} + C$$
$$= \frac{2}{3}(1+\ln(x))^{\frac{3}{2}} + C$$

Check:
$$\frac{d}{dx}\left[\frac{2}{3}(1+\ln(x))^{\frac{3}{2}} + C\right] = \frac{2}{3} \cdot \frac{3}{2}(1+\ln(x))^{\frac{1}{2}} \cdot \frac{1}{x}$$
$$= \frac{\sqrt{1+\ln(x)}}{x}$$

Lesson Practice 23D

1. Let $u = \cot(3\theta)$; $\frac{du}{d\theta} = -\csc^2(3\theta) \cdot 3$; $du = -3\csc^2(3\theta)\,d\theta$

$$-\frac{1}{3}\int \cot(3\theta)\,\csc^2(3\theta)\,d\theta(-3)$$
$$= -\frac{1}{3}\int u\,du$$
$$= -\frac{1}{3}\cdot\frac{u^2}{2} + C$$
$$= -\frac{1}{6}u^2 + C$$
$$= -\frac{1}{6}(\cot(3\theta))^2 + C$$

Check:
$$\frac{d}{d\theta}\left(-\frac{1}{6}(\cot(3\theta))^2 + C\right)$$
$$= -\frac{1}{6}\cdot 2(\cot(3\theta))\cdot 3(-\csc^2(3\theta))$$
$$= \cot(3\theta)\,\csc^2(3\theta)$$

LESSON PRACTICE 23D - LESSON PRACTICE 23D

2. There is no substitution that works. Multiply through and integrate.

$$\int (x^2+1)(x^2-2)\,dx = \int (x^4 - 2x^2 + x^2 - 2)\,dx$$
$$= \int (x^4 - x^2 - 2)\,dx$$
$$= \frac{x^5}{5} - \frac{x^3}{3} - 2x + C$$
$$= \frac{1}{5}x^5 - \frac{1}{3}x^3 - 2x + C$$

Check:
$$\frac{d}{dx}\left(\frac{1}{5}x^5 - \frac{1}{3}x^3 - 2x\right)$$
$$= \frac{1}{5} \cdot 5x^4 - \frac{1}{3} \cdot 3x^2 - 2$$
$$= x^4 - x^2 - 2$$
$$= (x^2+1)(x^2-2)$$

3. Let $u = 2y + 1$; $\frac{du}{dy} = 2$; $du = 2dy$

$$\int \frac{4\,dy}{(2y+1)^3} = 2\int \frac{4\,dy \cdot \frac{1}{2}}{(2y+1)^3}$$
$$= 2\int u^{-3}\,du$$
$$= \frac{2u^{-2}}{-2} + C$$
$$= -u^{-2} + C$$
$$= -(2y+1)^{-2} + C$$

Check:
$$\frac{d}{dy}\left(-(2y+1)^{-2}\right) = -\left(-2(2y+1)^{-3}(2)\right)$$
$$= 4(2y+1)^{-3}$$

4. Let $u = \tan(\theta) - 2$; $\frac{du}{d\theta} = \sec^2(\theta)$; $du = \sec^2(\theta)\,d\theta$

$$\int \frac{\sec^2(\theta)\,d\theta}{\sqrt{\tan(\theta) - 2}} = \int u^{-\frac{1}{2}}\,du$$
$$= \frac{u^{\frac{1}{2}}}{\frac{1}{2}} + C$$
$$= 2\sqrt{u} + C$$
$$= 2\sqrt{\tan(\theta) - 2} + C$$

Check:
$$\frac{d}{d\theta}\left(2\sqrt{\tan(\theta) - 2}\right)$$
$$= 2 \cdot \frac{1}{2}(\tan(\theta) - 2)^{-\frac{1}{2}} \sec^2(\theta)$$
$$= \sec^2(\theta)(\tan(\theta) - 2)^{-\frac{1}{2}}$$

5. Let $u = -2x^2$; $\frac{du}{dx} = -4x$; $du = -4x\,dx$

$$-\frac{1}{4}\int xe^{-2x^2}\,dx(-4) = -\frac{1}{4}\int e^u\,du$$
$$= -\frac{1}{4}e^u + C$$
$$= -\frac{1}{4}e^{-2x^2} + C$$

Check:
$$\frac{d}{dy}\left(-\frac{1}{4}e^{-2x^2} + C\right) = -\frac{1}{4}e^{-2x^2}(-4x)$$
$$= xe^{-2x^2}$$

Lesson Practice 24A

1. $\int_1^2 (x^2+2)dx = \int_1^2 x^2 dx + 2\int_1^2 dx$
    $= \left(\dfrac{x^3}{3}+2x\right)\Big|_1^2$
    $= \left(\dfrac{8}{3}+4\right)-\left(\dfrac{1}{3}+2\right)$
    $= \dfrac{7}{3}+2$
    $= \dfrac{13}{3}=4\dfrac{1}{3}$

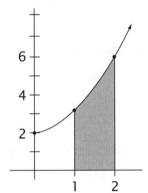

2. $\int_1^3 \dfrac{1}{x^2}dx = \int_1^3 x^{-2}dx$
    $= \dfrac{x^{-1}}{-1}\Big|_1^3$
    $= \left(-\dfrac{1}{x}\right)\Big|_1^3$
    $= -\dfrac{1}{3}+1 = \dfrac{2}{3}$

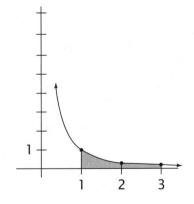

3. $\int_0^2 (x^3+3)dx = \int_0^2 x^3 dx + 3\int_0^2 dx$
    $= \left(\dfrac{x^4}{4}+3x\right)\Big|_0^2$
    $= (4+6)-(0) = 10$

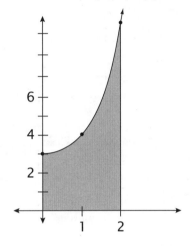

4. $\int_1^2 \left(\dfrac{x^3+3x}{x}\right)dx = \int_1^2 x^2 dx + 3\int_1^2 dx$
    $= \left(\dfrac{x^3}{3}+3x\right)\Big|_1^2$
    $= \left(\dfrac{8}{3}+6\right)-\left(\dfrac{1}{3}+3\right)$
    $= \dfrac{7}{3}+3 = \dfrac{16}{3} = 5\dfrac{1}{3}$

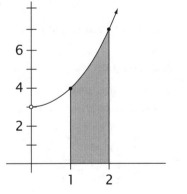

LESSON PRACTICE 24A - LESSON PRACTICE 24B

5. $y = x^3 - x$

   To find out where this graph hits the x-axis, we set $y = 0$.

   $0 = x(x^2 - 1)$
   $= x(x+1)(x-1)$
   $x = 0, 1, -1$

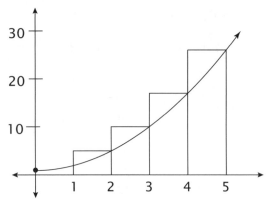

x	y
2	6
-2	-6
$\frac{1}{2}$	$-\frac{3}{8}$
$-\frac{1}{2}$	$\frac{3}{8}$

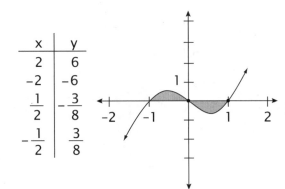

$\int_{-1}^{0} (x^3 - x)dx = \left(\frac{x^4}{4} - \frac{x^2}{2}\right)\Big|_{-1}^{0}$

$= 0 - \left(\frac{1}{4} - \frac{1}{2}\right) = \frac{1}{4}$

This graph is symmetrical so $\int_{0}^{1} (x^3 - x)dx$ will give us an area of $\frac{1}{4}$, but the answer will be $-\frac{1}{4}$ because the area is below the x-axis.

Our question was to find the area between the graph $y = x^3 - x$ and the x-axis. Since the area is positive, we have:

$\frac{1}{4} + \left|-\frac{1}{4}\right| =$

$\frac{1}{4} + \frac{1}{4} = \frac{1}{2}$

6. $y = x^2 + 1$ on [1, 5] using 4 intervals
   Area $= 1(5) + 1(10) + 1(17) + 1(26)$
   $= 32 + 26 = 58$

Lesson Practice 24B

1. $\int_{2}^{3}(x^3 - 1)dx = \left(\frac{x^4}{4} - x\right)\Big|_{2}^{3}$

   $= \left(\frac{81}{4} - 3\right) - \left(\frac{16}{4} - 2\right)$

   $= \frac{65}{4} - 1 = \frac{61}{4} = 15\frac{1}{4}$

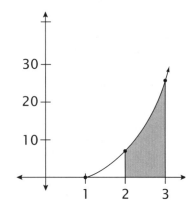

2. $\int_{-2}^{0}(3 + x^2)dx = \left(3x + \frac{x^3}{3}\right)\Big|_{-2}^{0}$

   $= 0 - \left(-6 + \frac{-8}{3}\right)$

   $= -\left(\frac{-18}{3} + \frac{-8}{3}\right)$

   $= \frac{26}{3} = 8\frac{2}{3}$

LESSON PRACTICE 24B - LESSON PRACTICE 24B

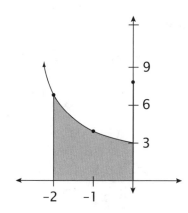

3. $\dfrac{1}{2}\int_2^3 \dfrac{dx}{\sqrt{2x}} \cdot 2$

Let $u = 2x$; $du = 2dx$

$= \dfrac{1}{2}\int_2^3 \dfrac{du}{\sqrt{u}}$

$= \dfrac{1}{2}\int_2^3 u^{-\frac{1}{2}} du$

$= \dfrac{1}{2} \dfrac{u^{\frac{1}{2}}}{\frac{1}{2}}\Big|_2^3$

$= \sqrt{u}\,\Big|_2^3$

$= \sqrt{2x}\,\Big|_2^3$

$= \sqrt{6} - 2 \approx .45$

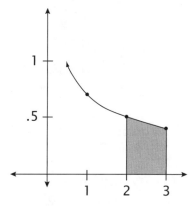

4. $\int_0^1 (3x+1)dx = \left(3\dfrac{x^2}{2} + x\right)\Big|_0^1$

$= \left(\dfrac{3}{2} + 1\right) = 2\dfrac{1}{2}$

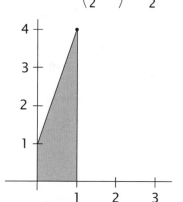

5. This graph crosses the x-axis when $y = 0$.

$x^3 - 4x = 0$
$x(x^2 - 4) = 0$
$x(x+2)(x-2) = 0$
$x = 0, 2, -2$

$\int_{-2}^0 (x^3 - 4x)dx = \left(\dfrac{x^4}{4} - 4\dfrac{x^2}{2}\right)\Big|_{-2}^0$

$= \left(\dfrac{x^4}{4} - 2x^2\right)\Big|_{-2}^0$

$= 0 - \left(\dfrac{16}{4} - 8\right) = 4$

The area between the curve $x^3 - 4x$ and the x-axis contains the area from $x = -2$ and $x = 0$ as well as the area from $x = 0$ to $x = 2$. Since the graph is symmetrical, the total area equals $2(4) = 8$ sq units.

x	y
1	−3
−1	3
3	15
−3	−15

LESSON PRACTICE 24B - LESSON PRACTICE 24C

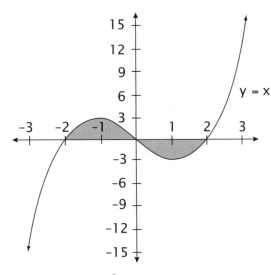

y = x³ − 4x

Lesson Practice 24C

1. $\int_0^1 (2-x^2)\,dx = \left(2x - \frac{x^3}{3}\right)\Big|_0^1$

   $= 2 - \frac{1}{3} = 1\frac{2}{3}$

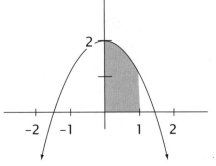

6. $y = x^2 + 2$ on [2, 6] using 4 intervals
   Area $= 1(6) + 1(11) + 1(18) + 1(27)$
   $= 35 + 27 = 62$

2. $\int_1^4 \sqrt{x}\,dx = \int_1^4 x^{\frac{1}{2}}\,dx$

   $= \dfrac{x^{\frac{3}{2}}}{\frac{3}{2}}\Big|_1^4$

   $= \dfrac{2}{3}(\sqrt{x})^3\Big|_1^4$

   $= \dfrac{2}{3}(8-1)$

   $= \dfrac{2}{3}(7) = \dfrac{14}{3}$

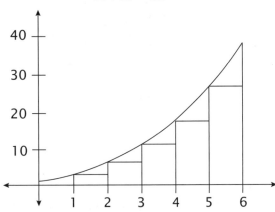

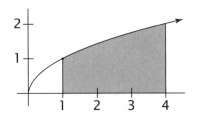

3. $y = x^3 - 6x^2 + 8x$
   This graph crosses the x-axis when y = 0.

   $0 = x(x^2 - 6x + 8)$
   $= x(x-4)(x-2)$
   $x = 0, 2, 4$

x	y
1	3
3	−3
−1	−15

382 SOLUTIONS

CALCULUS

LESSON PRACTICE 24C - LESSON PRACTICE 24D

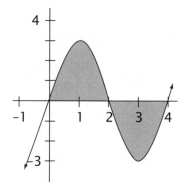

The area we desire is from $x = 0$ to $x = 2$ and from $x = 2$ to $x = 4$.

$$\int_0^2 (x^3 - 6x^2 + 8x)\,dx$$

$$= \left(\frac{x^4}{4} - 6\frac{x^3}{3} + 8\frac{x^2}{2}\right)\Big|_0^2$$

$$= \left(\frac{1}{4}x^4 - 2x^3 + 4x^2\right)\Big|_0^2$$

$$= \left(\frac{1}{4}(16) - 16 + 16\right) - 0 = 4$$

Since this graph is symmetrical, the area between the curve $y = x^3 - 6x^2 + 8x$ and the x-axis is $2(4) = 8$ sq units.

4. $\int_0^3 (9 - x^2)\,dx = \left(9x - \frac{x^3}{3}\right)\Big|_0^3$

    $= (27 - 9) - 0 = 18$

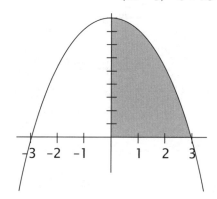

5. $\int_{-1}^0 (x^3 + 1)\,dx = \left(\frac{x^4}{4} + x\right)\Big|_{-1}^0$

    $= 0 - \left(\frac{1}{4} - 1\right) = \frac{3}{4}$

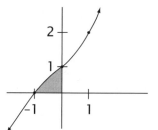

Lesson Practice 24D

1. $\int_{-2}^1 (4 - x^2)\,dx = \left(4x - \frac{x^3}{3}\right)\Big|_{-2}^1$

    $= \left(4 - \frac{1}{3}\right) - \left(-8 + \frac{8}{3}\right)$

    $= 12 - \frac{9}{3} = 9$

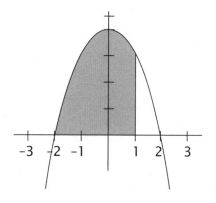

LESSON PRACTICE 24D - LESSON PRACTICE 24D

2. $\int_1^6 \dfrac{1}{\sqrt{x+3}} = \int_1^6 (x+3)^{-\frac{1}{2}}$

    $= \dfrac{(x+3)^{\frac{1}{2}}}{\frac{1}{2}}\bigg|_1^6$

    $= 2(\sqrt{x+3})\bigg|_1^6$

    $= 2(\sqrt{9} - \sqrt{4}) = 2(1) = 2$

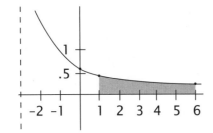

$\int_{-1}^0 (x^3 - x^2 - 2x)\,dx = \left(\dfrac{x^4}{4} - \dfrac{x^3}{3} - 2\dfrac{x^2}{2}\right)\bigg|_{-1}^0$

$= \left(\dfrac{1}{4}x^4 - \dfrac{1}{3}x^3 - x^2\right)\bigg|_{-1}^0$

$= 0 - \left(\dfrac{1}{4} + \dfrac{1}{3} - 1\right)$

$= -\left(\dfrac{7}{12} - 1\right) = \dfrac{5}{12}$

$\int_0^2 (x^3 - x^2 - 2x)\,dx = \left(\dfrac{1}{4}x^4 - \dfrac{1}{3}x^3 - x^2\right)\bigg|_0^2$

$= \left(4 - \dfrac{8}{3} - 4\right) - 0 = -\dfrac{8}{3}$

$\dfrac{5}{12} + \dfrac{8}{3} = \dfrac{37}{12} = 3\dfrac{1}{12}$

3. $y = x^3 - x^2 - 2x$
    $y = x(x^2 - x - 2)$
    $= x(x+1)(x-2)$

    This graph crosses the x-axis when y = 0.
    $0 = x(x+1)(x-2)$
    $x = 0, -1, 2$

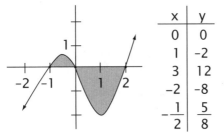

x	y
0	0
1	-2
3	12
-2	-8
$-\frac{1}{2}$	$\frac{5}{8}$

This graph is not symmetrical. It must be divided into 2 pieces. Integrate from -1 to 0 and again from 0 to 2.

4. $y = x^3 - 2x^2 - 3x$
    $y = x(x^2 - 2x - 3)$
    $= x(x-3)(x+1)$

    When y = 0, x = 0, or x = 3, or x = -1.
    Break this graph up into two pieces. Integrate from -1 to 0 and again from 0 to 3.

x	y
1	-4
2	-6
4	20
$-\frac{1}{2}$	$\frac{7}{8}$
-2	-10

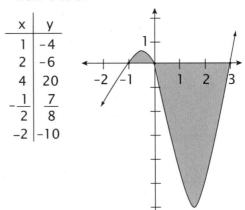

384 SOLUTIONS

CALCULUS

$$\int_{-1}^{0}(x^3-2x^2-3x)dx$$
$$=\left(\frac{x^4}{4}-2\frac{x^3}{3}-3\frac{x^2}{2}\right)\Big|_{-1}^{0}$$
$$=0-\left(\frac{1}{4}+\frac{2}{3}-\frac{3}{2}\right)$$
$$=-\left(\frac{3+8-18}{12}\right)=+\frac{7}{12}$$

$$\int_{0}^{3}(x^3-2x^2-3x)dx$$
$$=\left(\frac{x^4}{4}-\frac{2}{3}x^3-\frac{3}{2}x^2\right)\Big|_{0}^{3}$$
$$=\left[\frac{81}{4}-\frac{2}{3}(27)-\frac{3}{2}(9)\right]-0$$
$$=20\frac{1}{4}-18-\frac{27}{2}$$
$$=-11\frac{1}{4}$$

Total area $=\frac{7}{12}+11\frac{1}{4}=11\frac{5}{6}$

5. $2y^2=x^3$
$y^2=\frac{1}{2}x^3$
$y=\sqrt{\frac{x^3}{2}}$

x	y
0	0
1	≈ .7
2	2
3	≈ 3.7

$$\int_{0}^{2}\sqrt{\frac{x^3}{2}}dx=\int_{0}^{2}\left(\frac{1}{2}x^3\right)^{\frac{1}{2}}dx$$
$$=\frac{\sqrt{2}}{2}\int_{0}^{2}x^{\frac{3}{2}}dx$$
$$=\frac{\sqrt{2}}{2}\left[\frac{x^{\frac{5}{2}}}{\frac{5}{2}}\right]\Big|_{0}^{2}$$
$$=\frac{\sqrt{2}}{2}\cdot\frac{2}{5}\left(x^{\frac{5}{2}}\right)\Big|_{0}^{2}$$
$$=\frac{\sqrt{2}}{5}\left(2^{\frac{5}{2}}\right)-0$$
$$=\frac{\sqrt{2}}{5}\left(\sqrt{2}\right)^5=\frac{8}{5}=1\frac{3}{5}$$

Lesson Practice 25A

1. Let $u=25-x^2$; $\frac{du}{dx}=-2x$; $du=-2xdx$
$$-\frac{1}{2}\int_{0}^{4}x\sqrt{25-x^2}dx(-2)$$
$$=-\frac{1}{2}\int_{0}^{4}u^{\frac{1}{2}}du$$
$$=-\frac{1}{2}\left[\frac{u^{\frac{3}{2}}}{\frac{3}{2}}\right]\Big|_{0}^{4}$$
$$=-\frac{1}{2}\cdot\frac{2}{3}(25-x^2)^{\frac{3}{2}}\Big|_{0}^{4}$$
$$=-\frac{1}{3}\left(\sqrt{(25-x^2)}\right)^3\Big|_{0}^{4}$$
$$=-\frac{1}{3}(27-125)=\frac{98}{3}$$

LESSON PRACTICE 25A - LESSON PRACTICE 25A

2. Let $u = \dfrac{x^2}{-2}$; $\dfrac{du}{dx} = -\dfrac{1}{2}(2x)$; $du = -x\,dx$

$$-\int_0^2 xe^{\frac{x^2}{-2}}dx(-1) = -\int_0^2 e^u du$$

$$= -e^u \Big|_0^2$$

$$= -e^{-\frac{x^2}{2}} \Big|_0^2$$

$$= \left(-e^{-2} - (-e^0)\right) = 1 - \dfrac{1}{e^2}$$

3. Let $u = \sin(\theta)$; $\dfrac{du}{d\theta} = \cos(\theta)$; $du = \cos(\theta)\,d\theta$

$$\int_{\frac{\pi}{6}}^{\frac{\pi}{2}} \sin^2(\theta)\cos(\theta)\,d\theta$$

$$= \int_{\frac{\pi}{6}}^{\frac{\pi}{2}} u^2 du$$

$$= \dfrac{u^3}{3} \Big|_{\frac{\pi}{6}}^{\frac{\pi}{2}}$$

$$= \dfrac{1}{3}(\sin(\theta))^3 \Big|_{\frac{\pi}{6}}^{\frac{\pi}{2}}$$

$$= \dfrac{1}{3}\left(\sin\left(\dfrac{\pi}{2}\right)\right)^3 - \dfrac{1}{3}\left(\sin\left(\dfrac{\pi}{6}\right)\right)^3$$

$$= \dfrac{1}{3} - \dfrac{1}{3}\left(\dfrac{1}{2}\right)^3$$

$$= \dfrac{1}{3} - \dfrac{1}{24} = \dfrac{7}{24}$$

4. Let $u = \tan(x)$; $\dfrac{du}{dx} = \sec^2(x)$; $du = \sec^2(x)\,dx$

$$\int_{-\frac{\pi}{4}}^{0} \tan(x)\sec^2(x)\,dx$$

$$= \int_{-\frac{\pi}{4}}^{0} u\,du$$

$$= \dfrac{u^2}{2} \Big|_{-\frac{\pi}{4}}^{0}$$

$$= \dfrac{1}{2}(\tan(x))^2 \Big|_{-\frac{\pi}{4}}^{0}$$

$$= \dfrac{1}{2}(\tan(0))^2 - \dfrac{1}{2}\left(\tan\left(-\dfrac{\pi}{4}\right)\right)^2$$

$$= 0 - \dfrac{1}{2} = -\dfrac{1}{2}$$

5. Let $u = x^2 + 1$; $\dfrac{du}{dx} = 2x$; $du = 2x\,dx$

$$\dfrac{1}{2}\int_{-\sqrt{2}}^{\sqrt{2}} \dfrac{x}{\sqrt{x^2+1}}\,dx \cdot 2 = \dfrac{1}{2}\int_{-\sqrt{2}}^{\sqrt{2}} u^{-\frac{1}{2}}\,du$$

$$= \dfrac{1}{2} \dfrac{u^{\frac{1}{2}}}{\frac{1}{2}} \Big|_{-\sqrt{2}}^{\sqrt{2}}$$

$$= \left(\sqrt{x^2+1}\right)\Big|_{-\sqrt{2}}^{\sqrt{2}}$$

$$= \sqrt{3} - \sqrt{3} = 0$$

6. Let $u = 2x$; $\dfrac{du}{dx} = 2$; $du = 2dx$

$$\int_0^{\frac{\pi}{2}} 3\sin(2x)\,dx = 3 \cdot \dfrac{1}{2}\int_0^{\frac{\pi}{2}} \sin(2x)\,dx \cdot 2$$

$$= \dfrac{3}{2}\int_0^{\frac{\pi}{2}} \sin(u)\,du$$

$$= -\dfrac{3}{2}\cos(2x)\Big|_0^{\frac{\pi}{2}}$$

$$= -\dfrac{3}{2}[\cos(\pi) - \cos(0)] = 3$$

Lesson Practice 25B

1. Let $u = 2x$; $\dfrac{du}{dx} = 2$; $du = 2dx$

$$\dfrac{1}{2}\int_0^{\ln(3)} e^{2x}dx \cdot 2 = \dfrac{1}{2}\int_0^{\ln(3)} e^u du$$
$$= \dfrac{1}{2}e^{2x}\Big|_0^{\ln(3)}$$
$$= \dfrac{1}{2}\left(e^{2\ln(3)} - e^0\right)$$
$$= \dfrac{1}{2}(9 - 1) = 4$$

2. Let $u = \ln(x)$; $\dfrac{du}{dx} = \dfrac{1}{x}$; $du = \dfrac{1}{x}dx$

$$\int_1^2 \dfrac{\ln(x)}{x}dx = \int_1^2 u\, du$$
$$= \left(\dfrac{u^2}{2}\right)\Big|_1^2$$
$$= \dfrac{1}{2}(\ln(x))^2\Big|_1^2$$
$$= \dfrac{1}{2}\left[(\ln(2))^2 - (\ln(1))^2\right]$$
$$= \dfrac{1}{2}(\ln(2))^2$$

3. Let $u = \cos(\theta)$; $\dfrac{du}{d\theta} = -\sin(\theta)$; $du = -\sin(\theta)\, d\theta$

$$-\int_0^{\pi/2} \sin(\theta)\cos^2(\theta)\, d\theta\, (-1)$$
$$= -\int_0^{\pi/2} u^2 du$$
$$= -\dfrac{u^3}{3}\Big|_0^{\pi/2}$$
$$= -\dfrac{1}{3}(\cos^3(\theta))\Big|_0^{\pi/2}$$
$$= -\dfrac{1}{3}(0 - 1) = \dfrac{1}{3}$$

4. Let $u = 2x - 1$; $\dfrac{du}{dx} = 2$; $du = 2dx$

$$\dfrac{1}{2}\int_1^5 \dfrac{dx}{\sqrt{2x-1}}\cdot 2 = \dfrac{1}{2}\int_1^5 u^{-\frac{1}{2}}du$$
$$= \dfrac{1}{2}\cdot \dfrac{u^{\frac{1}{2}}}{\frac{1}{2}}\Big|_1^5$$
$$= \left(\sqrt{2x-1}\right)\Big|_1^5$$
$$= \sqrt{9} - \sqrt{1} = 2$$

5. Let $u = 4x$; $\dfrac{du}{dx} = 4$; $du = 4dx$

$$\dfrac{1}{4}\int_0^{\pi/16} \sec^2(4x)dx \cdot 4 = \dfrac{1}{4}\int_0^{\pi/16} \sec^2(u)du$$
$$= \dfrac{1}{4}\tan(4x)\Big|_0^{\pi/16}$$
$$= \dfrac{1}{4}\tan\left(\dfrac{\pi}{4}\right) = \dfrac{1}{4}$$

6. $y = \sin\left(\dfrac{1}{2}\right)x$

Let $u = \dfrac{1}{2}x$; $\dfrac{du}{dx} = \dfrac{1}{2}$; $du = \dfrac{1}{2}dx$

$$2\int_0^{2\pi} \sin\left(\dfrac{1}{2}(x)\right)dx\left(\dfrac{1}{2}\right) = 2\int_0^{2\pi}\sin(u)\,du$$
$$= -2\cos\left(\dfrac{1}{2}(x)\right)\Big|_0^{2\pi}$$
$$= -2(-1 - 1) = 4$$

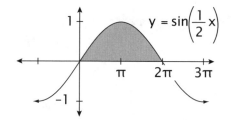

Lesson Practice 25C

1. Let $u = 1+\cos(\theta)$; $\frac{du}{d\theta} = -\sin(\theta)$;
$du = -\sin(\theta)d\theta$

$$-\int_0^{\frac{\pi}{2}} \frac{\sin(\theta)d\theta}{1+\cos(\theta)}(-1) = -\int_0^{\frac{\pi}{2}} \frac{du}{u}$$

$$= -\ln(1+\cos(\theta))\Big|_0^{\frac{\pi}{2}}$$

$$= -[\ln(1) - \ln(2)] = \ln(2)$$

2. $\int_{-\frac{3\pi}{4}}^0 \tan\left(\frac{x}{3}\right)dx = \int_{-\frac{3\pi}{4}}^0 \frac{\sin\left(\frac{x}{3}\right)}{\cos\left(\frac{x}{3}\right)}dx$

Let $u = \cos\left(\frac{x}{3}\right)$; $\frac{du}{d\theta} = \frac{-1}{3}\sin\left(\frac{x}{3}\right)$;
$du = -\frac{1}{3}\sin\left(\frac{x}{3}\right)dx$.

$$= -3\int_{-\frac{3\pi}{4}}^0 \frac{\sin\left(\frac{x}{3}\right)}{\cos\left(\frac{x}{3}\right)}dx\left(-\frac{1}{3}\right)$$

$$= -3\int_{-\frac{3\pi}{4}}^0 \frac{du}{u}$$

$$= -3\ln\left(\cos\left(\frac{x}{3}\right)\right)\Big|_{-\frac{3\pi}{4}}^0$$

$$= -3\left[\ln(\cos(0)) - \ln\left(\cos\left(-\frac{\pi}{4}\right)\right)\right]$$

$$= -3\left[0 - \ln\left(\frac{\sqrt{2}}{2}\right)\right]$$

$$= 3\ln\left(\frac{\sqrt{2}}{2}\right)$$

3. Let $u = 1+2\tan(\theta)$; $\frac{du}{d\theta} = 2\sec^2(\theta)$;
$du = 2\sec^2(\theta)d\theta$

$$\frac{1}{2}\int_0^{\frac{\pi}{4}} \frac{\sec^2(\theta)d\theta \cdot 2}{\sqrt{1+2\tan(\theta)}} = \frac{1}{2}\int_0^{\frac{\pi}{4}} \frac{du}{\sqrt{u}}$$

$$= \frac{1}{2}\int_0^{\frac{\pi}{4}} u^{-\frac{1}{2}}du$$

$$= \frac{1}{2}\cdot\frac{u^{\frac{1}{2}}}{\frac{1}{2}}\Big|_0^{\frac{\pi}{4}}$$

$$= \sqrt{u}\Big|_0^{\frac{\pi}{4}}$$

$$= \sqrt{1+2\tan(\theta)}\Big|_0^{\frac{\pi}{4}}$$

$$= \sqrt{3} - 1$$

4. Let $u = e^x + \cos(x)$; $\dfrac{du}{dx} = e^x - \sin(x)$; $du = (e^x - \sin(x))dx$

$\displaystyle\int_0^{\frac{\pi}{2}} \dfrac{(e^x - \sin(x))}{\sqrt{e^x + \cos(x)}} dx$

$= \displaystyle\int_0^{\frac{\pi}{2}} \dfrac{du}{\sqrt{u}}$

$= \displaystyle\int_0^{\frac{\pi}{2}} u^{-\frac{1}{2}}$

$= \dfrac{u^{\frac{1}{2}}}{\frac{1}{2}} \Big|_0^{\frac{\pi}{2}}$

$= 2\sqrt{u} \, \Big|_0^{\frac{\pi}{2}}$

$= 2\left[\sqrt{e^{\frac{\pi}{2}} + \cos\left(\dfrac{\pi}{2}\right)} - \sqrt{e^0 + \cos(0)}\right]$

$= 2\left[\sqrt{e^{\frac{\pi}{2}}} - \sqrt{1+1}\right]$

$= 2\sqrt{e^{\frac{\pi}{2}}} - 2\sqrt{2}$

5. Let $u = e^{2x} + 1$; $\dfrac{du}{dx} = e^{2x} \cdot 2$; $du = 2e^{2x}dx$

$\dfrac{1}{2}\displaystyle\int_0^1 \dfrac{e^{2x}dx \cdot 2}{e^{2x}+1} = \dfrac{1}{2}\int_0^1 \dfrac{du}{u}$

$= \dfrac{1}{2}\ln(u)\Big|_0^1$

$= \dfrac{1}{2}\ln(e^{2x}+1)\Big|_0^1$

$= \dfrac{1}{2}\left[\ln(e^2+1) - \ln(2)\right]$

$= \dfrac{1}{2}\ln\left(\dfrac{e^2+1}{2}\right)$

6. Let $u = \ln(x) + 2$; $\dfrac{du}{dx} = \dfrac{1}{x}$; $du = \dfrac{dx}{x}$

$\displaystyle\int_1^2 \dfrac{(\ln(x)+2)^2}{x} dx$

$= \displaystyle\int_1^2 u^2 du$

$= \dfrac{u^3}{3}\Big|_1^2$

$= \dfrac{1}{3}(\ln(x)+2)^3 \Big|_1^2$

$= \dfrac{1}{3}\left[(\ln(2)+2)^3 - (\ln(1)+2)^3\right]$

$= \dfrac{1}{3}\left[(\ln(2)+2)^3 - 8\right]$

Lesson Practice 25D

1. Let $u = \cot(4\theta)$; $\dfrac{du}{d\theta} = -4\csc^2(4\theta)$; $du = -4\csc^2(4\theta)d\theta$

$-\dfrac{1}{4}\displaystyle\int_{\frac{\pi}{24}}^{\frac{\pi}{16}} \cot(4\theta)\csc^2(4\theta)\, d\theta \cdot (-4)$

$= -\dfrac{1}{4}\displaystyle\int_{\frac{\pi}{24}}^{\frac{\pi}{16}} u\, du$

$= -\dfrac{1}{4} \cdot \dfrac{u^2}{2}\Big|_{\frac{\pi}{24}}^{\frac{\pi}{16}}$

$= -\dfrac{1}{8}(\cot(4\theta))^2 \Big|_{\frac{\pi}{24}}^{\frac{\pi}{16}}$

$= -\dfrac{1}{8}\left(1 - \left(\sqrt{3}\right)^2\right)$

$= -\dfrac{1}{8}(1-3) = \dfrac{2}{8} = \dfrac{1}{4}$

2. Let $u = \cos(x)$; $\dfrac{du}{dx} = -\sin(x)$;
   $du = -\sin(x)\,dx$

   $-\int_0^{\pi/2} \sin(x) e^{\cos(x)} dx\,(-1) = -\int_0^{\pi/2} e^u du$

   $= -e^u \Big|_0^{\pi/2}$

   $= -e^{\cos(x)} \Big|_0^{\pi/2}$

   $= -[e^0 - e^1]$

   $= e - 1$

3. Let $u = x^2 - 6x + 8$; $\dfrac{du}{dx} = 2x - 6$;
   $du = (2x - 6)\,dx$

   $\dfrac{1}{2}\int_0^1 \dfrac{(x-3)\,dx}{\sqrt{x^2 - 6x + 8}} \cdot 2 = \dfrac{1}{2}\int_0^1 u^{-\frac{1}{2}} du$

   $= \dfrac{1}{2} \dfrac{u^{\frac{1}{2}}}{\frac{1}{2}} \Big|_0^1$

   $= \sqrt{u}\Big|_0^1$

   $= \sqrt{x^2 - 6x + 8}\Big|_0^1$

   $= \sqrt{3} - \sqrt{8}$

4. Let $u = 1 - 2\cos(2x)$; $\dfrac{du}{dx} = 4\sin(2x)$;
   $du = 4\sin(2x)\,dx$

   $\dfrac{1}{4}\int_{\pi/2}^{5\pi/8} \dfrac{\sin(2x)\,dx}{(1 - 2\cos(2x))} \cdot 4$

   $= \dfrac{1}{4}\int_{\pi/2}^{5\pi/8} \dfrac{du}{u}$

   $= \dfrac{1}{4}\ln(u)\Big|_{\pi/2}^{5\pi/8}$

   $= \dfrac{1}{4}\ln(1 - 2\cos(2x))\Big|_{\pi/2}^{5\pi/8}$

   $= \dfrac{1}{4}\left[\ln\left(1 - 2\cos\left(\dfrac{5\pi}{4}\right)\right) - \ln(1 - 2\cos(\pi))\right]$

   $= \dfrac{1}{4}\left[\ln(1 + \sqrt{2}) - \ln(3)\right]$

   $= \dfrac{1}{4}\ln\left(\dfrac{1 + \sqrt{2}}{3}\right)$

5. Let $u = 1 + e^{2x}$; $\dfrac{du}{dx} = e^{2x} \cdot 2$; $du = 2e^{2x}\,dx$

   $\dfrac{1}{2}\int_0^1 \dfrac{e^{2x}\,dx}{\sqrt{1 + e^{2x}}} (2) = \dfrac{1}{2}\int_0^1 u^{-\frac{1}{2}} du$

   $= \dfrac{1}{2} \dfrac{u^{\frac{1}{2}}}{\frac{1}{2}} \Big|_0^1$

   $= \sqrt{u}\Big|_0^1$

   $= \sqrt{1 + e^{2x}}\Big|_0^1$

   $= \sqrt{1 + e^2} - \sqrt{2}$

6. Substitution does not work.
Divide through instead.

$$\int_0^{\frac{\pi}{4}} \frac{(\sin(x)-\cos(x))}{\cos(x)} dx$$

$$= -\int_0^{\frac{\pi}{4}} \frac{-\sin(x)}{\cos(x)} dx - \int_0^{\frac{\pi}{4}} dx$$

Now use substitution for the each integral.

Let $u = \cos(x); \frac{du}{dx} = -\sin(x);$
$du = -\sin(x) dx$

$$= -\int_0^{\frac{\pi}{4}} \frac{du}{u} - (x)\Big|_0^{\frac{\pi}{4}}$$

$$= -\ln(u)\Big|_0^{\frac{\pi}{4}} - x\Big|_0^{\frac{\pi}{4}}$$

$$= -\ln(\cos(x))\Big|_0^{\frac{\pi}{4}} - x\Big|_0^{\frac{\pi}{4}}$$

$$= -\left[\ln\left(\frac{\sqrt{2}}{2}\right) - 0\right] - \frac{\pi}{4}$$

$$= -\ln\left(\frac{\sqrt{2}}{2}\right) - \frac{\pi}{4}$$

7. Substitution does not work.
Divide through instead.

$$\int_0^1 \frac{(1+e^x)}{e^{2x}} dx = \int_0^1 e^{-2x} dx + \int_0^1 e^{-x} dx$$

Now use substitution for each integral.

Let $u = -2x; \frac{du}{dx} = -2; du = -2dx;$
$V = -x; \frac{dV}{dx} = -1; dV = -dx$

$$= -\frac{1}{2}\int_0^1 -2e^{-2x} dx - \int_0^1 e^{-x} dx (-1)$$

$$= -\frac{1}{2}\int_0^1 e^u du - \int_0^1 e^V dV$$

$$= -\frac{1}{2} e^u\Big|_0^1 - e^V\Big|_0^1$$

$$= -\frac{1}{2} e^{-2x}\Big|_0^1 - e^{-x}\Big|_0^1$$

$$= -\frac{1}{2}(e^{-2} - e^0) - (e^{-1} - e^0)$$

$$= -\frac{1}{2}e^{-2} + \frac{1}{2} - \frac{1}{e} + 1$$

$$= \frac{3}{2} - \frac{1}{e} - \frac{1}{2e^2}$$

Lesson Practice 26A

1. $\int_{-2}^{1} (-x+1) dx + \int_{1}^{3} (x-1) dx$

$$= \left(-\frac{x^2}{2} + x\right)\Big|_{-2}^{1} + \left(\frac{x^2}{2} - x\right)\Big|_{1}^{3}$$

$$= \left[\left(-\frac{1}{2}+1\right) - (-2-2)\right] + \left[\left(\frac{9}{2}-3\right) - \left(\frac{1}{2}-1\right)\right]$$

$$= \frac{7}{2} + 3 = 6\frac{1}{2}$$

Checking with geometry:

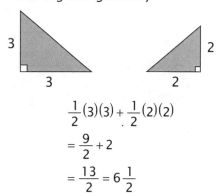

$$\frac{1}{2}(3)(3) + \frac{1}{2}(2)(2)$$

$$= \frac{9}{2} + 2$$

$$= \frac{13}{2} = 6\frac{1}{2}$$

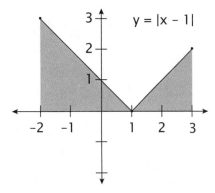

2. $y = \dfrac{8}{x}$ and $y = 1$ intersect at $(8, 1)$.

$y = \dfrac{8}{x}$ and $y = 4$ intersect at $(2, 4)$.

This requires two integrals:

$\int_0^2 (4-1)\,dx + \int_2^8 \left(\dfrac{8}{x} - 1\right)dx$

$= 3x \Big|_0^2 + \int_2^8 \dfrac{dx}{x} - x \Big|_2^8$

$= 6 + 8\ln(x)\Big|_2^8 - (8-2)$

$= 8(\ln(8) - \ln(2))$

$= 8\ln(4)$

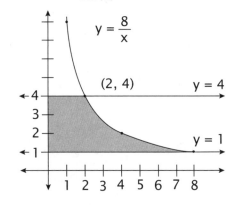

3. $y = x^2 + x + 1$ and $y = 4 - x^2$ meet when:

$x^2 + x + 1 = 4 - x^2$
$0 = 2x^2 + x - 3$
$0 = (2x + 3)(x - 1)$
$x = -\dfrac{3}{2}, 1$

Intersection points are $(1, 3)$ & $\left(-\dfrac{3}{2}, \dfrac{7}{4}\right)$

$\int_{-\frac{3}{2}}^{1} \left[(4 - x^2) - (x^2 + x + 1)\right] dx$

$= \int_{-\frac{3}{2}}^{1} (-2x^2 - x + 3) dx$

$= \left(-2\dfrac{x^3}{3} - \dfrac{x^2}{2} + 3x\right)\Big|_{-\frac{3}{2}}^{1}$

$= \left(-\dfrac{2}{3} - \dfrac{1}{2} + 3\right) - \left(\dfrac{9}{4} - \dfrac{9}{8} - \dfrac{9}{2}\right)$

$= 5\dfrac{5}{24}$

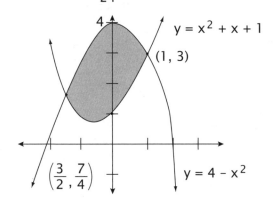

4. $y = x^2 - 1$ meets $y = 1 - x^2$ when:

$$x^2 - 1 = 1 - x^2$$
$$2x^2 = 2$$
$$x^2 = 1$$
$$x = \pm 1$$

Intersection points are $(1, 0)(-1, 0)$.

$$\int_{-1}^{1} [(1-x^2) - (x^2-1)] dx$$
$$= \int_{-1}^{1} (2 - 2x^2) dx$$
$$= \left(2x - 2\frac{x^3}{3}\right)\Big|_{-1}^{1}$$
$$= \left(2 - \frac{2}{3}\right) - \left(-2 + \frac{2}{3}\right)$$
$$= 4 - \frac{4}{3} = 2\frac{2}{3}$$

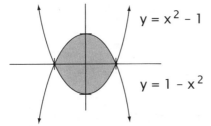

5. Let's work this problem from right to left
$x = y^2 \quad x = y + 6$

$$y^2 = y + 6$$
$$y^2 - y - 6 = 0$$
$$(y+2)(y-3) = 0 \quad y = -2, y = 3$$

Points of intersection are:
$(4, -2)$ and $(9, 3)$

$$\int_c^d [\text{right-left}] \, dy$$
$$\int_{-2}^{3} [(y+6) - y^2] dy$$
$$= \left(\frac{y^2}{2} + 6y - \frac{y^3}{3}\right)\Big|_{-2}^{3}$$
$$= \left(\frac{9}{2} + 18 - 9\right) - \left(2 - 12 + \frac{8}{3}\right)$$
$$= \frac{9}{2} + 9 + 10 - \frac{8}{3} = 20\frac{5}{6}$$

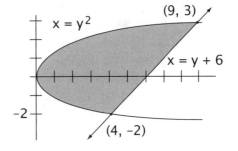

Lesson Practice 26B

1.
$$2\sqrt{x} = 2x^2$$
$$x = x^4$$
$$x^4 - x = 0$$
$$x(x^3 - 1) = 0$$
$$x = 0, 1$$

$$\int_0^1 (2\sqrt{x} - 2x^2) dx = \left(2\frac{x^{\frac{3}{2}}}{\frac{3}{2}} - 2\frac{x^3}{3}\right)\Big|_0^1$$
$$= \frac{4}{3} - \frac{2}{3} = \frac{2}{3}$$

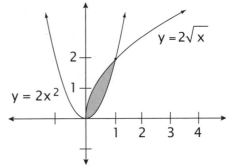

2. The intersection of $xy = 4$ and $y = -x + 5$ is:

$$\left(\frac{4}{x} = -x + 5\right)x$$
$$4 = -x^2 + 5x$$
$$x^2 - 5x + 4 = 0$$
$$(x-1)(x-4) = 0$$
$$x = 1, 4$$

Points of intersection are: $(1, 4)(4, 1)$.

$$\int_1^4 \left[(-x+5) - \frac{4}{x}\right]dx$$
$$= \left(-\frac{x^2}{2} + 5x - (4\ln(x))\right)\Big|_1^4$$
$$= (-8 + 20 - (4\ln(4))) - \left(-\frac{1}{2} + 5 - (4\ln(1))\right)$$
$$= 7 - (4\ln(4)) + \frac{1}{2}$$
$$= 7\frac{1}{2} - (4\ln(4))$$

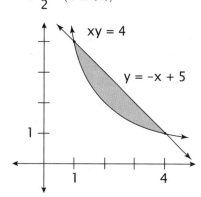

3. The intersection of $x = y^2$ and $x = -y^2 + 8$ is:
$$y^2 = -y^2 + 8$$
$$2y^2 = 8$$
$$y = \pm 2$$

Points of intersection are: $(4, -2)(4, 2)$.

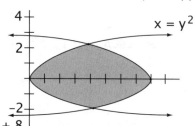

$$\int_{-2}^2 \left[(-y^2 + 8) - y^2\right]dy$$
$$= \int_{-2}^2 (-2y^2 + 8)dy$$
$$= \left(-2\frac{y^3}{3} + 8y\right)\Big|_{-2}^2$$
$$= \left(-\frac{16}{3} + 16\right) - \left(\frac{16}{3} - 16\right)$$
$$= -\frac{32}{3} + 32$$
$$= \frac{64}{3} = 21\frac{1}{3}$$

4. $\sin(x) = \cos(x)$ when $x = \frac{\pi}{4}$

$$\int_0^{\frac{\pi}{4}} (\cos(x) - \sin(x))dx + \int_{\frac{\pi}{4}}^{\frac{\pi}{2}} (\sin(x) - \cos(x))dx$$

$$= (\sin(x) + \cos(x))\Big|_0^{\frac{\pi}{4}} + (-\cos(x) - \sin(x))\Big|_{\frac{\pi}{4}}^{\frac{\pi}{2}}$$

$$= \left(\frac{\sqrt{2}}{2} + \frac{\sqrt{2}}{2}\right) - (0+1) + (0-1) - \left(-\frac{\sqrt{2}}{2} - \frac{\sqrt{2}}{2}\right)$$

$$= \sqrt{2} - 1 - 1 + \sqrt{2}$$
$$= 2\sqrt{2} - 2$$

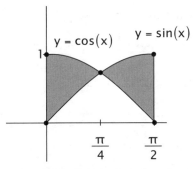

5. Find the area bounded by $y = x^2$ and $x = y^2$.

Intersection:
$$x^2 = \sqrt{x}$$
$$x^4 - x = 0$$
$$x(x^3 - 1) = 0$$
$$x = 0, 1$$

$$\int_0^1 (\sqrt{x} - x^2)dx = \left(\frac{x^{\frac{3}{2}}}{\frac{3}{2}} - \frac{x^3}{3}\right)\Big|_0^1$$
$$= \left(\frac{2}{3}x^{\frac{3}{2}} - \frac{1}{3}x^3\right)\Big|_0^1$$
$$= \frac{2}{3} - \frac{1}{3} = \frac{1}{3}$$

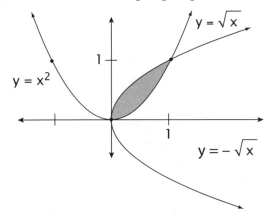

Lesson Practice 26C

1. These graphs meet when $x + 1 = x^2 - 1$.
$$x^2 - x - 2 = 0$$
$$(x-2)(x+1) = 0$$
$$x = 2, -1$$

Intersection points: $(2, 3)$, $(-1, 0)$

$$\int_{-1}^{2} \left[(x+1) - (x^2-1)\right]dx$$
$$= \int_{-1}^{2} (-x^2 + x + 2)dx$$
$$= \left(-\frac{x^3}{3} + \frac{x^2}{2} + 2x\right)\Big|_{-1}^{2}$$
$$= \left(-\frac{8}{3} + 2 + 4\right) - \left(\frac{1}{3} + \frac{1}{2} - 2\right)$$
$$= 4\frac{1}{2}$$

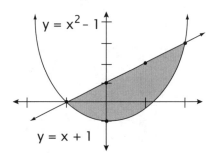

2. $y = 8x$ and $y = x$ intersect when $8x = x$ or $x = 0$

Intersection point: $(0, 0)$

$y = 8x$ and $y = \frac{1}{x^2}$

intersect when $8x = \frac{1}{x^2}$.

$8x^3 = 1$; $x^3 = \frac{1}{8}$; $x = \frac{1}{2}$

intersection $\left(\frac{1}{2}, 4\right)$

$y = x$ and $y = \frac{1}{x^2}$ intersect

when $x = \frac{1}{x^2}$ or $x^3 = 1$.

intersection point $(1, 1)$

This means we have two regions.

$$\int_0^{\frac{1}{2}} (8x - x)dx + \int_{\frac{1}{2}}^1 \left(\frac{1}{x^2} - x\right)dx$$

$$= \int_0^{\frac{1}{2}} 7x\,dx + \int_{\frac{1}{2}}^1 (x^{-2} - x)dx$$

$$= 7\frac{x^2}{2}\bigg|_0^{\frac{1}{2}} + \left(\frac{x^{-1}}{-1} - \frac{x^2}{2}\right)\bigg|_{\frac{1}{2}}^1$$

$$= \frac{7}{8} + \left(-\frac{1}{x} - \frac{1}{2}x^2\right)\bigg|_{\frac{1}{2}}^1$$

$$= \frac{7}{8} + \left(-1 - \frac{1}{2}\right) - \left(-2 - \frac{1}{8}\right)$$

$$= 1 + 1 - \frac{1}{2} = 1\frac{1}{2}$$

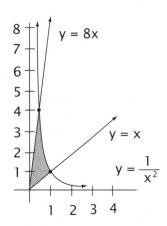

3. $y = -1$ and $y = x^3$ intersect at $(-1, -1)$.
$y = -1$ and $y = 2 - x$ intersect at $(3, -1)$.
$y = x^3$ and $y = 2 - x$ intersect when
$x^3 = 2 - x$ at $(1, 1)$ by inspection.

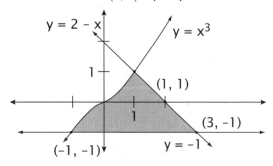

There are two integrals:

$$\int_{-1}^1 [x^3 - (-1)]dx + \int_1^3 [(2-x) - (-1)]dx$$

$$= \left(\frac{x^4}{4} + x\right)\bigg|_{-1}^1 + \left(3x - \frac{x^2}{2}\right)\bigg|_1^3$$

$$= \left(\frac{1}{4} + 1\right) - \left(\frac{1}{4} - 1\right) + \left(9 - \frac{9}{2}\right) - \left(3 - \frac{1}{2}\right) = 4$$

Second way: Right curve − Left curve
Right curve is $y = 2 - x$ or $x = 2 - y$.
Left curve is $y = x^3$ or $x = \sqrt[3]{y}$.

$$\int_{-1}^1 \left[(2-y) - y^{\frac{1}{3}}\right]dy$$

$$= \left[2y - \frac{y^2}{2} - \frac{y^{\frac{4}{3}}}{\frac{4}{3}}\right]\bigg|_{-1}^{+1}$$

$$= \left(2 - \frac{1}{2} - \frac{3}{4}\right) - \left(-2 - \frac{1}{2} - \frac{3}{4}\right) = 4$$

4. $y = 3\cos(\theta)$ intersects $y = 1 + \cos(\theta)$ when:
$$3\cos(\theta) = 1 + \cos(\theta)$$
$$1 = 2\cos(\theta)$$
$$\cos(\theta) = \frac{1}{2}$$

$(\theta) = 60° = \frac{\pi}{3}$ intersection point $\left(\frac{\pi}{3}, \frac{3}{2}\right)$

$y = 3\cos(\theta)$ $y = 1 + \cos(\theta)$

θ	y
0	3
$\frac{\pi}{2}$	0
π	−3
$\frac{\pi}{3}$	$\frac{3}{2}$

θ	y
0	2
$\frac{\pi}{2}$	1
π	0
$\frac{\pi}{3}$	$\frac{3}{2}$

Breaking this region into two separate integrals, we get:

$$\int_0^{\frac{\pi}{3}} (1+\cos(\theta))d\theta + \int_{\frac{\pi}{3}}^{\frac{\pi}{2}} 3\cos(\theta)d\theta$$

$$= (\theta + \sin(\theta))\Big|_0^{\pi/3} + 3\sin(\theta)\Big|_{\pi/3}^{\pi/2}$$

$$= \frac{\pi}{3} + \frac{\sqrt{3}}{2} + 3\left(1 - \frac{\sqrt{3}}{2}\right)$$

$$= \frac{\pi}{3} + \frac{\sqrt{3}}{2} + 3 - \frac{3\sqrt{3}}{2}$$

$$= 3 + \frac{\pi}{3} - \sqrt{3}$$

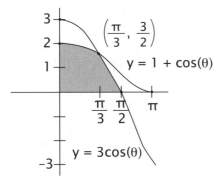

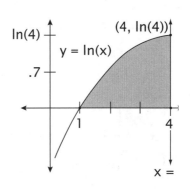

5. If we choose to integrate this with respect to x, we will have $\int_1^4 \ln(x)\,dx$ which we do not know how to integrate. Let's try integrating with respect to y. The intersection of $x = 4$ and $x = e^y$ (same as $y = \ln(x)$) is $4 = e^y$.

$\ln(4) = y$
$(4, \ln(4))$ is the intersection point.

$$\int_0^{\ln(4)} (4 - e^y)\,dy$$
$$= (4y - e^y)\Big|_0^{\ln(4)}$$
$$= \left(4\ln(4) - e^{\ln(4)}\right) - (0 - 1)$$
$$= 4\ln(4) - 4 + 1$$
$$= 4\ln(4) - 3$$

Lesson Practice 26D

1. These graphs intersect when:
$$y = 2(1) + \frac{1}{1^2} = 3$$
Intersection point: $(1, 3)$
for $y = 2x + \frac{1}{x^2}$

$$\int_1^4 \left(2x + \frac{1}{x^2}\right)dx$$
$$= \left(2\frac{x^2}{2} + \frac{x^{-1}}{-1}\right)\Big|_1^4$$
$$= \left(x^2 - \frac{1}{x}\right)\Big|_1^4$$
$$= \left(16 - \frac{1}{4}\right) - (1 - 1) = 15\frac{3}{4}$$

x	y
1	3
2	$4\frac{1}{4}$
$\frac{1}{2}$	5
4	$8\frac{1}{16}$

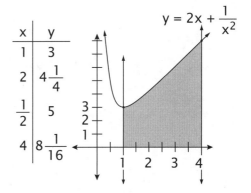

2. These graphs intersect when
$$y^2 = -2y^2 + 3$$
$$3y^2 = 3$$
$$y^2 = 1$$
$$y = \pm 1$$

Intersection points are $(1, 1)(+1, -1)$

$$\int_{-1}^{1} \left[(-2y^2 + 3) - y^2\right] dy = \int_{-1}^{1} (-3y^2 + 3) dy$$
$$= \left(-3 \frac{y^3}{3} + 3y\right)\Big|_{-1}^{1}$$
$$= \left(-y^3 + 3y\right)\Big|_{-1}^{1}$$
$$= (-1 + 3) - (1 - 3) = 4$$

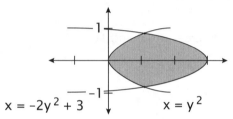

$x = -2y^2 + 3$ $x = y^2$

3. These graphs intersect when:
$$e^x = e^{-x}$$
$$x = -x$$
$$\text{or } x = 0$$
intersection: $(0, 1)$

$$\int_{-1}^{0} e^x dx + \int_{0}^{1} e^{-x} dx \cdot (-1)$$
$$= e^x \Big|_{-1}^{0} - e^{-x} \Big|_{0}^{1}$$
$$= (1 - e^{-1}) - (e^{-1} - 1)$$
$$= 2 - \frac{2}{e}$$

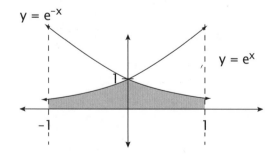

4. These graphs intersect when:
$$2x^2 = x^4 - 2x^2$$
$$4x^2 = x^4$$
$$x^4 - 4x^2 = 0$$
$$x^2(x^2 - 4) = 0$$
$$x = 0, \pm 2$$

Intersection points: $(0, 0)$, $(2, 8)$, $(-2, 8)$

$$y = x^4 - 2x^2$$
$$y' = 4x^3 - 4x$$
$$= 4x(x^2 - 1)$$

Critical points $x = 0, \pm 1$

$$y'' = 12x^2 - 4$$
$$= 4(3x^2 - 1) = 0$$
$$x = \pm\sqrt{\frac{1}{3}} = \frac{1}{\sqrt{3}} = \frac{\sqrt{3}}{3} \approx .6$$

Using our knowledge of the first and second derivatives as it pertains to max, min and inflection points, we find that $y = x^4 - 2x^2$ has the following extrema:

$x = 0$ max-local
$x = \pm 1$ min-absolute
$x = \frac{\sqrt{3}}{3}$ inflection point

$$\int_{-2}^{2} \left[2x^2 - (x^4 - 2x^2)\right] dx$$
$$= \int_{-2}^{2} (4x^2 - x^4) dx$$
$$= \left(4 \frac{x^3}{3} - \frac{x^5}{5}\right)\Big|_{-2}^{2}$$
$$= \left(\frac{32}{3} - \frac{32}{5}\right) - \left(-\frac{32}{3} + \frac{32}{5}\right)$$
$$= \frac{64}{3} - \frac{64}{5} = \frac{128}{15} = 8\frac{8}{15}$$

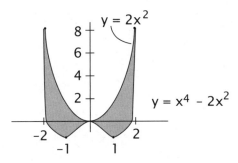

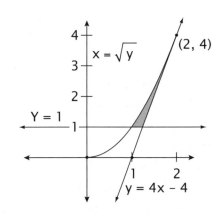

5. These graphs intersect when:

$x = \sqrt{y}$ meets $x = \dfrac{y+4}{4}$

$(\sqrt{y})^2 = \left(\dfrac{y+4}{4}\right)^2$

$y = \dfrac{y^2 + 8y + 16}{16}$

$16y = y^2 + 8y + 16$

$y^2 - 8y + 16 = 0$

$(y-4)(y-4) = 0$ Intersection point $(2, 4)$

$y = 4$

Using the right and left graphs and integrating with respect to y, we get:

$\displaystyle\int_1^4 \left(\dfrac{y+4}{4} - \sqrt{y}\right) dy$

$= \displaystyle\int_1^4 \left(\dfrac{1}{4}y + 1 - y^{\frac{1}{2}}\right) dy$

$= \left(\dfrac{1}{4}\dfrac{y^2}{2} + y - \dfrac{y^{\frac{3}{2}}}{\frac{3}{2}}\right)\Big|_1^4$

$= \left(\dfrac{y^2}{8} + y - \dfrac{2}{3}y^{\frac{3}{2}}\right)\Big|_1^4$

$= \left(2 + 4 - \dfrac{16}{3}\right) - \left(\dfrac{1}{8} + 1 - \dfrac{2}{3}\right)$

$= 5 - \dfrac{1}{8} - \dfrac{14}{3}$

$= 5 - 4\dfrac{19}{24} = \dfrac{5}{24}$

Lesson Practice 27A

1. Let $u = \dfrac{1}{x}$ or x^{-1}; $\dfrac{du}{dx} = -x^{-2}$

$y = \sec^{-1}\left(\dfrac{1}{x}\right)$

$\dfrac{dy}{dx} = \dfrac{-x^{-2}}{\left|\dfrac{1}{x}\right|\sqrt{\left(\dfrac{1}{x}\right)^2 - 1}} = \dfrac{-x^{-1}}{\sqrt{\dfrac{1}{x^2} - 1}}$

$= \dfrac{-1}{x\sqrt{\dfrac{1}{x^2} - 1}} = \dfrac{-1}{x\sqrt{\dfrac{1}{x^2} - 1}} \cdot \dfrac{\sqrt{x^2}}{\sqrt{x^2}}$

$= \dfrac{-x}{x\sqrt{1-x^2}} = \dfrac{-1}{\sqrt{1-x^2}}$

2. Let $u = \dfrac{1}{2}x^2$; $\dfrac{du}{dx} = x$

$y = \tan^{-1}\left(\dfrac{1}{2}x^2\right)$

$\dfrac{dy}{dx} = \dfrac{1}{1 + \left(\dfrac{1}{2}x^2\right)^2} \cdot x = \dfrac{x}{1 + \dfrac{1}{4}x^4}$

3. Let $u = 2x$; $\frac{du}{dx} = 2$

$$y = x \cdot \sin^{-1}(2x)$$
This function requires the product rule
$$\frac{dy}{dx} = x \cdot \frac{d}{dx}\left(\sin^{-1}(2x)\right) + \sin^{-1}(2x) \cdot \frac{d}{dx}(x)$$
$$= x \cdot \frac{1}{\sqrt{1-(2x)^2}}(2) + \sin^{-1}(2x)(1)$$
$$= \frac{2x}{\sqrt{1-4x^2}} + \sin^{-1}(2x)$$

4. Let $u = 3x$; $\frac{du}{dx} = 3$; $du = 3dx$

$$\frac{1}{3}\int \frac{dx}{1+9x^2} \cdot 3 = \frac{1}{3}\int \frac{du}{1+u^2}$$
$$= \frac{1}{3}\tan^{-1}(u) + C$$
$$= \frac{1}{3}\tan^{-1}(3x) + C$$

5. Let $u = \frac{y}{5}$; $\frac{du}{dy} = \frac{1}{5}$; $du = \frac{1}{5}dy$

$$\int \frac{dy}{\sqrt{25-y^2}} = \int \frac{dy}{\sqrt{25\left(1-\frac{y^2}{25}\right)}}$$
$$= \int \frac{dy}{5\sqrt{1-\frac{y^2}{25}}}$$
$$= \int \frac{du}{\sqrt{1-u^2}}$$
$$= \sin^{-1}(u)$$
$$= \sin^{-1}\left(\frac{y}{5}\right) + C$$

6. Let $u = e^x$; $\frac{du}{dy} = e^x$; $du = e^x dx$

$$\int \frac{e^x dx}{1+e^{2x}} = \int \frac{du}{1+u^2}$$
$$= \tan^{-1}(u) + C$$
$$= \tan^{-1}(e^x) + C$$

7. Let $u = \frac{2}{5}y$; $\frac{du}{dy} = \frac{2}{5}$; $du = \frac{2}{5}dy$

$$\int_0^{\frac{5}{4}} \frac{dy}{\sqrt{25-4y^2}} = \int_0^{\frac{5}{4}} \frac{dy}{\sqrt{25\left(1-\frac{4}{25}y^2\right)}}$$
$$= \frac{1}{2}\int_0^{\frac{5}{4}} \frac{dy}{5\sqrt{1-\left(\frac{2}{5}y\right)^2}} \cdot 2$$
$$= \frac{1}{2}\int_0^{\frac{5}{4}} \frac{du}{\sqrt{1-u^2}}$$
$$= \frac{1}{2}\sin^{-1}\left(\frac{2}{5}y\right)\Big|_0^{\frac{5}{4}}$$
$$= \frac{1}{2}\left(\sin^{-1}\left(\frac{1}{2}\right) - \sin^{-1}(0)\right)$$
$$= \frac{1}{2}\left(\frac{\pi}{6}\right) = \frac{\pi}{12}$$

8. Let $u = 4t$; $\frac{du}{dt} = 4$; $du = 4dt$

$$\frac{1}{4}\int_0^{\frac{1}{4}} \frac{dt}{16t^2+1} \cdot 4 = \frac{1}{4}\int_0^{\frac{1}{4}} \frac{du}{1+u^2}$$
$$= \frac{1}{4}\tan^{-1}(4t)\Big|_0^{\frac{1}{4}}$$
$$= \frac{1}{4}\left(\tan^{-1}(1) - \tan^{-1}(0)\right)$$
$$= \frac{1}{4}\left(\frac{\pi}{4} - 0\right) = \frac{\pi}{16}$$

Lesson Practice 27B

1. Let $u = \dfrac{x}{a}$; $\dfrac{du}{dx} = \dfrac{1}{a}$

   $y = \cos^{-1}\left(\dfrac{x}{a}\right)$

   $\dfrac{dy}{dx} = \dfrac{-\dfrac{1}{a}}{\sqrt{1-\left(\dfrac{x}{a}\right)^2}} = \dfrac{-1}{a\sqrt{1-\dfrac{x^2}{a^2}}}$

   $= \dfrac{-1}{\sqrt{a^2\left(1-\dfrac{x^2}{a^2}\right)}} = \dfrac{-1}{\sqrt{a^2-x^2}}$

2. $y = \tan^{-1}\left(\dfrac{2+r}{1-2r}\right)$

   Let $u = \dfrac{2+r}{1-2r}$;

   $\dfrac{du}{dr} = \dfrac{(1-2r)(1)-(2+r)(-2)}{(1-2r)^2}$

   $= \dfrac{1-2r+4+2r}{(1-2r)^2}$

   $= \dfrac{5}{(1-2r)^2}$

   $\dfrac{dy}{dr} = \dfrac{\dfrac{5}{(1-2r)^2}}{1+\left(\dfrac{2+r}{1-2r}\right)^2}$

   $= \dfrac{\dfrac{5}{(1-2r)^2}}{\dfrac{(1-2r)^2+(2+r)^2}{(1-2r)^2}}$

   $= \dfrac{5}{1-4r+4r^2+4+4r+r^2}$

   $= \dfrac{5}{5+5r^2} = \dfrac{1}{1+r^2}$

3. $y = e^x \cot^{-1}(x)$

   $\dfrac{dy}{dx} = e^x \cdot \dfrac{d}{dx}\left(\cot^{-1}(x)\right) + \cot^{-1}(x) \cdot \dfrac{d}{dx}\left(e^x\right)$

   $= e^x\left(-\dfrac{1}{1+x^2}\right) + \left(\cot^{-1}(x)\right)\left(e^x\right)$

   $= e^x\left[\cot^{-1}(x) - \dfrac{1}{1+x^2}\right]$

4. Let $u = 3x$; $\dfrac{du}{dx} = 3$; $du = 3\,dx$

   $\displaystyle\int \dfrac{dx}{|x|\sqrt{9x^2-1}} = \int \dfrac{3\,dx}{3|x|\sqrt{9x^2-1}}$

   $= \displaystyle\int \dfrac{du}{|u|\sqrt{u^2-1}}$

   $= \sec^{-1}(3x) + C$

5. Let $u = e^{2x}$; $\dfrac{du}{dx} = e^{2x} \cdot (2)$; $du = 2e^{2x}dx$

   $\dfrac{1}{2}\displaystyle\int \dfrac{e^{2x}dx}{1+e^{4x}} \cdot 2 = \dfrac{1}{2}\int \dfrac{du}{1+u^2}$

   $= \dfrac{1}{2}\tan^{-1}(e^{2x}) + C$

6. $u = \ln(x)$; $\dfrac{du}{dx} = \dfrac{1}{x}$; $du = \dfrac{dx}{x}$

   $\displaystyle\int \dfrac{dx}{x\sqrt{1-(\ln(x))^2}} = \int \dfrac{du}{\sqrt{1-u^2}}$

   $= \sin^{-1}(\ln(x)) + C$

7. Let $u = \sqrt{x} = x^{\frac{1}{2}}$;

$\dfrac{du}{dx} = \dfrac{1}{2}x^{-\frac{1}{2}}$; $du = \dfrac{dx}{2\sqrt{x}}$

$= \dfrac{1}{2\sqrt{x}}$

$2\displaystyle\int_0^1 \dfrac{dx}{\sqrt{x}(1+x)} \cdot \dfrac{1}{2}$

$= 2\displaystyle\int_0^1 \dfrac{du}{1+u^2}$

$= 2\tan^{-1}(u)\Big|_0^1$

$= 2\tan^{-1}(\sqrt{x})\Big|_0^1$

$= 2\tan^{-1}(1) - 2\tan^{-1}(0) = \dfrac{\pi}{2}$

8. Let $u = \dfrac{x}{3}$; $\dfrac{du}{dx} = \dfrac{1}{3}$; $du = \dfrac{1}{3}dx$

$\displaystyle\int_0^3 \dfrac{dx}{\sqrt{9-x^2}} = \int_0^3 \dfrac{dx}{\sqrt{9\left[1-\left(\dfrac{x}{3}\right)^2\right]}}$

$= \displaystyle\int_0^3 \dfrac{dx}{3\sqrt{1-\left(\dfrac{x}{3}\right)^2}}$

$= \displaystyle\int_0^3 \dfrac{du}{\sqrt{1-u^2}}$

$= \sin^{-1}(u)\Big|_0^3$

$= \sin^{-1}\left(\dfrac{x}{3}\right)\Big|_0^3$

$= \sin^{-1}(1) - \sin^{-1}(0) = \dfrac{\pi}{2}$

Lesson Practice 27C

1. $y = x\sin^{-1}(x)$

$\dfrac{dy}{dx} = x \cdot \dfrac{1}{\sqrt{1-x^2}} + \sin^{-1}(x)$

when $x = \dfrac{1}{2}$,

$\dfrac{1}{2} \cdot \dfrac{1}{\sqrt{1-\left(\dfrac{1}{2}\right)^2}} + \sin^{-1}\left(\dfrac{1}{2}\right)$

$= \dfrac{1}{2\sqrt{\dfrac{3}{4}}} + \dfrac{\pi}{6}$

$= \dfrac{1}{\sqrt{4 \cdot \dfrac{3}{4}}} + \dfrac{\pi}{6}$

$= \dfrac{1}{\sqrt{3}} + \dfrac{\pi}{6}$

$= \dfrac{\sqrt{3}}{3} + \dfrac{\pi}{6}$

$= \dfrac{2\sqrt{3} + \pi}{6}$

2. $y = \dfrac{\tan^{-1}(x)}{x}$

$\dfrac{dy}{dx} = \dfrac{x \cdot \dfrac{d}{dx}(\tan^{-1}(x)) - (\tan^{-1}(x))\dfrac{d}{dx}(x)}{x^2}$

$= \dfrac{x\left(\dfrac{1}{1+x^2}\right) - \tan^{-1}(x)}{x^2}$

when $x = 1$,

$\dfrac{\dfrac{1}{2} - \tan^{-1}(1)}{1} = \dfrac{1}{2} - \dfrac{\pi}{4} = -.285$

3. $y = \ln(\tan^{-1}(x))$

$\dfrac{dy}{dx} = \dfrac{1}{\tan^{-1}(x)} \cdot \dfrac{d}{dx}(\tan^{-1}(x))$

$= \dfrac{\dfrac{1}{1+x^2}}{\tan^{-1}(x)}$

$= \dfrac{1}{(1+x^2)\tan^{-1}(x)}$

when $x = 1$,

$\dfrac{1}{(2)\tan^{-1}(1)} = \dfrac{1}{2\left(\dfrac{\pi}{4}\right)} = \dfrac{2}{\pi}$

4. $\displaystyle\int_{-1}^{1} \dfrac{dx}{1+x^2} = \tan^{-1}(x)\Big|_{-1}^{1}$

$= \tan^{-1}(+1) - \tan^{-1}(-1)$

$= \dfrac{\pi}{4} - \left(-\dfrac{\pi}{4}\right)$

$= \dfrac{\pi}{2}$

5. Let $u = x - 1$; $\dfrac{du}{dx} = 1$; $du = dx$

$\displaystyle\int_{0}^{2} \dfrac{dx}{1+(x-1)^2} = \int_{0}^{2} \dfrac{du}{1+u^2}$

$= \tan^{-1}(x-1)\Big|_{0}^{2}$

$= \tan^{-1}(1) - \tan^{-1}(-1)$

$= \dfrac{\pi}{4} - \left(-\dfrac{\pi}{4}\right) = \dfrac{\pi}{2}$

6. $\displaystyle\lim_{x \to 0} \dfrac{\sin^{-1}(3x)}{x}$ has form $\dfrac{0}{0}$

$= \displaystyle\lim_{x \to 0} \dfrac{\dfrac{1}{\sqrt{1-9x^2}} \cdot (3)}{1}$

$= \displaystyle\lim_{x \to 0} \dfrac{3}{\sqrt{1-9x^2}} = 3$

7. $\displaystyle\lim_{x \to 0} \dfrac{2\pi \cdot \tan^{-1}(2x)}{3x}$ has form $\dfrac{0}{0}$

$= \displaystyle\lim_{x \to 0} \dfrac{2\pi\left(\dfrac{1}{1+4x^2}\right)}{3}$

$= \displaystyle\lim_{x \to 0} \dfrac{2\pi}{3(1+4x^2)} = \dfrac{2\pi}{3}$

Lesson Practice 27D

1. $y = x^2 \cos^{-1}(x)$

$\dfrac{dy}{dx} = x^2 \cdot \dfrac{d}{dx}(\cos^{-1}(x)) + \cos^{-1}(x) \cdot \dfrac{d}{dx}(x^2)$

$= x^2\left(\dfrac{-1}{\sqrt{1-x^2}}\right) + (\cos^{-1}(x))(2x)$

$= \dfrac{-x^2}{\sqrt{1-x^2}} + 2x\cos^{-1}(x)$

when $x = 0$, $\dfrac{dy}{dx} = 0$

2. $y = \cot^{-1}(3x)$

$\dfrac{dy}{dx} = \dfrac{-1}{1+(3x)^2} \cdot (3) = \dfrac{-3}{1+9x^2}$

when $x = 1$, $\dfrac{dy}{dx} = \dfrac{-3}{10}$

3. $y = \csc^{-1}(x^2+1)$

$\dfrac{dy}{dx} = \dfrac{-1}{|x^2+1|\sqrt{(x^2+1)^2 - 1}}(2x)$

$= \dfrac{-2x}{|x^2+1|\sqrt{x^4+2x^2+1-1}}$

$= \dfrac{-2x}{|x^2+1|\sqrt{x^4+2x^2}}$

$= \dfrac{-2x}{|x^2+1|x\sqrt{x^2+2}}$

$= \dfrac{-2}{(x^2+1)\sqrt{x^2+2}}$

4. Let $u = 2x$; $\frac{du}{dx} = 2$; $du = 2dx$

$$\frac{1}{2}\int_0^{\frac{1}{2}} \frac{dx}{1+4x^2} \cdot 2 = \frac{1}{2}\int_0^{\frac{1}{2}} \frac{du}{1+u^2}$$

$$= \frac{1}{2}\tan^{-1}(u)\Big|_0^{\frac{1}{2}}$$

$$= \frac{1}{2}\tan^{-1}(2x)\Big|_0^{\frac{1}{2}}$$

$$= \frac{1}{2}\left[\tan^{-1}(1) - \tan^{-1}(0)\right]$$

$$= \frac{1}{2}\left[\frac{\pi}{4} - 0\right] = \frac{\pi}{8}$$

5. Let $u = x^2$, $\frac{du}{dx} = 2x$; $du = 2x\,dx$

$$\int \frac{2x\,dx}{x^2\sqrt{x^4-1}} = \int \frac{du}{u\sqrt{u^2-1}}$$

$$= \sec^{-1}|u| + C$$

$$= \sec^{-1}|x^2| + C$$

6. $\lim_{x \to 0} \frac{3\tan^{-1}(2x)}{7x}$ has form: $\frac{0}{0}$

$$= \lim_{x \to 0} \frac{3 \cdot \left[\frac{1}{1+(2x)^2}\right]^{(2)}}{7}$$

$$= \lim_{x \to 0} \frac{6}{7(1+4x^2)} = \frac{6}{7}$$

7. $\lim_{x \to 0} \frac{x - \sin^{-1}(x)}{x^2}$ has form: $\frac{0}{0}$

$$= \lim_{x \to 0} \frac{1 - \frac{1}{\sqrt{1-x^2}}}{2x}$$

$$= \lim_{x \to 0} \frac{\frac{\sqrt{1-x^2} - 1}{\sqrt{1-x^2}}}{2x}$$

$$= \lim_{x \to 0} \frac{\sqrt{1-x^2} - 1}{2x\sqrt{1-x^2}} \text{ has form: } \frac{0}{0}$$

$$= \lim_{x \to 0} \frac{\frac{1}{2}(1-x^2)^{-\frac{1}{2}}(-2x)}{2\left[x \cdot \frac{1}{2}(1-x^2)^{-\frac{1}{2}} \cdot (-2x) + \sqrt{1-x^2}\right]}$$

$$= \lim_{x \to 0} \frac{-x}{2\sqrt{1-x^2}(1-x^2)^{-\frac{1}{2}}\left[-x^2 + 1 - x^2\right]}$$

$$= \lim_{x \to 0} \frac{-x}{2(-2x^2+1)} = 0$$

Lesson Practice 28A

1. $\int \frac{dx}{x^2(3+x)}$

Use #4: $\int \frac{du}{u^2(a+bu)}$

where $a = 3$, $b = 1$ and $u = x$

$$= -\frac{1}{au} + \frac{b}{a^2}\ln\left(\frac{a+bu}{u}\right) + C$$

$$= -\frac{1}{3x} + \frac{1}{9}\ln\left(\frac{3+x}{x}\right) + C$$

2. $\int \dfrac{dx}{x(4-3x)}$

use #3: $\int \dfrac{du}{u(a+bu)} = -\dfrac{1}{a}\ln\left(\dfrac{a+bu}{u}\right)+C$

where $u = x$, $a = 4$, and $b = -3$.

$= -\dfrac{1}{4}\ln\left(\dfrac{4-3x}{x}\right)+C$

3. $\int y\sqrt{2+3y}\,dy$

use #8: $\int u\sqrt{a+bu}\,du$

$= -\dfrac{2(2a-3bu)(a+bu)^{\frac{3}{2}}}{15b^2}+C$

where $u = y$, $a = 2$, and $b = 3$.

$= -\dfrac{2[2(2)-3(3)y](2+3y)^{\frac{3}{2}}}{15(3)^2}+C$

$= \dfrac{-2(4-9y)(2+3y)^{\frac{3}{2}}}{135}+C$

4. $\int \sqrt{x^2-4}\,dx$

use #9: $\int \sqrt{u^2-a^2}\,du$
$= \dfrac{u}{2}\sqrt{u^2-a^2} - \dfrac{a^2}{2}\ln\left(u+\sqrt{u^2-a^2}\right)$

where $u = x$ and $a = 2$.

$= \dfrac{x}{2}\sqrt{x^2-4} - 2\ln\left(x+\sqrt{x^2-4}\right)+C$

5. $\dfrac{1}{3}\int \ln(3x-1)\,dx \cdot 3$

use #18: $\int \ln(u)\,du = u\ln(u)-u+C$

where $u = 3x-1$; $\dfrac{du}{dx} = 3$; and $du = 3dx$.

$= \dfrac{1}{3}\int \ln(u)\,du$

$= \dfrac{1}{3}(3x-1)\ln(3x-1) - \dfrac{(3x-1)}{3}+C$

6. $\int \dfrac{2}{x\ln(x)}\,dx = 2\int \dfrac{dx}{x\ln(x)}$

Use #20: $\int \dfrac{du}{u\ln(u)} = \ln(\ln(u))+C$

where $u = x$, $\dfrac{du}{dx} = 1$, and $du = dx$.

$= 2\int \dfrac{du}{u\ln(u)}$

$= 2\ln(\ln(u))+C$

$= 2\ln(\ln(x))+C$

7. $\int \dfrac{x^2\,dx}{\sqrt{1-x^2}}$

use #13: $\int \dfrac{u^2\,du}{\sqrt{a^2-u^2}}$

$= -\dfrac{u}{2}\sqrt{a^2-u^2} + \dfrac{a^2}{2}\sin^{-1}\left(\dfrac{u}{a}\right)+C$

where $u = x$ and $a = 1$.

$\int \dfrac{x^2\,dx}{\sqrt{1-x^2}} = -\dfrac{x}{2}\sqrt{1-x^2} + \dfrac{1}{2}\sin^{-1}(x)+C$

8. $\int_2^3 \sqrt{x^2-4}\,dx$

use #9: $\int \sqrt{u^2-a^2}\,du$
$= \frac{u}{2}\sqrt{u^2-a^2} - \frac{a^2}{2}\ln\left(u+\sqrt{u^2-a^2}\right) + C$

where $u = x$ and $a = 2$.

$= \left[\frac{x}{2}\sqrt{x^2-4} - 2\ln\left(x+\sqrt{x^2-4}\right)\right]\Big|_2^3$

$= \left[\frac{3}{2}\sqrt{5} - 2\ln\left(3+\sqrt{5}\right)\right] - \left[-2\ln(2)\right]$

$= \frac{3}{2}\sqrt{5} - 2\ln\left(3+\sqrt{5}\right) + 2\ln(2)$

9. $\int \cos^3(x)\,dx$

Use #29: $\int \cos^n(u)\,du$
$= \frac{\cos^{n-1}(u)\sin(u)}{n} + \frac{n-1}{n}\int \cos^{n-2}(u)\,du$

$= \frac{\cos^2(x)\sin(x)}{3} + \frac{2}{3}\int \cos(x)\,dx$

$= \frac{\cos^2(x)\sin(x)}{3} + \frac{2}{3}\sin(x) + C$

Lesson Practice 28B

1. $2\int \frac{x\,dx}{(2+x)^3}$

Use #2: $\int \frac{u\,du}{(a+bu)^3}$
$= \frac{1}{b^2}\left[-\frac{1}{a+bu} + \frac{a}{2(a+bu)^2}\right] + C$

where $u = x$, $a = 2$, $b = 1$, and $du = dx$.

$= 2\left(-\frac{1}{2+x} + \frac{2}{2(2+x)^2}\right) + C$

2. use #15:
$\int \sqrt{\frac{1+u}{1-u}}\,du = -\sqrt{1-u^2} + \sin^{-1}(u) + C$

where $u = 3x$ and $du = 3dx$.

$\frac{1}{3}\int \sqrt{\frac{1+3x}{1-3x}}\,dx(3) = \frac{1}{3}\int \sqrt{\frac{1+u}{1-u}}\,du$

$= \frac{1}{3}\left[-\sqrt{1-(3x)^2} + \sin^{-1}(3x)\right] + C$

$= -\frac{1}{3}\sqrt{1-9x^2} + \frac{1}{3}\sin^{-1}(3x) + C$

LESSON PRACTICE 28B - LESSON PRACTICE 28B

3. use #19:

$$\int u^n \ln(u)\,du = u^{n+1}\left[\frac{\ln(u)}{n+1} - \frac{1}{(n+1)^2}\right] + C$$

where $u = 2x$, $du = 2dx$, and $n = 3$.

$$\frac{1}{2}\int (2x)^3 \ln(2x)\,dx(2) = \frac{1}{2}\int u^3 \ln(u)\,du$$

$$= \frac{1}{2}(2x)^4\left[\frac{\ln(2x)}{4} - \frac{1}{16}\right] + C$$

$$= 8x^4\left(\frac{\ln(2x)}{4} - \frac{1}{16}\right) + C$$

$$= 2x^4 \ln(2x) - \frac{x^4}{2} + C$$

4. $\frac{1}{2}\int x\sin(2x)\,dx = \frac{1}{2}\int 2x\sin(2x)\,dx$

use #23:

$$\int u\sin(u)\,du = \sin(u) - (u\cos(u)) + C$$

where $u = 2x$ and $du = 2dx$.

$$= \frac{1}{2}\int u\sin(u)\frac{du}{2}$$

$$= \frac{1}{4}\int u\sin(u)\,du$$

$$= \frac{1}{4}\left[\sin(u) - (u\cos(u))\right] + C$$

$$= \frac{1}{4}\left[\sin(2x) - (2x\cos(2x))\right] + C$$

$$= \frac{\sin(2x)}{4} - \frac{x\cos(2x)}{2} + C$$

5. $\int \frac{dx}{9x^2\sqrt{9-x^2}}$

use #14: $\int \frac{du}{u^2\sqrt{a^2-u^2}} = -\frac{\sqrt{a^2-u^2}}{a^2 u} + C$

where $u = x$ and $a = 3$.

$$= \frac{1}{9}\int \frac{dx}{x^2\sqrt{9-x^2}}$$

$$= -\frac{1}{9}\left[\frac{\sqrt{9-x^2}}{9x}\right] + C$$

$$= -\frac{1}{81x}\sqrt{9-x^2} + C$$

6. $\frac{1}{3}\int \frac{dx \cdot 3}{\sqrt{9x^2+3}}$

use #10:

$$\int \frac{du}{\sqrt{u^2+a^2}} = \ln\left(u + \sqrt{u^2+a^2}\right) + C$$

$u = 3x$, $a = \sqrt{3}$, $\frac{du}{dx} = 3$, $du = 3dx$

$$= \frac{1}{3}\int \frac{du}{\sqrt{u^2+a^2}}$$

$$= \frac{1}{3}\ln\left(3x + \sqrt{9x^2+3}\right) + C$$

7. $\int \frac{dx}{4+2x^2} = \int \frac{du}{a^2+b^2u^2}$

use #5: $\int \frac{du}{a^2+b^2u^2} = \frac{1}{ab}\tan^{-1}\left(\frac{bu}{a}\right) + C$

$u = x$, $du = dx$, $a = 2$, $b = \sqrt{2}$

$$= \frac{1}{2\sqrt{2}}\tan^{-1}\left(\frac{\sqrt{2}x}{2}\right) + C$$

$$= \frac{\sqrt{2}}{4}\tan^{-1}\left(\frac{\sqrt{2}x}{2}\right) + C$$

LESSON PRACTICE 28B - LESSON PRACTICE 28C

8. $\int_1^2 \dfrac{dx}{9x^2\sqrt{9-x^2}} = -\dfrac{1}{81x}\sqrt{9-x^2}\Big|_1^2$

   $= -\dfrac{1}{162}\sqrt{5} + \dfrac{1}{81}\sqrt{8}$

9. $\int x^2 e^{3x} dx$

   use # 17:

   $\int u^n e^{au} du = \dfrac{u^n e^{au}}{a} - \dfrac{n}{a}\int u^{n-1} e^{au} du$

   where $u = x$, $a = 3$, and $n = 2$.

   $= \dfrac{x^2 e^{3x}}{3} - \dfrac{2}{3}\int x e^{3x} dx + C$

   Now use this formula again
   where $u = x$, $n = 1$, and $a = 3$.

   $= \dfrac{x^2 e^{3x}}{3} - \dfrac{2}{3}\left[\dfrac{xe^{3x}}{3} - \dfrac{1}{3}\int x^0 e^{3x} dx\right] + C$

   $= \dfrac{x^2 e^{3x}}{3} - \dfrac{2xe^{3x}}{9} + \dfrac{2}{9}\cdot\dfrac{1}{3}\int e^{3x} dx \cdot 3 + C$

   $= \dfrac{x^2 e^{3x}}{3} - \dfrac{2xe^{3x}}{9} + \dfrac{2}{27} e^{3x} + C$

Lesson Practice 28C

1. $\int \sqrt{\dfrac{x^2}{4} + 2}\, dx$

   use #9: $\int (u^2 + a^2)^{\frac{1}{2}} du$

   $= \dfrac{u}{2}\sqrt{u^2 + a^2} + \dfrac{a^2}{2}\ln\left(u + \sqrt{u^2+a^2}\right) + C$

   $u = \dfrac{x}{2}$, $\dfrac{du}{dx} = \dfrac{1}{2}$, $du = \dfrac{1}{2} dx$, $a = \sqrt{2}$

   $= 2\int \sqrt{\dfrac{x^2}{4}+2}\, dx\left(\dfrac{1}{2}\right)$

   $= 2\left[\dfrac{x}{4}\sqrt{\dfrac{x^2}{4}+2} + \ln\left(\dfrac{x}{2} + \sqrt{\dfrac{x^2}{4}+2}\right)\right] + C$

   $= \dfrac{x}{2}\sqrt{\dfrac{x^2}{4}+2} + 2\ln\left(\dfrac{x}{2} + \sqrt{\dfrac{x^2}{4}+2}\right) + C$

2. $\int \dfrac{x\, dx}{(1+2x)^2}$

   use #1:

   $\int \dfrac{u\, du}{(a+bu)^2} = \dfrac{1}{b^2}\left[\dfrac{a}{a+bu} + \ln(a+bu)\right] + C$

   $u = x;\ a = 1;\ b = 2;\ du = dx$

   $= \dfrac{1}{4}\left[\dfrac{1}{1+2x} + \ln(1+2x)\right] + C$

3. $\int \dfrac{dx}{\sqrt{x^2-4}}$

   use #10:

   $\int \dfrac{du}{(u^2 \pm a^2)^{\frac{1}{2}}} = \ln\left(u + \sqrt{u^2 \pm a^2}\right) + C$

   Let $u = x;\ a = 2$

   $= \ln\left(x + \sqrt{x^2-4}\right) + C$

4. $\dfrac{1}{2}\int \sin^2(2x)\,dx \cdot 2$

use #21:
$$\int \sin^2(u)\,du = \dfrac{1}{2}u - \dfrac{1}{4}\sin(2u) + C$$

$u = 2x,\ du = 2dx$

$$= \dfrac{1}{2}\int \sin^2(u)\,du$$
$$= \dfrac{1}{2}\left[\dfrac{1}{2}(2x) - \dfrac{1}{4}\sin(4x)\right] + C$$
$$= \dfrac{x}{2} - \dfrac{1}{8}\sin(4x) + C$$

5. $\int x^2 e^{2x}\,dx$

use #17:
$$\int u^n e^{au}\,du = \dfrac{u^n e^{au}}{a} - \dfrac{n}{a}\int u^{n-1} e^{au}\,du$$

$u = x;\ du = dx;\ a = 2,\ n = 2$

$$= \dfrac{x^2 e^{2x}}{2} - \dfrac{2}{2}\int x^1 e^{2x}\,dx$$

This reduces the exponents and we can now use the same formula again.

$\int x e^{2x}\,dx$

$u = x;\ du = dx;\ a = 2;\ n = 1;$
$u = 2x;\ \dfrac{du}{dx} = 2;\ du = 2dx$

$$= \dfrac{x e^{2x}}{2} - \dfrac{1}{2}\int x^0 e^{2x}\,dx$$
$$= \dfrac{x e^{2x}}{2} - \dfrac{1}{2}\cdot\dfrac{1}{2}\int e^{2x}\,dx(2)$$
$$= \dfrac{x e^{2x}}{2} - \dfrac{1}{4}e^{2x} + C$$

Now substitute this back and you get:

$$= \dfrac{x^2 e^{2x}}{2} - \dfrac{x e^{2x}}{2} + \dfrac{1}{4}e^{2x} + C$$

6. $\int \tan^3(x)\,dx$

use # 24
$$\int \tan^n(u)\,du = \dfrac{\tan^{n-1}(u)}{n-1} - \int \tan^{n-2}(u)\,du$$
Let $n = 3;\ u = x;\ du = dx$

$$\int \tan^3(x)\,dx = \dfrac{\tan^2(x)}{2} - \int \tan(x)\,dx$$
$$= \dfrac{\tan^2(x)}{2} + \int -\dfrac{\sin(x)}{\cos(x)}\,dx$$

Let $u = \cos(x);\ du = -\sin(x)\,dx$

$$= \dfrac{\tan^2(x)}{2} + \int \dfrac{du}{u}$$
$$= \dfrac{\tan^2(x)}{2} + \ln(u) + C$$
$$= \dfrac{\tan^2(x)}{2} + \ln(\cos(x)) + C$$

7. $\displaystyle\int_0^{\pi/4} \sin^2(2x)\,dx = \left(\dfrac{x}{2} - \dfrac{1}{8}\sin(4x)\right)\bigg|_0^{\pi/4}$

$f(x) = \dfrac{x}{2} - \dfrac{1}{8}\sin(4x)$

$f\left(\dfrac{\pi}{4}\right) = \dfrac{\pi}{8} - \dfrac{1}{4}\sin(\pi) = \dfrac{\pi}{8}$

$f(0) = 0 - \dfrac{1}{4}\sin(0) = 0$

$\dfrac{\pi}{8} - 0 = \dfrac{\pi}{8}$

LESSON PRACTICE 28C - LESSON PRACTICE 28D

8. $\int_0^1 \frac{x\,dx}{(1+2x)^2} = \frac{1}{4}\left[\frac{1}{1+2x} + \ln(1+2x)\right]\Big|_0^1$

$f(x) = \frac{1}{4}\left[\frac{1}{1+2x} + \ln(1+2x)\right]$

$f(1) = \frac{1}{4}\left[\frac{1}{3} + \ln(3)\right]$

$f(0) = \frac{1}{4}(1 + \ln(1)) = \frac{1}{4}(1) = \frac{1}{4}$

$= -\frac{1}{6} + \frac{1}{4}\ln(3)$

$f(1) - f(0) = \frac{1}{12} + \frac{1}{4}\ln(3) - \frac{1}{4}$

$= -\frac{1}{6} + \frac{1}{4}\ln(3)$

5. Use #23 where $u = \frac{1}{2}x$ and $du = \frac{1}{2}dx$.

$\int x\sin\left(\frac{1}{2}x\right)dx$

$= 4\int \frac{1}{2}x\sin\left(\frac{1}{2}x\right)dx \cdot \frac{1}{2}$

$= 4\sin\left(\frac{1}{2}x\right) - 2x\cos\left(\frac{1}{2}x\right) + C$

6. Use #16 where $a = 1$, $b = 4$, $du = 2xdx$, and $u = x^2$.

$\int x \cdot 4^{x^2} dx = \frac{1}{2}\int 2x \cdot 4^{x^2} dx$

$= \frac{4^{x^2}}{2\ln(4)} + C$

Lesson Practice 28D

1. $\int \frac{dx}{\sqrt{x^2 - 4}}$

$= \ln\left(x + \sqrt{x^2 - 4}\right) + C$ Use #10

2. Use #6 where $a = 5$, $u = x$ and $b = 1$

$\int \frac{dx}{25 - x^2} = -\frac{1}{10}\ln\left(\frac{5+x}{5-x}\right) + C$

3. use #22 where $m = 3$, $n = 4$

$\int \sin(3x)\sin(4x)\,dx$

$= \frac{\sin(-x)}{-2} - \frac{\sin(7x)}{14} + C$

4. use #12 where $a = 10$ and $u = x$

$\int \frac{dx}{x^2\sqrt{x^2 - 100}} = \frac{\sqrt{x^2 - 100}}{100x} + C$

7. $\int_{12}^{13} \frac{dx}{x^2\sqrt{x^2 - 100}}$

$= \frac{\sqrt{x^2 - 100}}{100x}\Big|_{12}^{13}$

$= \frac{\sqrt{169 - 100}}{1300} - \frac{\sqrt{144 - 100}}{1200}$

$= \frac{\sqrt{69}}{1300} - \frac{\sqrt{44}}{1200}$

$= \frac{\sqrt{69}}{1300} - \frac{\sqrt{11}}{600}$

$= \frac{6\sqrt{69} - 13\sqrt{11}}{7800}$

8. $\int_{2\pi}^{4\pi} x \sin\left(\frac{1}{2}x\right) dx$

$= \left(4\sin\left(\frac{1}{2}x\right) - 2x\cos\left(\frac{1}{2}x\right)\right)\Big|_{2\pi}^{4\pi}$

$= (4\sin(2\pi) - 8\pi\cos(2\pi)) - (4\sin(\pi) - 4\pi\cos(\pi))$

$= -8\pi - 4\pi = -12\pi$

Lesson Practice 29A

1. $y(x) = -3x^2 + C$

   $\frac{dy}{dx} = -6x$ so C and i are a match.

   $y(x) = -6x + C$

   $\frac{dy}{dx} = -6$

   Plugging into A, you get $-6 + 6 = 0$ so A and ii are a match.
   Finally, B and iii are a match.

2. A and C are separable.

   $\frac{dy}{dx} = (\sin(x))(\cos(y))$

   $\frac{dy}{\cos(y)} = \sin(x)\, dx$

   $\frac{dy}{dx} = \frac{3}{y\sin(x)}$

   $y\, dy = \frac{3}{\sin(x)} dx$

3. $\frac{dy}{dx} = x^3 y$

   $\int \frac{dy}{y} = \int x^3 dx$

   $\ln(y) = \frac{x^4}{4} + C$

   $y = e^{\frac{x^4}{4} + C}$

4. $\frac{dy}{dt} = \frac{\sin(t)}{y}$

   $\int y\, dy = \int \sin(t)\, dt$

   $\frac{y^2}{2} = -\cos(t) + C$

   $y^2 = -2\cos(t) + 2C$

   $y = \pm\sqrt{2C - 2\cos(t)}$

5. $\frac{dy}{dx} - 2x = 1$

   $\frac{dy}{dx} = 2x + 1$

   $\int dy = \int (2x+1)\, dx$

   $y = x^2 + x + C$

6. $2\frac{dy}{dx} = \cos(x)$

   $dy = \frac{1}{2}\cos(x) dx$

   $\int dy = \int \frac{1}{2}\cos(x) dx$

   $y = \frac{1}{2}\sin(x) + C$

7. $2\frac{dy}{dx} = 6x \qquad y(2) = 9$

   $\frac{dy}{dx} = 3x$

   $\int dy = \int 3x\, dx$

   $y = 3\frac{x^2}{2} + C$

   Substituting $y(2) = 9$ we get:

   $9 = 3\left(\frac{4}{2}\right) + C$

   $C = 3$

   Our particular solution is $y = \frac{3}{2}x^2 + 3$.

8. $\dfrac{dP}{dt}(t^2+1) = Pt \qquad P(0) = 6$

$\int \dfrac{dP}{P} = \dfrac{1}{2}\int \dfrac{t\,dt}{t^2+1}(2)$

$\ln(P) = \dfrac{1}{2}\ln(t^2+1) + C$

$\ln(P) = \left(\ln\sqrt{t^2+1}\right) + C$

$P = \sqrt{t^2+1} \cdot e^C$

Substituting $P(0) = 6$ we get:

$6 = \sqrt{0+1} \cdot e^C$

$6 = e^C$

$\ln(6) = C$

Our particular solution is
$P = \sqrt{t^2+1} \cdot e^{\ln(6)}$ which reduces to
$P = 6\sqrt{t^2+1}$.

9. $\dfrac{dy}{dx}\cos(2x) = \sin(2x) \qquad y(0) = 1$

$\int dy = -\dfrac{1}{2}\int \dfrac{\sin(2x)}{\cos(2x)}dx \cdot (-2)$

$y = -\dfrac{1}{2}\ln(\cos(2x)) + C$

$1 = -\dfrac{1}{2}\ln(\cos(0)) + C$

$C = 1$

Our particular solution is:
$y = -\dfrac{1}{2}\ln(\cos(2x)) + 1$

10. $e^t \dfrac{dy}{dt} = e^{3t} \qquad y(0) = \dfrac{3}{2}$

$dy = \dfrac{e^{3t}}{e^t}\,dt$

$\int dy = \dfrac{1}{2}\int e^{2t}\,dt \cdot 2$

$y = \dfrac{1}{2}e^{2t} + C$

$\dfrac{3}{2} = \dfrac{1}{2} + C$

$C = 1$

Our particular solution is $y = \dfrac{1}{2}e^{2t} + 1$.

Lesson Practice 29B

1. $y = x^2 + C$

$\dfrac{dy}{dx} = 2x$

Substituting into C we get:
$2x + 2x = 4x$ so C and i are a match.

$y = 2x^3 + C$

$\dfrac{dy}{dx} = 6x^2$

Substituting into A we get:
$6x^2 - 5x^2 = x^2$, so A and ii are a match.

$y = \dfrac{1}{2}x^2 + C$

$\dfrac{dy}{dx} = x$ so B and iii are a match.

2. B and C are separable.

$dy = (e^x + \sin(x))dx$

$\dfrac{dy}{dx} = e^y(\sin(x+1))$

$\dfrac{dy}{e^y} = (\sin(x+1))dx$

3. $\dfrac{dy}{dx} = 2x^3 y$

$\displaystyle\int \dfrac{dy}{y} = \int 2x^3\, dx$

$\ln(y) = 2\dfrac{x^4}{4} + C$

$\ln(y) = \dfrac{1}{2}x^4 + C$

$y = e^{\frac{1}{2}x^4 + C}$

4. $\dfrac{dx}{dt} = e^x \sin(t)$

$\displaystyle\int \dfrac{dx}{e^x} = \int \sin(t)\, dt$

$-\displaystyle\int e^{-x}(-1)\, dx = \int \sin(t)\, dt$

$-e^{-x} = -\cos(t) + C$

$e^{-x} = \cos(t) + C$

$-x = \ln(\cos(t)) + C$

$x = -\ln(\cos(t)) + C$

5. $\dfrac{dy}{dx} - 4y = 2$

$\dfrac{dy}{dx} = 4y + 2$

$\dfrac{1}{4}\displaystyle\int \dfrac{4\,dy}{4y+2} = \int dx$

$\dfrac{1}{4}\ln(4y+2) = x + C$

$\ln(4y+2) = 4x + 4C$

$4y + 2 = e^{4x + 4C}$

$4y = e^{4x + 4C} - 2$

$y = \dfrac{e^{4x + 4C} - 2}{4}$

6. $\left[\csc(t) + \dfrac{2}{\sin(t)}\right]\dfrac{ds}{dt} = 6$

$\left[\dfrac{\csc(t)\sin(t) + 2}{\sin(t)}\right]\dfrac{ds}{dt} = 6$

$\left[\dfrac{3}{\sin(t)}\right]\dfrac{ds}{dt} = 6$

$\dfrac{\frac{ds}{dt}}{\sin(t)} = 2$

$\displaystyle\int ds = \int 2\sin(t)\, dt$

$s = -2\cos(t) + C$

7. $\dfrac{w^5}{5t^4} \cdot \dfrac{dw}{dt} = e^{t^5} \qquad w(0) = 2$

$\displaystyle\int w^5\, dw = \int e^{t^5}(5t^4)\, dt$

$\dfrac{w^6}{6} = e^{t^5} + C$

$w^6 = 6e^{t^5} + 6C$

$w = \sqrt[6]{6e^{t^5} + 6C}$

$w(0) = \sqrt[6]{6e^0 + 6C}$

$2 = \sqrt[6]{6 + 6C}$

$2^6 = 6C + 6$

$58 = 6C$

$C = 9\dfrac{2}{3}$

Our particular solution is:
$w = \sqrt[6]{6e^{t^5} + 58}$

LESSON PRACTICE 29B - LESSON PRACTICE 29C

8. $x^{-10}\dfrac{dz}{dx} + \dfrac{1}{2x^{10}}\dfrac{dz}{dx} = 11 \quad z(1) = \dfrac{4}{3}$

$\dfrac{1}{x^{10}}\dfrac{dz}{dx} + \dfrac{1}{2x^{10}} = 11$

$\dfrac{1}{x^{10}}\dfrac{dz}{dx}\left[1 + \dfrac{1}{2}\right] = 11$

$\int \dfrac{3}{2} dz = \int 11 x^{10} dx$

$\dfrac{3}{2} z = x^{11} + C$

$z = \dfrac{2}{3}x^{11} + \dfrac{2}{3}C$

$\dfrac{4}{3} = \dfrac{2}{3} + \dfrac{2}{3}C$

$C = 1$

Our particular solution is $z = \dfrac{2}{3}x^{11} + \dfrac{2}{3}$.

9. $\dfrac{dy}{dx} = \dfrac{e^y}{4x} \quad y(1) = 0$

$\int \dfrac{dy}{e^y} = \int \dfrac{dx}{4x}$

$-\int e^{-y}(-1) = \dfrac{1}{4}\int \dfrac{dx}{x}$

$-e^{-y} = \dfrac{1}{4}\ln(4x) + C$

$y = \ln\left[\dfrac{1}{4}\ln(4x) + C\right]$

$y = \ln\left[\dfrac{1}{4}\ln(4x) + C\right]$

$0 = \ln\left[\dfrac{1}{4}\ln(1) + C\right]$

$0 = \ln[C]$

$C = 1$

Our particular solution is:

$y = -\ln\left[\dfrac{1}{4}\ln(x) + 1\right]$

10. $\dfrac{dw}{dt} = .2w + 2 \quad w(0) = 10$

$\dfrac{1}{.2}\int \dfrac{dw(.2)}{.2w + 2} = \int dt$

$5\ln(.2w + 2) = t + C$

$\ln(.2w + 2) = \dfrac{1}{5}t + \dfrac{1}{5}C$

$.2w = e^{\frac{1}{5}t + \frac{1}{5}C} - 2$

$w = 5\left[e^{\frac{1}{5}t + \frac{1}{5}C} - 2\right]$

$10 = 5\left[e^{\frac{1}{5}C} - 2\right]$

$2 = e^{\frac{1}{5}C} - 2$

$4 = e^{\frac{1}{5}C}$

$\ln(4) = \dfrac{1}{5}C$

$C = 5\ln(4) = \ln(1024)$

Our particular solution is:

$w = 5\left[e^{\frac{1}{5}t + \ln(4)} - 2\right]$

Lesson Practice 29C

1. $\dfrac{dx}{dt} = e^{2x}\cos(2t)$

$-\dfrac{1}{2}\int \dfrac{dx}{e^{2x}}(-2) = \dfrac{1}{2}\int \cos(2t)\, dt$

$-\dfrac{1}{2}e^{-2x} = \dfrac{1}{2}\sin(2t) + C$

$e^{-2x} = -\sin(2t) + C$

$-2x = \ln[-\sin(2t) + C]$

$x = -\dfrac{1}{2}\ln[-\sin(2t) + C]$

LESSON PRACTICE 29C

2. $\dfrac{dy}{dx} e^{-x} - e^{3x} = 0$

$\dfrac{\frac{dy}{dx} e^{-x}}{e^{-x}} = \dfrac{e^{3x}}{e^{-x}}$

$\dfrac{dy}{dx} = e^{4x}$

$\int dy = \dfrac{1}{4} \int e^{4x} dx \, (4)$

$y = \dfrac{1}{4} e^{4x} + C$

3. $\dfrac{dy}{dx} \sec(2x) = -6$

$dy = \dfrac{-6 \, dx}{\sec(2x)}$

$\int dy = -6 \int \cos(2x) \, dx$

$y = \left(\dfrac{1}{2}\right)(-6) \int \cos(2x) \, dx \,(2)$

$y = -3 \sin(2x) + C$

4. $\int dy = \int x(x+2)$ \quad $y(1) = \dfrac{1}{3}$

$\int dy = \int (x^2 + 2x) \, dx$

$y = \dfrac{x^3}{3} + 2 \dfrac{x^2}{2} + C$

$y = \dfrac{1}{3} x^3 + x^2 + C$

$\dfrac{1}{3} = \dfrac{1}{3} + 1 + C$

$\dfrac{1}{3} = \dfrac{4}{3} + C$

$C = -1$

$y = \dfrac{1}{3} x^3 + x^2 - 1$

5. $\dfrac{dy}{dx} = xe^{x^2 - \ln(y)}$ \quad $y(0) = 3$

$\dfrac{dy}{dx} = \dfrac{xe^{x^2}}{e^{\ln(y)}}$

$\int y \, dy = \dfrac{1}{2} \int xe^{x^2} dx \,(2)$

$\dfrac{y^2}{2} = \dfrac{1}{2} e^{x^2} + C$

$y^2 = e^{x^2} + 2C$

$y = \pm \sqrt{e^{x^2} + 2C}$

$3 = \pm \sqrt{1 + 2C}$

$9 = 1 + 2C$

$C = 4$

$y = \pm \sqrt{e^{x^2} + 8}$ \quad is the particular solution.

6. $\dfrac{dy}{d\theta} \cos(\theta) \cot(\theta) = 2$ \quad $y(0) = 2$

$dy = \dfrac{2 \, d\theta}{\cos(\theta) \cot(\theta)}$

$\int dy = 2 \int \sec(\theta) \tan(\theta) \, d\theta$

$y = 2(\sec(\theta)) + C$

$C = 0$

$y = 2 \sec(\theta)$ is the particular solution.

Lesson Practice 29D

1. $x^{-3}\dfrac{dy}{dx} - 4 = -\dfrac{1}{x^3}$

   $\dfrac{\frac{dy}{dx}}{x^3} + \dfrac{1}{x^3} = 4$

   $\dfrac{1}{x^3}\left[\dfrac{dy}{dx} + 1\right] = 4$

   $\dfrac{dy}{dx} + 1 = 4x^3$

   $\dfrac{dy}{dx} = 4x^3 - 1$

   $\int dy = \int(4x^3 - 1)dx$

   $y = 4\dfrac{x^4}{4} - x + C$

   $y = x^4 - x + C$

2. $\dfrac{dy}{dx}\csc(3x) = 5$

   $\dfrac{dy}{dx} = \dfrac{5}{\csc(3x)}$

   $\dfrac{dy}{dx} = 5\sin(3x)$

   $dy = 5\sin(3x)\,dx$

   $\int dy = \dfrac{5}{3}\int \sin(3x)\cdot 3\,dx$

   $y = \dfrac{5}{3}(-\cos(3x)) + C$

   $y = -\dfrac{5}{3}\cos(3x) + C$

3. $\dfrac{dy}{dt}\cdot y = 3\cos(3t)$

   $\int y\,dy = \int 3\cos(3t)\,dt$

   $\dfrac{y^2}{2} = \sin(3t) + C$

   $y^2 = 2(\sin(3t)) + 2C$

   $y = \pm\sqrt{2(\sin(3t)) + 2C}$

4. $(t^2 + 3)\dfrac{dR}{dt} = Rt \qquad R(1) = 2$

   $\int\dfrac{dR}{R} = \dfrac{1}{2}\int\dfrac{t\,dt}{t^2 + 3}(2)$

   $\ln(R) = \dfrac{1}{2}\ln(t^2 + 3) + C$

   $R = e^{\frac{1}{2}\ln(t^2+3)+C}$

   $R = \left[e^{\ln(t^2+3)^{\frac{1}{2}}}\right]\cdot e^C$

   $R = \left[\sqrt{(t^2+3)}\right]e^C$

   $2 = (\sqrt{4})e^C$

   $2 = 2(e^C)$

   $1 = e^C \qquad C = 0$

   Our particular solution is $R = \sqrt{t^2 + 3}$

5. $\dfrac{dy}{d\theta}\cdot\csc(2\theta) = -y\sec(2\theta) \qquad y(0) = 2$

   $\dfrac{dy}{y} = -\dfrac{\sec(2\theta)}{\csc(2\theta)}d\theta$

   $\int\dfrac{dy}{y} = -\dfrac{1}{2}\int\dfrac{\sin(2\theta)}{\cos(2\theta)}d\theta\cdot 2$

   $\ln(y) = \dfrac{1}{2}\ln(\cos(2\theta)) + C$

   $\ln(y) = \ln\sqrt{\cos(2\theta)} + C$

   $y = \sqrt{\cos(2\theta)}\cdot e^C$

   $2 = 1\cdot e^C$

   $\ln(2) = C$

   $y = \sqrt{\cos(2\theta)}\cdot e^{\ln(2)}$

   $y = 2\sqrt{\cos(2\theta)}$ is our particular solution.

6. $\dfrac{dM}{dt} - t = Mt \qquad M(0) = 1$

$\dfrac{dM}{dt} = Mt + t$

$dM = t(M+1)dt$

$\int \dfrac{dM}{M+1} = \int t\, dt$

$\ln(M+1) = \dfrac{t^2}{2} + C$

$M + 1 = e^{\frac{t^2}{2} + C}$

$M = e^{\frac{t^2}{2} + C} - 1$

$1 = e^C - 1$

$2 = e^C$

$C = \ln(2)$

$M = e^{\frac{t^2}{2} + \ln(2)} - 1$

$M = 2e^{\frac{t^2}{2}} - 1$ is the particular solution.

Lesson Practice 30A

1. $P = Ce^{kt}$

$P(0) = 20$

$20 = Ce^0$

so $C = 20$

Our model is $P = 20e^{kt}$

$P(10) = 22$

$22 = 20e^{10k}$

$\dfrac{11}{10} = e^{10k}$

$\dfrac{1}{10}\ln\left(\dfrac{11}{10}\right) = k = .009531$

Answer the question.

$P(30) = 20e^{.009531 \cdot (30)}$

≈ 26.6 million people

2. $\dfrac{dN}{dt} = .16N$

$\int \dfrac{dN}{N} = \int .16\, dt$

$\ln(N) = .16t + C$

$N = e^{.16t + C}$

$N = Ce^{.16t}$

$3000 = Ce^0$

Our model becomes $N = 3000e^{.16t}$

$N(1) = 3000e^{.16(1)} \approx 3521$

3521 bacteria were present after 1 hour.

$9000 = 3000e^{.16t}$

$3 = e^{.16t}$

$\ln(3) = .16t$

$t \approx 6.9$

The bacteria will triple after 6.9 hours.

3. $T = (T_0 - T_s)e^{kt} + T_s$

$T_0 = 25°C$

$T_s = 2°C$

$T(5) = 20°C$

$T = (25 - 2)e^{kt} + 2$

$T = 23e^{kt} + 2$

$20 = 23e^{5k} + 2$

$18 = 23e^{5k}$

$\ln\left(\dfrac{18}{23}\right) = 5k$

$k \approx -.0490$

$T = 23e^{-.0490t} + 2$
$38 = 23e^{-.0490t} + 2$
$36 = 23e^{-.0490t}$
$\ln\left(\frac{36}{23}\right) = -.0490t$
$t \approx -9.1$

The deer died 9.1 minutes earlier or at approximately 8:51 am.

Lesson Practice 30B

1. a) $\frac{dy}{dt} = ky$

$y = Ce^{kt}$

$y = 100e^{kt}$ at $t = 0$, $y = 100$

When $t = 1$, $y = 420$.

$420 = 100e^k$
$4.2 = e^k$
$k = \ln(4.2) \approx 1.4351$

b) When $t = 3$:

$y = 100e^{1.4351t}$
$y = 100e^{1.4351(3)}$
$= 7409.1$ bacteria cells after 3 hours

2. $\frac{dy}{dt} = ky$

$\int \frac{dy}{y} = \int k\,dt$

$\ln(y) = kt + C$

$y = e^{kt+C}$

$y = Ce^{kt}$

If the population doubles every 10 years, then when $y = 2C$, $t = 10$.

$2C = Ce^{10k}$
$2 = e^{10k}$
$\frac{\ln(2)}{10} = k$
$k \approx .06931$

3. $T = (T_0 - T_s)e^{kt} + T_s$

$T_0 = 170°F \quad T_s = 76°F \quad T(2) = 125°F$

$T = (170 - 76)e^{kt} + 76$
$T = 94e^{kt} + 76$

$125 = 94e^{2k} + 76$
$49 = 94e^{2k}$
$\ln\frac{49}{94} = 2k$
$k = \frac{1}{2}\ln\left(\frac{49}{94}\right)$
$\approx -.3257$

$80 = 94e^{-.3257t} + 76$
$4 = 94e^{-.3257t}$
$\ln\left(\frac{4}{94}\right) = -.3257t$
$t \approx 9.7$

It will take approximately 9.7 minutes for the tea to reach a temperature of 80°F.

4. a) $\dfrac{dv}{dt} = -9.8$

$\int dv = \int -9.8\, dt$

$v(t) = -9.8t + C$
$v(0) = 10$ so $C = 10$

$v(t) = -9.8t + 10$

b) $\dfrac{ds}{dt} = -9.8t + 10$

$\int ds = \int (-9.8t + 10)\, dt$

$s(t) = -9.8\dfrac{t^2}{2} + 10t + C$
$s(0) = 3$ so $C = 3$

$s(t) = -4.9t^2 + 10t + 3$

c) When the rock is at its peak, the velocity will be zero.

$-9.8t + 10 = 0$
$-9.8t = -10$
$t \approx 1.02$ seconds

d) $s(1.02) = -4.9(1.02)^2 + 10(1.02) + 3$
$-5.09796 + 10.2 + 3 = 8.1$ meters

The peak will be reached in 1.02 seconds and the height will be 8.1 meters. These are approximate values.

Lesson Practice 30C

1. $\dfrac{di}{dt} = ki$

$\int \dfrac{di}{i} = \int k\, dt$ \qquad $i_0 = 1{,}000$

$\ln(i) = kt + C$ \qquad $i_7 = 1{,}200$

$i = e^{kt+C}$ \qquad $i_{12} = ?$

$i = Ce^{kt}$
$1{,}000 = Ce^0$
$1{,}000 = C$

$i = 1{,}000e^{kt}$

$1200 = 1{,}000e^{k(7)}$
$1.2 = e^{7k}$
$k \approx .02605$

$i_{12} = 1{,}000e^{.02605(12)}$
$\approx 1{,}367$ people infected in 12 days

$2{,}000 = 1{,}000e^{.02605t}$
$2 = e^{.02605t}$
$t = \ln(2) \div .02605 \approx 26.6$ days

It wil take approximately 27 days for 2,000 people to become infected.

2. $\dfrac{dF}{dt} = -110e^{-.4t}$ \qquad $F(0) = 375$

$\int dF = -\dfrac{1}{.4}\int -110e^{-.4t}\, dt\,(-.4)$

$F = -\dfrac{110}{-.4}e^{-.4t} + C$

$F = 275e^{-.4t} + C$

When $t = 0$, the pizza is 375°F.

$375 = 275e^0 + C$
$100 = C$
$F = 275e^{-.4t} + 100$

$F(5) = 275e^{-.4(5)} + 100$
$= 275e^{-2} + 100$
≈ 137 degrees F

LESSON PRACTICE 30C - LESSON PRACTICE 30D

3. $T = (T_0 - T_s)e^{kt} + T_s$

$T_0 = 1{,}500°F; \ T_s = 80°F; \ T(1) = 1{,}120°F$

$$T = (1{,}500 - 80)e^{kt} + 80$$
$$T = 1420e^{kt} + 80$$
$$1120 = 1420e^{k(1)} + 80$$
$$\frac{1{,}040}{1{,}420} = e^k$$
$$k = \ln\left(\frac{1{,}040}{1{,}420}\right) \approx -.3114$$
$$T = 1{,}420e^{-.3114t} + 80$$
$$T = 1{,}420e^{-.3114(5)} + 80$$
$$\approx 379.3°F$$

After 5 hours, the core temperature would be approximately 379.3°F.

4. $\int \frac{dM}{M} = \int .05 \, dt$

$\ln(M) = .05t + C$

$M = e^{.05t + C}$

$M = Ce^{.05t}$

$M(0) = 1{,}000$

$1{,}000 = C$; so our model is $M = 1{,}000e^{.05t}$

a) $M(10) = 1{,}000e^{.05(10)}$
$\approx 1{,}649$

Approximately \$1,649 would be available after 10 years.

b) $2{,}000 = 1{,}000e^{.05t}$
$2 = e^{.05t}$
$\ln(2) = .05t$
$t \approx 13.86$

It would take approximately 13.86 years to double a \$1,000 investment at 5% compounded continuously.

Lesson Practice 30D

1. $T_0 = 2 \quad T_2 = 3.6 \quad T_3 = ?$

$2 = Ce^0; \ 2 = C$ so our model is $W = 2e^{kt}$

First use the information at 2 months to determine k.

$$3.6 = 2e^{2k}$$
$$1.8 = e^{2k}$$
$$k \approx .2939$$

Now determine weight at 3 months.

$$W = 2e^{.2939t}$$
$$W_3 = 2e^{(.2939)3} \approx 4.8 \text{ lbs}$$

The puppy should weigh approximately 4.8 lbs at three months of age.

$$6 = 2e^{.2939t}$$
$$3 = e^{.2939t}$$
$$t = 3.7 \text{ months}$$

The puppy should weigh 6 lbs at approximately the age of 3.7 months.

2. $T = (T_0 - T_s)e^{kt} + T_s$

$T_0 = 100, \ T_s = 30, \ T_3 = 70, \ T_? = 40$

$$T = (100 - 30)e^{kt} + 30$$
$$T = 70e^{kt} + 30$$

$$70 = 70e^{3k} + 30$$
$$40 = 70e^{3k}$$
$$\ln\left(\frac{4}{7}\right) = 3k$$
$$k \approx -.1865$$

LESSON PRACTICE 30D - LESSON PRACTICE 30D

$$T = 70e^{-.1865t} + 30$$
$$40 = 70e^{-.1865t} + 30$$
$$10 = 70e^{-.1865t}$$
$$\ln\left(\frac{1}{7}\right) = -.1865t$$
$$t \approx 10.4 \text{ minutes}$$

The copper ball will reach a temperature of 40°C in approximately 10.4 minutes.

3. $T = (T_0 - T_s)e^{kt} + T_s$

We know $T_0 = 98°C$, $T_s = 18°C$, and $T(5) = 48°C$.

$$T = (98 - 18)e^{kt} + 18$$
$$T = 80e^{kt} + 18$$

$$48 = 80e^{5k} + 18$$
$$30 = 80e^{5k}$$
$$\frac{3}{8} = e^{5k}$$
$$5k = \ln\left(\frac{3}{8}\right)$$
$$k = \frac{1}{5}\ln\left(\frac{3}{8}\right)$$
$$k = -.1962$$

$$T = 80e^{-.1962t} + 18$$
$$28 = 80e^{-.1962t} + 18$$
$$\ln\left(\frac{1}{8}\right) = -.1962t$$
$$t \approx 10.6 \text{ min}$$

Since it will take approximately 10.6 minutes to cool to 28°C, the time it takes past the first 5 minutes is 5.6 minutes.

4. $T = (T_0 - T_s)e^{kt} + T_s$

We know $T_0 = 90°C$, $T_s = 30°C$ and $T(10) = 60°C$.

$$T = (90 - 30)e^{kt} + 30$$
$$T = 60e^{kt} + 30$$
$$60 = 60e^{10k} + 30$$
$$30 = 60e^{10k}$$
$$\frac{1}{2} = e^{10k}$$
$$\ln\left(\frac{1}{2}\right) = 10k$$
$$k \approx -.06931$$

$$T = 60e^{(-.06931)t} + 30$$
$$40 = 60e^{(-.06931)t} + 30$$
$$10 = 60e^{(-.06931)t}$$
$$t \approx 25.85$$

It will take about 26 minutes to cool from 90°C to 40°C so it will take about 16 minutes longer to reach that temperature once it has cooled to 60°C.

Test Solutions

Test 1
1. D: constant
2. C: variable
3. B: coefficient
4. E: dependent
5. G: tangent
6. D: $|x+1| > 2$

    $\begin{array}{ll} x+1 > 2 & -(x+1) > 2 \\ x > 1 & -x-1 > 2 \\ & -x > 3 \\ & x < -3 \end{array}$

7. A: The discontinuity is at $(-1, -2)$.
8. D: The $\frac{1}{3}$ dilates the graph.

    The -3 translates the graph downwards.
9. D: $y = -\sqrt{x+2}$

    $y = 0$ when $x = -2$, so this graph begins at $(-2, 0)$. It has the shape of $y = \sqrt{x}$, but it is negative.
10. B: The $\frac{1}{2}$ dilates the graph.

Test 2
1. A: line
2. C: different coefficients on x^2 and y^2, but signs are the same
3. A: horizontal line: $y = 3$
4. D: one squared term
5. F: cubic, which is not quadratic
6. B: circle of radius 2

7. B:

$$3x^2 + 3y^2 + 6x + 12y = 1$$
$$3(x^2 + 2x + 1) + 3(y^2 + 4y + 4) = 1 + 3 + 12$$
$$3(x+1)^2 + 3(y+2)^2 = 16$$
$$\frac{3(x+1)^2}{16} + \frac{3(y+2)^2}{16} = 1$$
$$\frac{(x+1)^2}{\frac{16}{3}} + \frac{(y+2)^2}{\frac{16}{3}} = 1$$

center: $(-1, -2)$

8. C: $\sqrt{\frac{16}{3}} = \frac{4}{\sqrt{3}} = \frac{4\sqrt{3}}{3}$

    (see answer to question 7)

9. B: $x = -y^2 + 10y - 25 + 2$

    $x = -(y^2 - 10y) - 23$

    $x = -(y^2 - 10y + 25) - 23 + 25$

    $x = -(y-5)^2 + 2$

    Vertex is $(2, 5)$.

10. A: $y = 5$

Test 3
1. C All have the form $xy = $ constant or there are two squared terms with opposite signs.

2. D: $y = \frac{1}{2}x^2 - 2$

$y - x = -2 \Rightarrow y = x - 2$

Two things that are both equal to y are equal to each other, so:

$\frac{1}{2}x^2 - 2 = x - 2$

$\frac{1}{2}x^2 = x$

$x^2 = 2x$

$x^2 - 2x = 0$

$(x)(x-2) = 0$

$x = 0 \qquad x - 2 = 0$

$\qquad\qquad x = 2$

$y = x - 2 \Rightarrow y = (0) - 2$

$y = -2$

$y = (2) - 2$

$y = 0$

intersection: $(0, -2); (2, 0)$

3. A: $y = x$

$x^2 + y^2 = 4 \Rightarrow x^2 + (x)^2 = 4$

$2x^2 = 4$

$x^2 = 2$

$x = \pm\sqrt{2}$

$y = x \Rightarrow y = (\pm\sqrt{2})$

intersection: $(\sqrt{2}, \sqrt{2}); (-\sqrt{2}, -\sqrt{2})$

4. B   Try $(0, 0)$. It doesn't work, so the solution is outside the circle.

5. D   Quadrants 2 and 4 have opposite signs for x and y.

6. B   Because the equality is present C & D are incorrect. Try $(0, 3)$. It doesn't work so B is correct.

7. D   Putting into standard form, we get $\frac{y^2}{4} - \frac{x^2}{16} = 1$.

Asymptotes are $y = \pm\frac{2}{4}x$ or $y = \pm\frac{1}{2}x$

8. A   Because y^2 is positive, the hyperbola is an up-down hyperbola.

9. C   Substituting we get $x^2 = 1$, $x = \pm 1$ so it intersects at $(1, 1)$ and $(-1, 1)$.

10. D   Substituting we get $x^2 = -1$ and there is no solution.

Test 4

1. $f(x) = x^2 - 2x$

   B: $f(-1) = (-1)^2 - 2(-1)$

   $= 1 + 2$

   $= 3$

2. $f(x) = x^2 - 2x$

   C: $f(-x^2) = (-x^2)^2 - 2(-x^2)$

   $= x^4 + 2x^2$

3. $f(x) = x^2 - 2x$

   D: $f(x+h) = (x+h)^2 - 2(x+h)$

   $= x^2 + 2xh + h^2 - 2x - 2h$

4. $f(x) = x^2 - 2x;\ g(x) = 3x$

   C: $g(f(x)) = g(x^2 - 2x)$

   $= 3(x^2 - 2x)$

   $= 3x^2 - 6x$

5. A: $f(x) = 3 - x$

   $y = 3 - x$

   $x = 3 - y$ (switch variables)

   $x - 3 = -y$

   $-x + 3 = y$

   $y = 3 - x$

   $f^{-1}(x) = 3 - x$

6. C: A is the sine function.
   B is a non-vertical line, so it is a function.
   C is a parabola with the "C" shape so it fails the vertical line test.
   D is the natural log function.

7. B: definition of inverse

8. C: $f(x) = 2x^2$
$y = 2x^2$
$x = 2y^2$ (switch variables)
$\frac{x}{2} = y^2$
$y = \pm\sqrt{\frac{x}{2}}$
$f^{-1}(x) = \pm\sqrt{\frac{x}{2}}$

This is a graph of a parabola with the "C" shape. It is not a function, but its graph is in quadrants I and IV.

9. A. $C = \pi d$
$\frac{C}{\pi} = d$
$d(C) = \frac{C}{\pi}$

5. A: $\tan\left(\frac{3\pi}{4}\right) = \tan 135°$
quadrant II, tangent is negative
$\tan\left(\frac{3\pi}{4}\right) = -1$

6. C Definition of amplitude

7. B $-\pi$

8. B 2 is the vertical shift

9. C $\csc\left(\frac{7\pi}{4}\right) = \csc 315° = -\sqrt{2}$

10. D $\sin(2x) = 0$ when $x = 0, \frac{\pi}{2}, \pi, \frac{3\pi}{2}, 2\pi, \frac{5\pi}{2}, \ldots \frac{19\pi}{2}, 10\pi$
There are 21 solutions.

Test 5

1. B: The period of $y = \cos(x)$ is 2π, so the period of $y = \cos(2x)$ will be half as large.

2. C: $\sec\left(\frac{\pi}{4}\right) = \frac{\text{hyp}}{\text{adj}} = \frac{\sqrt{2}}{1} = \sqrt{2}$

3. C $\sin(y) = 0$ When $y = 0, \pi, 2\pi, 3\pi, 4\pi$ etc

$2\theta = 0 \quad 2\theta = \pi \quad 2\theta = 2\pi$
$\theta = 0 \quad \theta = \frac{\pi}{2} \quad \theta = \pi$
$\theta = 0, \frac{\pi}{2}, \pi$

4. A definition of frequency

Test 6

1. C $\frac{\ln(9)}{2} = \frac{1}{2}\ln(9) = \ln\left(9^{\frac{1}{2}}\right) = \ln\left(\sqrt{9}\right) = \ln(3)$

2. D $\ln(x) - \ln(4) = 2$
$\ln\left(\frac{x}{4}\right) = 2$
$e^{\ln\left(\frac{x}{4}\right)} = e^2$
$\frac{x}{4} = e^2$
$x = 4e^2$

3. D $\ln\left(\frac{6}{3}\right) = \ln(6) - \ln(3)$

4. B $y = \ln(x - 2)$
$x = \ln(y - 2)$ (switch variables)
$e^x = e^{\ln(y-2)}$
$e^x = y - 2$
$e^x + 2 = y$
$f^{-1}(x) = e^x + 2$

5. C $\ln(\sqrt{2}) + \ln(\sqrt{10}) = \ln(\sqrt{2})(\sqrt{10})$
       $= \ln(\sqrt{20}) = \ln(2\sqrt{5})$

6. A: $\ln^2(x) - 5\ln(x) = -4$
       $y^2 - 5y = -4$
       $y^2 - 5y + 4 = 0$
       $(y-1)(y-4) = 0$

   $y - 1 = 0$ $y - 4 = 0$
   $y = 1$ $y = 4$
   $\ln(x) = 1$ $\ln(x) = 4$
   $e^{\ln(x)} = e^1$ $e^{\ln(x)} = e^4$
   $x = e$ $x = e^4$

7. D definition of an inverse from lesson 4

8. B $e^{2x} = 3e^x$
      $\ln(e^{2x}) = \ln(3e^x)$
      $2x = \ln(3) + \ln(e^x)$
      $2x = \ln(3) + x$
      $x = \ln(3)$

9. D $\ln(2) + \ln(x) = 7$
      $\ln(2x) = 7$
      $e^{\ln(2x)} = e^7$
      $2x = e^7$
      $x = \dfrac{e^7}{2}$

10. A $e^{3x-1} = 1$
       $\ln(e^{3x-1}) = \ln(1)$
       $3x - 1 = 0$
       $3x = 1$
       $x = \dfrac{1}{3}$

Test 7

1. D Limit from the left = 1 and limit from the right = 2. They are not equal, so the limit does not exist.

2. C

   limit from the left and right = 0

3. C $\lim\limits_{x \to \pi} \sin(x) = 0$

4. B $\lim\limits_{x \to 2} \dfrac{x+2}{x+3} = \dfrac{2+2}{2+3} = \dfrac{4}{5}$

5. D $\lim\limits_{x \to 1} \dfrac{3+x}{2-\ln(x)} = \dfrac{3+1}{2-\ln(1)} = \dfrac{4}{2} = 2$

6. B $\lim\limits_{\theta \to 0} \dfrac{2\sec(\theta)}{\theta - 2} = \dfrac{2(1)}{0-2} = \dfrac{2}{-2} = -1$

7. C $\lim\limits_{x \to 2} \dfrac{x^2 + x - 6}{x - 2} =$

   $\lim\limits_{x \to 2} \dfrac{(x+3)(x-2)}{x-2} =$

   $\lim\limits_{x \to 2} x + 3 = 5$

8. D $\lim\limits_{x \to 2} \dfrac{-3}{x-2}$ can't be factored.

   A vertical asymptote occurs at $x = 2$.

9. A $\lim\limits_{x \to 0} \dfrac{e^{2x+1}}{e^x} = \dfrac{e^1}{e^0} = e$

10. A $\lim\limits_{t\to-2} \dfrac{t^3-4t}{t+2} = \lim\limits_{t\to-2} \dfrac{t(t^2-4)}{t+2} =$

$\lim\limits_{t\to-2} \dfrac{t(t+2)(t-2)}{(t+2)} =$

$\lim\limits_{t\to-2} t(t-2) = -2(-2-2) = -2(-4) = 8$

Unit Test I

I. 1.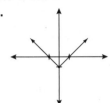

2.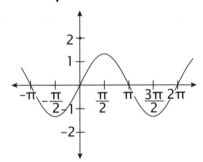

3. $4(x^2+2x)+y^2=12$
$4(x^2+2x+1)+y^2=12+4$
$4(x+1)^2+y^2=16$
$\dfrac{4(x+1)^2}{16}+\dfrac{y^2}{16}=1$
$\dfrac{(x+1)^2}{4}+\dfrac{y^2}{16}=1$

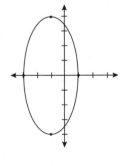

4.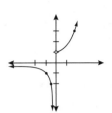

5.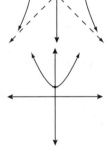

6.

II. No. Problems 3 & 5 are not functions.

III. Yes. Problem 4 has a discontinuity at x = 0.

IV. $|x-2| \geq 1$

When x = 2; 0 ≥ 1 is false, so shade outside of 1 & 3.

V.
1. $\lim\limits_{x\to\infty} 3^x = 3^\infty = \infty$

2. $\lim\limits_{x\to 2} \dfrac{x^2+x-6}{x-2} = \lim\limits_{x\to 2} \dfrac{(x-2)(x+3)}{(x-2)}$
$= \lim\limits_{x\to 2} (x+3) = 5$

3. $\lim\limits_{x\to 0} (\cos(x)-3\sin(x)) = \cos(0)-3\sin(0)$
$= 1$

UNIT TEST I -

VI. $f(x) = 3x + 2$
$y = 3x + 2$
$x = 3y + 2$
$x - 2 = 3y$
$y = \dfrac{x-2}{3}$
$f^{-1}(x) = \dfrac{x-2}{3}$

VII.
1. $\cos\left(\dfrac{2\pi}{3}\right) = \cos 120° = -\dfrac{1}{2}$
2. $\cot\left(\dfrac{3\pi}{4}\right) = \cot 135° = -1$

VIII. $e^{3x+1} = 2$
$3x + 1 = \ln(2)$
$3x = \ln(2) - 1$
$x = \dfrac{\ln(2) - 1}{3}$

Test 8
1. B definition of continuous
2. D
3. A
4. A $\lim\limits_{x\to\infty} \dfrac{-4}{x} =$
   $\lim\limits_{x\to\infty} \dfrac{-4}{\infty} = 0$

5. A $\lim\limits_{\theta\to\frac{\pi}{2}} \dfrac{\cos(\theta)}{\sec(\theta)} =$
   $\lim\limits_{\theta\to\frac{\pi}{2}} \dfrac{\cos(\theta)}{\frac{1}{\cos(\theta)}} = \lim\limits_{\theta\to\frac{\pi}{2}} \cos^2(\theta) = 0$

6. D $\lim\limits_{\theta\to 0} \dfrac{2}{\sin(\theta)}$
   $\sin(\theta) = 0$ so this limit does not exist

7. C $\lim\limits_{x\to\infty} \dfrac{x}{\sqrt{3}} = \infty$

8. C $\lim\limits_{x\to\infty} \dfrac{4 - 3x^3}{7x^3 + 2x^2 - 3} =$
   $\lim\limits_{x\to\infty} \dfrac{\frac{4}{x^3} - 3}{7 + \frac{2}{x} - \frac{3}{x^3}} = -\dfrac{3}{7}$

9. B $\lim\limits_{x\to 4} \dfrac{4-x}{2-\sqrt{x}} \cdot \dfrac{2+\sqrt{x}}{2+\sqrt{x}} =$
   $\lim\limits_{x\to 4} \dfrac{(4-x)(2+\sqrt{x})}{(4-x)} =$
   $\lim\limits_{x\to 4} 2 + \sqrt{x} = 4$

10. B All of these are stated as reasons for the failure of the existence of a limit except when there is a removable discontinuity.

Test 9
1. C
2. B
3. A $\lim\limits_{h\to 0} \dfrac{f(x+h)-f(x)}{h} =$

   $\lim\limits_{h\to 0} \dfrac{4-4}{h} =$

   $\lim\limits_{h\to 0} 0 = 0$

4. A $\lim\limits_{h\to 0} \dfrac{f(x+h)-f(x)}{h} =$

   $\lim\limits_{h\to 0} \dfrac{4x+4h-4x}{h} =$

   $\lim\limits_{h\to 0} \dfrac{4h}{h} =$

   $\lim\limits_{h\to 0} 4 = 4$

5. C
6. B
7. B $\lim\limits_{h\to 0} \dfrac{f(x+h)-f(x)}{h} =$

   $\lim\limits_{h\to 0} \dfrac{[3(x+h)^2-2]-(3x^2-2)}{h} =$

   $\lim\limits_{h\to 0} \dfrac{3x^2+6xh+3h^2-2-3x^2+2}{h} =$

   $\lim\limits_{h\to 0} \dfrac{6xh+3h^2}{h} = \lim\limits_{h\to 0} (6x+3h) = 6x$

8. D $\lim\limits_{h\to 0} \dfrac{f(x+h)-f(x)}{h} =$

   $\lim\limits_{h\to 0} \dfrac{[6-2(x+h)^2]-(6-2x^2)}{h} =$

   $\lim\limits_{h\to 0} \dfrac{6-2x^2-4xh-2h^2-6+2x^2}{h} =$

   $\lim\limits_{h\to 0} \dfrac{-4xh-2h^2}{h} =$

   $\lim\limits_{h\to 0} -4x-2h = -4x$

9. C $\lim\limits_{h\to 0} \dfrac{f(x+h)-f(x)}{h} =$

   $\lim\limits_{h\to 0} \dfrac{[3-5(x+h)]-(3-5x)}{h} =$

   $\lim\limits_{h\to 0} \dfrac{3-5x-5h-3+5x}{h} =$

   $\lim\limits_{h\to 0} \dfrac{-5h}{h} =$

   $\lim\limits_{h\to 0} -5 = -5$

10. D $\lim_{h\to 0} \dfrac{f(x+h)-f(x)}{h} =$

$\lim_{h\to 0} \dfrac{\dfrac{5}{x+h} - \dfrac{5}{x}}{h} =$

$\lim_{h\to 0} \dfrac{\dfrac{5x - 5x - 5h}{x(x+h)}}{h} =$

$\lim_{h\to 0} \dfrac{\dfrac{-5h}{x(x+h)}}{h} =$

$\lim_{h\to 0} \dfrac{-5}{x(x+h)} = \dfrac{-5}{x^2}$

Test 10

1. A. 0
2. B. Power rule
3. B. 1
4. D. $y = 6 - 5x$
   $y' = -5$
5. A. $y = (x^3 + 1)^2$
   $y' = 2(x^3 + 1)(3x^2)$
   $= 6x^2(x^3 + 1)$
6. C. $y = (2x - 3)(5x^4)$
   $y' = (2x - 3)(20x^3) + 5x^4(2)$
   $= 40x^4 - 60x^3 + 10x^4$
   $= 50x^4 - 60x^3$

7. C. $y = \dfrac{x^2}{2x+1}$
   $y' = \dfrac{(2x+1)(2x) - x^2(2)}{(2x+1)^2}$
   $= \dfrac{4x^2 + 2x - 2x^2}{(2x+1)^2}$
   $= \dfrac{2x^2 + 2x}{(2x+1)^2} = \dfrac{2x(x+1)}{(2x+1)^2}$

8. A. $y = (4x)^{\frac{1}{2}}$
   $y' = \dfrac{1}{2}(4x)^{-\frac{1}{2}}(4)$
   $= \dfrac{2}{\sqrt{4x}} = \dfrac{2}{2\sqrt{x}} = \dfrac{1}{\sqrt{x}} = \dfrac{\sqrt{x}}{x}$

9. B. $y = -3x^2$
   $y' = -6x$

10. C. $y = \dfrac{1}{2}x^2 - 4x + 7$
    $Y' = x - 4$

Test 11

1. A In order for the derivative to exist, at x = a, the function must be continuous at x = a. iii is the definition of the derivative.
2. D All of these problems were discussed in the instruction manual as places where the derivative does not exist.
3. D See instruction manual to verify.
4. C $\dfrac{dy}{dx} = \dfrac{dy}{du} \cdot \dfrac{du}{dx}$
   $= (-6u)(3) = -18u$
   $= -18(3x) = -54x$
5. A $\dfrac{dy}{dx} = \dfrac{dy}{du} \cdot \dfrac{du}{dx}$
   $= (2u)\left[\dfrac{1}{2}(4x)^{-\frac{1}{2}}(4)\right]$
   $= \dfrac{4u}{\sqrt{4x}} = \dfrac{4\sqrt{4x}}{\sqrt{4x}} = 4$

6. C $\quad \dfrac{dy}{dx} = \dfrac{dy}{du} \cdot \dfrac{du}{dr} \cdot \dfrac{dr}{dx}$

$= (7)(4r)\left(\dfrac{1}{2}\right)$

$= 14r = 14\left(\dfrac{1}{2}x + 1\right)$

$= 7x + 14$

7. B $\quad y = -4x^3 + 8x^2 + 17$

$y' = -12x^2 + 16x$

$y'' = -24x + 16$

8. D $\quad f(x) = \dfrac{(x-2)^{\frac{2}{3}}}{(x+1)}$

$f'(x) = \dfrac{(x+1) \cdot \frac{d}{dx}(x-2)^{\frac{2}{3}} - (x-2)^{\frac{2}{3}} \cdot \frac{d}{dx}(x+1)}{(x+1)^2}$

$f'(x) = \dfrac{(x+1) \cdot \frac{2}{3}(x-2)^{-\frac{1}{3}} - (x-2)^{\frac{2}{3}}}{(x+1)^2}$

$f'(x) = \dfrac{(x-2)^{-\frac{1}{3}}\left[\frac{2}{3}(x+1) - (x-2)\right]}{(x+1)^2}$

$f'(x) = \dfrac{\frac{2}{3}x + \frac{2}{3} - x + 2}{(x-2)^{\frac{1}{3}}(x+1)^2}$

$f'(x) = \dfrac{-\frac{1}{3}x + \frac{8}{3}}{(x-2)^{\frac{1}{3}}(x+1)^2}$

$x = 2$ and $x = -1$ are not differentiable.

9. A There are corners at $x = -4$ and $x = 2$.
There is a discontinuity at $x - 3$.

10. B $\quad y = \sqrt{x} = x^{\frac{1}{2}}$

$y' = \dfrac{1}{2}x^{-\frac{1}{2}} = \dfrac{1}{2\sqrt{x}}$

$y'' = \dfrac{1}{2}\left(-\dfrac{1}{2}x^{-\frac{3}{2}}\right) = -\dfrac{1}{4\left(x^{\frac{3}{2}}\right)}$

The derivative does not exist when $x = 0$ for either the first or second derivative. Since square roots must be positive in this course, answer D will not work.

Test 12

$f(x) = \sin(x) \qquad f(x) = \cos(x)$
$f'(x) = \cos(x) \qquad f'(x) = -\sin(x)$
$f''(x) = -\sin(x) \qquad f''(x) = -\cos(x)$
$f'''(x) = -\cos(x) \qquad f'''(x) = \sin(x)$
$f^4(x) = \sin(x) \qquad f^4(x) = \cos(x)$

1. C You must first rewrite the problems.

$f(x) = [2x \cdot \sin(x+1)]^{\frac{1}{2}}$

and apply the power rule

2. B $\quad f(x) = \sin(x)$

$f'(x) = \cos(x)$

$\cos(x)$ is 0 when $x = \dfrac{\pi}{2}$ in $[0, \pi]$

3. D $\quad f(x) = \cos x^2$

$f'(x) = -\sin x^2 \cdot 2x$

$= -2x \sin(x^2)$

4. B $\quad f(x) = 2x \tan(x^2)$

$f'(x) = 2x(\sec^2(x^2))(2x) + (\tan(x^2))(2)$

$= 2[2x^2 \sec^2(x^2) + \tan(x^2)]$

5. A $\quad f(x) = \sin(3x)$

$f'(x) = (\cos(3x))3 = 3\cos(3x)$

$f''(x) = 3(-\sin(3x) \cdot 3) = -9\sin(3x)$

6. B $f(x) = 2\cot(x) - \csc(2x)$
$f'(x) = -2\csc^2(x) + (\csc(2x) \cdot \cot(2x))2$
$= -2[\csc^2(x) - \csc(2x) \cot(2x)]$

7. A $f(x) = 2\csc(2x)$
$f'(x) = 2(-\csc(2x) \cot(2x))(2)$
$= -4\csc(2x) \cot(2x)$

8. D $f(x) = (\csc(3x))^3$
$f'(x) = 3(\csc(3x))^2(-\csc(3x) \cot(3x))(3)$
$= -9(\csc(3x))^2 \csc(3x) \cot(3x)$

9. C $f(x) = \dfrac{\sin(x)+1}{1-\cos(x)}$
$f'(x) = \dfrac{(1-\cos(x))(\cos(x)) - (\sin x + 1)(\sin(x))}{(1-\cos(x))^2}$
$= \dfrac{\cos(x) - \cos^2(x) - \sin^2(x) - \sin(x)}{(1-\cos(x))^2}$
$= \dfrac{\cos(x) - \sin(x) - (\cos^2(x) + \sin^2(x))}{(1-\cos(x))^2}$
$= \dfrac{\cos(x) - \sin(x) - 1}{(1-\cos(x))^2}$

10. A $f(x) = \dfrac{1}{2}\sec(x^2+2)$
$f'(x) = \dfrac{1}{2}\sec(x^2+2)\tan(x^2+2)(2x)$
$= x \sec(x^2+2) \tan(x^2+2)$

Test 13

1. C
$f(x) = \sin(x) \qquad f(x) = \cos(x)$
$f'(x) = \cos(x) \qquad f'(x) = -\sin(x)$
$f''(x) = -\sin(x) \qquad f''(x) = -\cos(x) \qquad f(x) = e^x$
$f'''(x) = -\cos(x) \qquad f'''(x) = \sin(x) \qquad f''''(x) = e^x$
$f''''(x) = \sin(x) \qquad f''''(x) = \cos(x)$

2. D $f'(x) = e^{\sqrt{2x}} \cdot \dfrac{d}{dx}(\sqrt{2x})$
$f'(x) = e^{\sqrt{2x}} \cdot \dfrac{1}{2}(2x)^{-\frac{1}{2}} \cdot 2 = \dfrac{e^{\sqrt{2x}}}{\sqrt{2x}}$

3. A $f'(x) = e^x(\sec(x)\tan(x)) + \sec(x) \cdot e^x$
$= e^x \sec(x)(\tan(x)+1)$

4. B $f'(x) = \dfrac{\cos(x) \cdot e^x - e^x(-\sin(x))}{\cos^2(x)}$
$= \dfrac{e^x(\cos(x) + \sin(x))}{\cos^2(x)}$

5. B $f'(x) = e^{2x} \cdot 2 - \cos 2x \cdot 2 - \sin 2x \cdot 2$
$= 2(e^{2x} - \cos 2x - \sin 2x)$

6. B

7. B $f(x) = \ln(2x)$
$f'(x) = \dfrac{1}{2x} \cdot 2 = \dfrac{1}{x}$

8. A $f'(x) = \dfrac{1}{\cos(2x)} \cdot \dfrac{(-\sin(2x))(2)}{1}$
$= \dfrac{-2\sin(2x)}{\cos(2x)}$
$= -2\tan(2x)$

9. D $f'(x) = \dfrac{x \cdot \dfrac{1}{3x}(3) - \ln(3x)(1)}{x^2}$
$= \dfrac{1 - \ln 3x}{x^2}$

10. D $f'(x) = \dfrac{1}{e^{2x} + e^{4x}} \cdot \dfrac{d}{dx}(e^{2x} + e^{4x})$
$f'(x) = \dfrac{2e^{2x} + 4e^{4x}}{e^{2x} + e^{4x}}$
$= \dfrac{e^{2x}(2 + 4e^{2x})}{e^{2x}(1 + e^{2x})}$
$= \dfrac{2 + 4e^{2x}}{1 + e^{2x}}$

Test 14

1. **B** $y^2 = x$
   $2yy' = 1$
   $y' = \dfrac{1}{2y}$

2. **D** $2 \cdot \dfrac{d}{dx} y = 2y'$

3. **C** a line which is perpendicular to the tangent line

4. **A** $2y^3 - 3y^2 + y = 2x$
   $6y^2 y' - 6yy' + y' = 2$
   $y'(6y^2 - 6y + 1) = 2$
   $y' = \dfrac{2}{6y^2 - 6y + 1}$

5. **C** $\tan(2y) = 2x^2$
   $\sec^2(2y)(2)y' = 4x$
   $2\sec^2(2y) \cdot y' = 4x$
   $y' = \dfrac{4x}{2\sec^2(2y)}$
   $y' = \dfrac{2x}{\sec^2(2y)}$

6. **D** $y^3 - 2y = 2x^2$
   $3y^2 y' - 2y' = 4x$
   $y'(3y^2 - 2) = 4x$
   $y' = \dfrac{4x}{3y^2 - 2}$
   At $(1, 2)$: $y' = \dfrac{4(1)}{3(2)^2 - 2}$
   $= \dfrac{4}{12 - 2} = \dfrac{4}{10} = \dfrac{2}{5}$

7. **A** $x^3 + 2y^2 = 5$
   $3x^2 + 4yy' = 0$
   $4yy' = -3x^2$
   $y' = \dfrac{-3x^2}{4y}$

8. **B** $\ln(y) + \ln(x) = x$
   $\dfrac{1}{y} y' + \dfrac{1}{x} = 1$
   $\dfrac{1}{y} y' = 1 - \dfrac{1}{x}$
   $y' = y - \dfrac{y}{x}$
   $y' = \dfrac{xy}{x} - \dfrac{y}{x} = \dfrac{xy - y}{x}$

9. **D** $y^2 - 3y^3 + y = 2x$
   $2yy' - 9y^2 y' + y' = 2$
   $y'(2y - 9y^2 + 1) = 2$
   $y' = \dfrac{2}{2y - 9y^2 + 1}$

10. **C**
    $x^3 - Axy + 3Ay^2 = 3A^3$ at (A, A)
    $3x^2 - A(xy' + y) + 6Ayy' = 0$
    $3x^2 - Axy' - Ay + 6Ayy' = 0$
    $y'(6Ay - Ax) = Ay - 3x^2$
    $y' = \dfrac{Ay - 3x^2}{6Ay - Ax}$
    $y'(A, A) = \dfrac{A(A) - 3(A)^2}{6A(A) - A(A)}$
    $= \dfrac{A^2 - 3A^2}{6A^2 - A^2}$
    $= \dfrac{-2A^2}{5A^2}$
    $= -\dfrac{2}{5}$

Unit Test II

I.

1. $y = 10x^3 - 4x^2 + 7$
   $y' = 30x^2 - 8x$

2. $y = [\sin(5x)]^3$
   $y' = 3(\sin(5x))^2 \cdot \frac{d}{dx}(\sin(5x))$
   $y' = 3(\sin(5x))^2 \cdot [\cos(5x)] \cdot 5$
   $y' = 15(\sin^2(5x))(\cos(5x))$

3. $y = e^{4x} + \sec(4x)$
   $y' = (e^{4x}) \cdot 4 + [\sec(4x)\tan(4x)] \cdot 4$
   $y' = 4[e^{4x} + \sec(4x)\tan(4x)]$

4. $y = \dfrac{e^{2x}}{\ln(2x)}$
   $y' = \dfrac{\ln(2x) \cdot (e^{2x}) \cdot 2 - e^{2x} \cdot \frac{1}{2x} \cdot 2}{(\ln(2x))^2}$
   $y' = \dfrac{e^{2x}\left[2\ln(2x) - \frac{1}{x}\right]}{\ln^2(2x)}$

5. $2x + 3y = xy$
   $2 + 3y' = xy' + y$
   $3y' - xy' = y - 2$
   $y' = \dfrac{y-2}{3-x}$

II.

1. $y = \dfrac{2}{x}$
   $y' = \lim\limits_{h \to 0} \dfrac{f(x+h) - f(x)}{h}$
   $y' = \lim\limits_{h \to 0} \dfrac{\frac{2}{x+h} - \frac{2}{x}}{h}$
   $y' = \lim\limits_{h \to 0} \dfrac{\frac{2x - 2(x+h)}{x(x+h)}}{h}$
   $y' = \lim\limits_{h \to 0} \dfrac{\frac{-2h}{x(x+h)}}{h}$
   $y' = \lim\limits_{h \to 0} \dfrac{-2}{x(x+h)} = -\dfrac{2}{x^2}$

III.

1. $\lim\limits_{x \to \infty} 3x^{-1} = \lim\limits_{x \to \infty} \dfrac{3}{x} = \dfrac{3}{\infty} = 0$

2. $\lim\limits_{x \to \infty} \dfrac{x^2 - 4}{3x^2 + 1} = \lim\limits_{x \to \infty} \dfrac{1 - \frac{4}{x^2}}{3 + \frac{1}{x^2}} = \dfrac{1}{3}$

3. $\lim\limits_{x \to \frac{\pi}{2}} \dfrac{2(\csc(x) + \cot(x))}{1 + \cos(x)}$
   $= \lim\limits_{x \to \frac{\pi}{2}} 2\left(\dfrac{\frac{1}{\sin(x)} + \frac{\cos(x)}{\sin(x)}}{1 + \cos(x)}\right)$
   $= \lim\limits_{x \to \frac{\pi}{2}} 2\left(\dfrac{\frac{1 + \cos(x)}{\sin(x)}}{1 + \cos(x)}\right)$
   $= \lim\limits_{x \to \frac{\pi}{2}} 2\left(\dfrac{1}{\sin(x)}\right) = 2$

4. $\lim\limits_{x \to 4} \dfrac{x-4}{\sqrt{x} - 2} = \lim\limits_{x \to 4} \dfrac{(x-4)(\sqrt{x}+2)}{(\sqrt{x}-2)(\sqrt{x}+2)}$
   $= \lim\limits_{x \to 4} \dfrac{(x-4)(\sqrt{x}+2)}{(x-4)}$
   $= \lim\limits_{x \to 4} (\sqrt{x} + 2) = 4$

IV. $f(x) = \dfrac{x^2+3x+2}{x^2-x-6} = \dfrac{(x+1)(x+2)}{(x-3)(x+2)}$

There is a hole at x = -2 because there is a removal discontinuity there. There will be a vertical asymptote at x = 3.

$$\lim_{x\to\infty} \dfrac{x^2+3x+2}{x^2-x-6} = \lim_{x\to\infty} \dfrac{1+\dfrac{3}{x}+\dfrac{2}{x^2}}{1-\dfrac{1}{x}-\dfrac{6}{x^2}} = 1$$

Therefore there is an horizontal asymptote at y = 1.

V. $y = 3u^2;\ u = -2t;\ t = x+1$

$$\dfrac{dy}{dx} = (6u)(-2)(1)$$
$$= -12u$$
$$= -12(-2t)$$
$$= 24t$$
$$= 24(x+1)$$

VI. $f(x) = x^6 - 5x^4 - 3$
$f'(x) = 6x^5 - 20x^3$
$f''(x) = 30x^4 - 60x^2$
$= 30x^2(x^2-2)$

$f''(x) = 0$ when $x = 0,\ \pm\sqrt{2}$

VII. Graphs a, b, d, & e are not differentiable at x = 0. Graph a is not continuous at x = 0. Graphs b and d come to a point, and graph e does not exist when x = 0.

a. $y = \sqrt{x}$

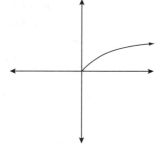

b. $y = \sqrt{|x|}$

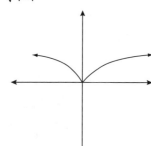

c. $y = x$

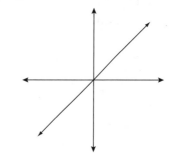

d. $y = |x|$

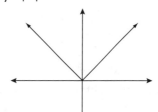

e. $y = \dfrac{1}{x}$

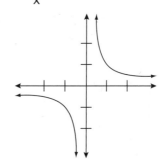

Test 15
1. A
2. B
3. C
4. D
5. B

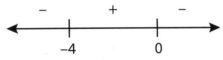

minimum when x = −4
maximum when x = 0

6. B: $f(x) = x^{\frac{2}{5}}$

$f'(x) = \frac{2}{5}x^{-\frac{3}{5}} = \frac{2}{x^{\frac{3}{5}}}$

Critical point at x = 0

7. D: $f(x) = (9-x^2)^{\frac{1}{2}}$

$f'(x) = \frac{1}{2}(9-x^2)^{-\frac{1}{2}}(-2x)$

$= \frac{-x}{\sqrt{9-x^2}}$

Critical points: $x = 0$, $9-x^2 = 0$
$x = 0, 3, -3$

8. B: $f(x) = \sin(x) [0, \pi]$
   From the graph we have
   a global or absolute max at $x = \frac{\pi}{2}$
   and two global mins at $x = 0$, $x = \pi$.

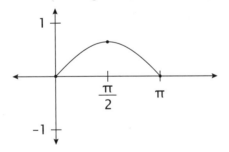

9. C: When $x^2 - 1 = 0$ there are vertical asymptotes $x = -1$, +1 are vertical asymptotes.

$\lim_{x \to \infty} \frac{2x^2 + 2x}{x^2 - 1} = \lim_{x \to \infty} \frac{2 + \frac{2}{x}}{1 - \frac{1}{x}} = 2$

There is a horizontal asymptote at y = 2.

10. A: $f(x) = 3x^2 + bx + c$
$f'(x) = 6x + b$
$6x + b = 0$
$6x = -b$
$x = \frac{-b}{6}$

$(1) = \frac{-b}{6}$
$6 = -b$
$b = -6$

$f'(x)$ becomes $6x - 6$, and there is a minimum at $x = 1$ because to the left of 1 the derivative is negative, and to the right it is positive.

Test 16
1. B. The inflection has to change on either side of x = a in order to be an inflection point.
2. B. When the second derivative is less than 0 at a critical point, we have a maximum.
3. C

```
           x + 1
     ┌─────────────
x − 1 │ x² + 0x + 1
        x² −  x
        ─────────
             x + 1
             x − 1
             ─────
               2 R
```

4. B

$$8x + b = 0$$
$$8x = -b$$
$$x = -\frac{b}{8}$$
$$\frac{1}{2} = -\frac{b}{8}$$
$$b = -4$$

Check: $f''(x) = 8$
Therefore when $b = -4$ there is a local minimum at $x = \frac{1}{2}$.

5. C $f(x) = \dfrac{1}{2-x}$ on $[0, 3]$

$$f'(x) = -(2-x)^{-2}(-1) = \frac{1}{(2-x)^2}$$
$$f''(x) = -2(2-x)^{-3}(-1) = \frac{2}{(2-x)^3}$$

critical pt: $x = 2$

x	$f(x)$	$f'(x)$	Conclusions
0	$\frac{1}{2}$	+	
2	und	und	vertical asymptote
3	−1	+	

There are no absolute max or min values.

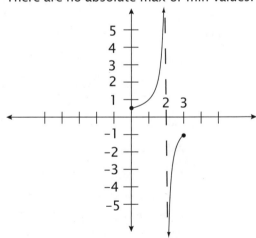

It can be seen from the graph that there is no absolute max or min.

6. B $f(x) = 3x^4 - 4x^3 + 2$
$$f'(x) = 12x^3 - 12x^2$$
$$= 12x^2(x-1)$$

critical pts $x = 0, 1$

The minimum must exist at one of the two critical points.

$$f''(x) = 36x^2 - 24x$$
$$= 12x(3x - 2)$$

$f''(0) = 0$
That means $x = 0$ is a possible inflection pt.

$f''(1) =$ positive number, so $x = 1$ is a minimum

7. A $f(x) = \dfrac{x^4}{12} - \dfrac{x^3}{3}$

$$f'(x) = \frac{1}{12} \cdot 4x^3 - \frac{1}{3} \cdot 3x^2$$
$$= \frac{1}{3}x^3 - x^2$$
$$= x^2\left(\frac{1}{3}x - 1\right) \quad \text{critical pts } x = 0, 3$$
$$f''(x) = \frac{1}{3} \cdot 3x^2 - 2x$$
$$= x^2 - 2x$$
$$= x(x-2)$$

possible inflection pts $x = 0, 2$

x	$f(x)$	$f'(x)$	$f''(x)$	
−1	$\frac{5}{12}$	−	+	
0	0	0	0	Inflection point
1	$-\frac{1}{4}$	−	−	
2	$-\frac{4}{3}$	−	0	Inflection point
3	$-\frac{9}{4}$	0	+	

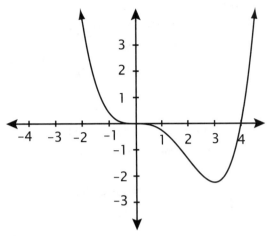

8. D

$f(x) = 5x - x^5$
$f'(x) = 5 - 5x^4$
$\quad = 5(1 - x^4)$
$\quad = 5(1-x)(1+x)(1+x^2)$ critical pts $x = \pm 1$
$f''(x) = -20x^3$
$f''(1) = -20 \quad x = 1$ is a maximum
$f''(-1) = +20 \quad x = -1$ is a minimum

As $x \to \infty$, $f(x)$ grows infinitely larger and negative.

As $x \to -\infty$, $f(x)$ grows infinitely larger and positive.

Therefore, $x = 1$ is a local maximum and $x = -1$ is a local minimum.

9. A $\quad f(x) = \dfrac{6x}{x^2+3}$

$f'(x) = \dfrac{(x^2+3)(6) - 6x(2x)}{(x^2+3)^2}$

$\quad = \dfrac{6x^2 + 18 - 12x^2}{(x^2+3)^2}$

$\quad = \dfrac{18 - 6x^2}{(x^2+3)^2}$

$\quad = \dfrac{6(3 - x^2)}{(x^2+3)^2}$

critical points $x = \pm\sqrt{3}$

$\lim\limits_{x \to \infty} \dfrac{6x}{x^2+3} = 0$ so $y = 0$
is an horizontal asymptote.

x	f(x)	f'(x)	f''(x)	conclusion
-2	$-\dfrac{12}{7} \approx 1.71$	$-$		
$-\sqrt{3}$	$-\sqrt{3} \approx 1.73$	0	$+$	min-absolute
0	0	$+$	0	
$\sqrt{3}$	$\sqrt{3} \approx 1.73$	0	$-$	max-absolute
2	$\dfrac{12}{7} \approx 1.71$	$-$		

$f''(x) = \dfrac{(x^2+3)^2(-12x) - 6(3-x^2)2(x^2+3)(2x)}{(x^2+3)^4}$

$\quad = \dfrac{(-12x)(x^2+3)[x^2+3+12(3-x^2)]}{(x^2+3)^4}$

$f''(\sqrt{3}) = -$
$f''(-\sqrt{3}) = +$
$x = -\sqrt{3}$ would be a local minimum and
$x = +\sqrt{3}$ would be a local maximum.
By plotting a few points it can be shown that the local min and local max are both indeed absolute.

10. D $f(x) = 3x - x^3$
         $f'(x) = 3 - 3x^2$
              $= 3(1 - x^2)$
              $= 3(1-x)(1+x)$
         critical points $x = \pm 1$
         $f''(x) = -6x$
         possible inflection point at $x = 0$

x	$f(x)$	$f'(x)$	$f''(x)$	
-1	-2	0	+	min-local
0	0	+	0	inflection point
1	2	0	-	max-local

min at $x = -1$
max at $x = 1$
inflection point at $x = 0$
D is the correct shape.

Test 17
1. B   All the exponents are even.å
2. D $f(-x) = \dfrac{3}{-2x+1}$
        $-f(x) = -\dfrac{3}{2x+1}$
        These are not equal, so the function is neither even nor odd.
3. D
4. A
5. D $x^2 - 3x$ [0, 1] Slope of segment
        $f(0) = 0$ between $(0, 0)$ and $(1, -2)$
        $f(1) = 1 - 3 = -2$ $m = \dfrac{-2-0}{1-0} = \dfrac{-2}{1} = -2$
        $f'(x) = 2x - 3$
        $2x - 3 = -2$
        $2x = 1$
        $x = \dfrac{1}{2}$

6. B $\lim\limits_{x \to 1} \dfrac{x^4 - 1}{x - 1}$ Has form $\dfrac{0}{0}$ Use LR
        $= \lim\limits_{x \to 1} \dfrac{4x^3}{1} = 4$

7. C $\lim\limits_{\theta \to 0} \dfrac{2\sin 2\theta}{\tan 3\theta}$ Has form $\dfrac{0}{0}$ Use LR
        $= \lim\limits_{\theta \to 0} \dfrac{4\cos 2\theta}{3\sec^2 3\theta} = \dfrac{4}{3}$

8. B $\lim\limits_{x \to 3} \dfrac{2x^2 - x - 15}{x - 3}$ Has form $\dfrac{0}{0}$ Use LR
        $= \lim\limits_{x \to 3} \dfrac{4x - 1}{1} = 11$

9. D $\lim\limits_{x \to \infty} \dfrac{x^3 + 1}{2\sqrt{x}}$ Has form $\dfrac{\infty}{\infty}$ Use LR
        $= \lim\limits_{x \to \infty} \dfrac{3x^2}{2\left(\dfrac{1}{2} x^{-\tfrac{1}{2}}\right)}$
        $= \lim\limits_{x \to \infty} \dfrac{3x^2 \sqrt{x}}{1} = \infty$

10. A $\lim\limits_{x \to \infty} \dfrac{\sin 3x}{e^{3x}}$
        Has form $\dfrac{\text{small number}}{\infty} = 0$

Test 18
1. B
2. C $v(t) = d'(t)$
               $= d''(t)$
        $a(t)$
3. D $0 = 400 - 16t^2$
        $400 = 16t^2$
        $t = 5$
4. C $v(t) = -32t$
        $v(5) = -32(5) = -160$
5. C $d(3) = 400 - 16(3)^2$
              $= 256$

6. A The initial velocity in problem 3 was 0 ft/sec.

7. D Since a = 32 ft/sec^2 and v_0 = 48 ft/sec we have:

$$d = 48t - \left(\frac{1}{2}\right)32t^2$$
$$= 48t - 16t^2$$

The 32 is negative because the force of gravity is downward.

8. B $v(t) = 64 - 32t = 0$
$64 = 32t$
$t = 2$ sec.

9. B $s(t) = \frac{3}{5}t^5 - 100$
$v(t) = 3t^4$
$a(t) = 12t^3$

10. D $s(t) = \frac{3}{5}t^5 - 100$
$s(10) = \frac{3}{5}(100,000) - 100$
$= 59,900$ m
$= 59.9$ km

Test 19

1. A $C'(x) = .004x + 25$

2. A $P(x) = R(x) - C(x)$

$= (75x - .008x^2) - (40,000 + 25x + .002x^2)$
$= -40,000 + 50x - .01x^2$

3. C $P'(x) = 50 - .02x = 0$
$x = 2,500$

4. A The cost of the 1,001st item is $C'(1,000)$
$C'(x) = 25 + .004x$
$C'(1,000) = 25 + .004(1,000) = 29$

5. A Profit = Revenue - Cost
$P(x) = 200x - \frac{1}{30}x^2 - 72,000 - 60x$
$P(x) = 140x - \frac{1}{30}x^2 - 72,000$

$P(100) = 14,000 - \left(\frac{1}{30}\right)10,000 - 72,000$
$= -58,333$

6. D $P'(x) = 140 - \left(\frac{1}{15}\right)x = 0$
$x = 2,100$

7. C The cost function is the fixed costs plus the variable costs.
$C(x) = 6,000 + 2x$

8. C The break-even point is where the cost equals the revenue.
$R(x) = 5x$
$C(x) = 6,000 + 2x$
$6,000 + 2x = 5x$
$x = 2,000$

9. B $C(x) = 20x + 10,000$
$C(100) = 20(100) + 10,000 = \$12,000$

10. A $R(x) = 80x$
$P(x) = 80x - 20x - 10,000$
$P(x) = 60x - 10,000$

$P(200) = 60(200) - 10,000 = \$2,000$

Test 20

1. C: $2x + 2y = 40$

The constraint equation equals a real number.
Max area = xy is the optimization equation.

2. A Max $A = xy$ $2y = 40 - 2x$
   $= x(20 - x)$ $y = 20 - x$
   $= 20x - x^2$
   $A' = 20 - 2x$
   $x = 10$
   $A'' = -2$
   $A''(10) =$ negative, so it is a max.

   $2(10) + 2y = 40$, so $y = 10$ also.

3. B $x + y = 8$ $x \geq 0$; $y \geq 0$
   $P = x^2 y^2$
   $= x^2(8-x)^2$
   $= x^2(64 - 16x + x^2)$
   $= 64x^2 - 16x^3 + x^4$
   $P' = 128x - 48x^2 + 4x^3$
   $= 4x(32 - 12x + x^2)$
   $= 4x(8-x)(4-x)$
   $x = 0, 4, 8$

4. C When $x = 0$ or $x = 8$ the product is 0.
   When $x = 4$, the product is 256.
   Therefore $x = 4$ yields the largest product.
   Since $x = 4$, then $y = 4$.
   $P = 16 \cdot 16 = 256$

   ```
   -     +     -     +    P'
   ―――+―――――+―――――+―――
      0     4     8
   ```

5. C $(0, 3)(x, x^2 + 1)$
   $= \sqrt{(x-0)^2 + [(x^2+1)-3]^2}$
   $= \sqrt{x^2 + (x^2 - 2)^2}$
   $= \sqrt{x^2 + x^4 - 4x^2 + 4}$
   $= \sqrt{x^4 - 3x^2 + 4}$

6. D $d = \sqrt{x^4 - 3x^2 + 4}$
   The x value for a minimum d would be the same as the x value for a minimum d^2, so we can square both sides:
   Min $d^2 = x^4 - 3x^2 + 4$
   $(d^2)' = 4x^3 - 6x$
   $= 2x(2x^2 - 3)$
   $x = 0$ $2x^2 = 3$
                  $x^2 = \frac{3}{2}$
                  $x = \pm\sqrt{\frac{3}{2}}$
                  $= \pm\frac{\sqrt{6}}{2}$

   So critical points are $0, \pm\frac{\sqrt{6}}{2}$

7. D $d'' = 12x^2 - 6$
   $d''\left(\sqrt{\frac{3}{2}}\right) = 12\left(\sqrt{\frac{3}{2}}\right)^2 - 6$
   $= 12\left(\frac{3}{2}\right) - 6$
   $= 18 - 6 = 12$
   so $\sqrt{\frac{3}{2}} = \frac{\sqrt{6}}{2}$ is a minimum
   $d'' = \left(-\sqrt{\frac{3}{2}}\right)$
   $= 12\left(-\sqrt{\frac{3}{2}}\right)^2 - 6 = 12$
   so $-\sqrt{\frac{3}{2}} = -\frac{\sqrt{6}}{2}$ is also a minimum
   Points:
   $\left(\frac{\sqrt{6}}{2}, \left(\frac{\sqrt{6}}{2}\right)^2 + 1\right)$
   $\left(\frac{\sqrt{6}}{2}, \frac{3}{2} + 1\right) = \left(\frac{\sqrt{6}}{2}, 2.5\right)$
   $\left(-\frac{\sqrt{6}}{2}, \left(-\frac{\sqrt{6}}{2}\right)^2 + 1\right) = \left(-\frac{\sqrt{6}}{2}, 2.5\right)$

8. B   V = Area of base · height
    $= \pi r^2 h$
    $20\pi = \pi r^2 h$
    $20 = r^2 h$

9. D $h = \dfrac{20}{r^2}$

    C = Cost of bottom + cost of top + cost of side
    $= 10(\pi r^2) + 10(\pi r^2) + 6(2\pi rh)$
    $= 20\pi r^2 + 12\pi rh$
    $= 20\pi r^2 + 12\pi r\left(\dfrac{20}{r^2}\right)$
    $= 20\pi r^2 + \dfrac{240\pi}{r}$

10. D $C' = 40\pi r + \dfrac{240\pi}{-r^2} = 0$
    $40\pi r = \dfrac{240\pi}{r^2}$
    $r = \sqrt[3]{6}$
    $h = 20 \cdot 6^{-\frac{2}{3}}$

Test 21

1. C
2. A
3. D $A = \pi r^2$
    $\dfrac{dA}{dt} = 2\pi r \dfrac{dr}{dt}$
    $\dfrac{dA}{dt} = 2\pi(20)\left(\dfrac{2}{3}\right) = \dfrac{80\pi}{3} \dfrac{in^2}{sec}$
4. B

5. A $\dfrac{1}{R} = \dfrac{1}{50} + \dfrac{1}{60}$
    $\dfrac{1}{R} = \left(\dfrac{6}{6}\right)\dfrac{1}{50} + \left(\dfrac{5}{5}\right)\dfrac{1}{60}$
    $\dfrac{1}{R} = \left(\dfrac{6+5}{300}\right) = \dfrac{11}{300}$
    $R = \dfrac{300}{11}$

    $-\dfrac{1}{R^2}R' = -\dfrac{1}{(R_1)^2}(R_1') - \dfrac{1}{(R_2)^2}(R_2')$

    $R' = R^2\left[\dfrac{R_1'}{(R_1)^2} + \dfrac{R_2'}{(R_2)^2}\right]$

    $R' = \left(\dfrac{300}{11}\right)^2\left[\dfrac{(-.4)}{50^2} + \dfrac{(.2)}{60^2}\right]$

    $R' = (743.80)(-.00016 + .000056)$

    $R' \approx -.08 \dfrac{ohm}{min}$

    Therefore R is decreasing at a rate of .08 ohm/min.

6. A $V = \dfrac{4}{3}\pi r^3$
    $\dfrac{dV}{dt} = \dfrac{4}{3}\pi(3r^2)\dfrac{dr}{dt}$
    $= 4\pi r^2 \dfrac{dr}{dt}$
    $10 = 4\pi(5)^2 \dfrac{dr}{dt}$
    $10 = 100\pi \dfrac{dr}{dt}$
    $\dfrac{dr}{dt} = \dfrac{1}{10\pi}$

7. A   Surface Area $= 4\pi r^2$
    $\dfrac{ds}{dt} = 8\pi r \dfrac{dr}{dt}$
    $\dfrac{ds}{dt} = 8\pi(5)\left(\dfrac{1}{10\pi}\right)$
    $= 4 \dfrac{in^2}{min}$

8. A   y = distance from the top of ladder to ground
    x = distance from bottom of the ladder to the wall

$x^2 + y^2 = 100$

When $x = 2$, $y^2 = 100 - 4$
$y = \sqrt{96}$

$2x\dfrac{dx}{dt} + 2y\dfrac{dy}{dt} = 0$

$\dfrac{dx}{dt} = \dfrac{-2y\dfrac{dy}{dt}}{2x}$

$= \dfrac{-2\sqrt{96}(-5)}{2(2)}$

$= \dfrac{10\sqrt{96}}{4}$

$= \dfrac{40\sqrt{6}}{4} = 10\sqrt{6} \,\dfrac{\text{ft}}{\text{sec}}$

9. B
10. D

Unit Test III
I.
1. $f'(x) = 12x^3 - 12x^2 - 24x$
   $= 12x(x+1)(x-2)$
   critical points: $x = 0, -1, 2$

2.

   $f(x)$ is decreasing on $(-\infty, -1)$ and $(0, 2)$.
   $f(x)$ is increasing on $(-1, 0)$ and $(2, \infty)$.

3. From #2 we see that $f(x)$ has a minimum at $x = -1$ and $x = 2$. $f(x)$ has a maximum at $x = 0$.

4. $f''(x) = 36x^2 - 24x - 24$
   $0 = 3x^2 - 2x - 2$
   This cannot be factored.
   Using the quadratic formula we get:
   $x = \dfrac{2 \pm \sqrt{4 - 4(3)(-2)}}{2(3)}$
   $= \dfrac{2 \pm \sqrt{28}}{6}$
   $= \dfrac{1 \pm \sqrt{7}}{3}$

5. iii. The graph begins by decreasing from $-\infty$ to -1 and then alternates back and forth three more times.

II. Using the Pythagorean theorem:
$x^2 + y^2 = z^2$, $z = \sqrt{x^2 + y^2}$

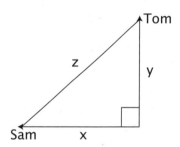

Differentiating with respect to t implicitly we have:

$2x\dfrac{dx}{dt} + 2y\dfrac{dy}{dt} = 2z\dfrac{dz}{dt}$

We know that when:

$y = .6$; $\dfrac{dy}{dt} = 20$ mph

and $x = .8$, $\dfrac{dx}{dt} = 25$ mph

$2(.8)(25) + 2(.6)(20) = 2\left(\sqrt{.8^2 + .6^2}\right)\dfrac{dz}{dt}$

$40 + 24 = 2\dfrac{dz}{dt}$

$\dfrac{dz}{dt} = 32$

At the instant in question, the distance between the trucks is increasing at 32 mph.

III. $x = $ 1st number
$10 - x = $ 2nd number
Max $f(x) = x(10-x)$
$= 10x - x^2$
$f'(x) = 10 - 2x$
$x = 5$ is the critical point
$f''(x) = -2$ so $x = 5$ is a maximum.

The two numbers are 5 and 5 and the product is 25.

IV.
1. $R(x) = x \cdot p(x) = 50x - .01x^2$
2. $P(x) = R(x) - C(x)$
$= 50x - .01x^2 - 1,000 - 39x$
$P(x) = -.01x^2 + 11x - 1,000$
3. $P'(x) = -.02x + 11 = 0$
$x = 550$ games to maximize profit because $P'' = -.02$.
4. The break even points occur when $R(x) = C(x)$:
$50x - .01x^2 = 1,000 + 39x$
$-.01x^2 + 11x - 1,000 = 0$
$x^2 - 1,100x + 100,000 = 0$
$(x - 100)(x - 1,000) = 0$
$x = 100;\ x = 1,000$
5. $P(550) = -.01(550)^2 + 11(550) - 1,000$
$= \$2,025$
6. $C'(x) = 39$
7. $p(550) = 50 - .01(550) = \$44.50$

V.
1. $d(0) = 18$ meters
2. $v(t) = 60 - 10t$
3. $a(t) = -10$
4. When velocity $= 0$, the height is at its peak.
$60 - 10t = 0$ when $t = 6$ seconds
5. $d(6) = 60(6) - 5(6)^2 + 18 = 198$ meters

Test 22

1. C   The antiderivative of a constant K is $Kx + C$.
2. B $z + C$
3. D $\int (x^5 - 2)dx = \int x^5 dx - 2\int dx$
$= \frac{1}{6}x^6 - 2x + C$
4. D $\int (3\sqrt{r})dr = 3\int r^{\frac{1}{2}}dr$
$= 3\frac{r^{\frac{3}{2}}}{\frac{3}{2}} + C$
$= 2r^{\frac{3}{2}} + C$
5. A $\int \frac{dy}{2y^3} = \frac{1}{2}\int y^{-3}dy$
$= \frac{1}{2}\frac{y^{-2}}{-2} + C$
$= -\frac{1}{4}y^{-2} + C$
6. A $\int x^{-2}(x^3 - 2x^2)dx = \int x\,dx - 2\int x^0 dx$
$= \frac{x^2}{2} - 2x + C$
7. C $\int \sqrt{w}\left(3\sqrt{w} - w^{\frac{1}{3}}\right)dw$
$= 3\int w\,dw - \int w^{\frac{5}{6}}dw$
$= 3\frac{w^2}{2} - \frac{w^{\frac{11}{6}}}{\frac{11}{6}} + C$
$= \frac{3}{2}w^2 - \frac{6}{11}w^{\frac{11}{16}} + C$

8. B $\int \frac{2-x}{\sqrt{x}} dx = 2\int x^{-\frac{1}{2}} dx - \int x^{\frac{1}{2}} dx$

$= 2\frac{x^{\frac{1}{2}}}{\frac{1}{2}} - \frac{x^{\frac{3}{2}}}{\frac{3}{2}} + C$

$= 4\sqrt{x} - \frac{2}{3}x^{\frac{3}{2}} + C$

9. C $y = \int (12x) dx$

$= 12\frac{x^2}{2} + C_1$

$= 6x^2 + C_1$

$y' = 6x^2 + C_1 \qquad y' = 0$ when $x = 1$

$0 = 6 + C_1 \qquad$ so $C_1 = -6$

$y' = 6x^2 - 6$

$y = \int (6x^2 - 6) dx$

$= 6\int x^2 dx - 6\int dx$

$= 6\frac{x^3}{3} - 6x + C_2$

$= 2x^3 - 6x + C_2$

$y = 2x^3 - 6x + C_2$

At point $(1, -3)$ we get:

$-3 = 2 - 6 + C_2$

$C_2 = 1$

So $y = 2x^3 - 6x + 1$

10. B $y'' = 6x - 4$

$\int (6x - 4) dx = 6\int x \, dx - 4\int dx$

$y' = 6\frac{x^2}{2} - 4x + C_1$

$= 3x^2 - 4x + C_1$

$\int (3x^2 - 4x + C_1) dx$

$y = 3\int x^2 dx - 4\int x dx + C_1 \int dx$

$= 3\frac{x^3}{3} - 4\frac{x^2}{2} + C_1 x + C_2$

$= x^3 - 2x^2 + C_1 x + C_2$

using $(0, -2)$ and $(1, -2)$

$-2 = C_2 \qquad -2 = 1 - 2 + C_1 + C_2$

$-2 = -1 + C_1 - 2$

$1 = C_1$

$y = x^3 - 2x^2 + x - 2$

Test 23

1. A $\sec(x) + C$

2. C $-\frac{1}{2}\int e^{-2x}(-2) dx = -\frac{1}{2}e^{-2x} + C$

3. D $\int (x^2 - 2x) dx = \int x^2 dx - 2\int x dx$

$= \frac{x^3}{3} - 2\frac{x^2}{2} + C$

$= \frac{1}{3}x^3 - x^2 + C$

4. C Let $u = 3 - y$; $\dfrac{du}{dy} = -1$; $du = -dy$

$$\int \dfrac{dy}{(3-y)^3} = -\int \dfrac{dy(-1)}{(3-y)^3}$$
$$= -\int u^{-3} du$$
$$= -\dfrac{u^{-2}}{-2} + C$$
$$= \dfrac{1}{2}(3-y)^{-2} + C$$

5. A Let $u = 5 - 3x^2$; $\dfrac{du}{dx} = -6x$;
$du = -6x\,dx$

$$\int x(5-3x^2)^5 dx = -\dfrac{1}{6}\int -6x(5-3x^2)^5 dx$$
$$= -\dfrac{1}{6}\int u^5 du$$
$$= -\dfrac{1}{6}\dfrac{u^6}{6} + C$$
$$= -\dfrac{1}{36} u^6 + C$$
$$= -\dfrac{1}{36}(5-3x^2)^6 + C$$

6. B Let $u = 2 - \ln(x)$; $\dfrac{du}{dx} = -\dfrac{1}{x}$;
$du = -\dfrac{1}{x} dx$

$$-\int \dfrac{(2-\ln(x))^3}{x} dx(-1) = -\int u^3 du$$
$$= -\dfrac{u^4}{4} + C$$
$$= -\dfrac{1}{4}(2-\ln(x))^4 + C$$

7. D

$$\int (1+e^x)^2 dx = \int (1+2e^x + e^{2x}) dx$$
$$= \int dx + 2\int e^x dx + \dfrac{1}{2}\int e^{2x} dx(2)$$
$$= x + 2e^x + \dfrac{1}{2} e^{2x} + C$$

8. D Let $u = 1 + \cot(2\theta)$; $\dfrac{du}{d\theta} = -\csc^2(2\theta) \cdot 2$;
$du = -2\csc^2(2\theta) d\theta$

$$-\dfrac{1}{2}\int \dfrac{\csc^2(2\theta) d\theta}{\sqrt{1+\cot(2\theta)}}(-2) = -\dfrac{1}{2}\int \dfrac{du}{\sqrt{u}}$$
$$= -\dfrac{1}{2}\int u^{-\frac{1}{2}} du$$
$$= -\dfrac{1}{2}\dfrac{u^{\frac{1}{2}}}{\frac{1}{2}} + C$$
$$= -\sqrt{u} + C$$
$$= -\sqrt{1+\cot(2\theta)} + C$$

9. A Let $u = x^3 - 15x + 3$; $\dfrac{du}{dx} = 3x^2 - 15$;
$du = (3x^2 - 15) dx$

$$\dfrac{1}{3}\int \dfrac{(x^2-5) dx}{x^3 - 15x + 3}(3) = \dfrac{1}{3}\int u^{-1} du$$
$$= \dfrac{1}{3}\ln(x^3 - 15x + 3) + C$$

10. B Let $u = 1 - 2\cos(2\theta)$; $\dfrac{du}{dx} = -2(-\sin(2\theta))2$;
$du = 4\sin(2\theta) d\theta$

$$\dfrac{1}{4}\int \dfrac{\sin(2\theta) d\theta \cdot 4}{(1-2\cos(2\theta))^3} = \dfrac{1}{4}\int u^{-3} du$$
$$= \dfrac{1}{4}\dfrac{u^{-2}}{-2} + C$$
$$= -\dfrac{1}{8}(1-2\cos(2\theta))^{-2} + C$$

Test 24

1. **C** This is the first integral property.
$$\int_a^a f(x)dx = 0$$

2. **C** This is the fourth integral property.
$$\int_a^b f(x)dx = -\int_b^a f(x)dx$$

3. **D** The middle term's value would be negative, so the absolute value would need to be taken.

4. **C** $\int_0^4 x^3 dx = \dfrac{x^4}{4}\Big|_0^4 = 64$

5. **A** $\int_2^3 (x^2 + x + 1)dx$

$= \left(\dfrac{x^3}{3} + \dfrac{x^2}{2} + x\right)\Big|_2^3$

$= \left(\dfrac{27}{3} + \dfrac{9}{2} + 3\right) - \left(\dfrac{8}{3} + \dfrac{4}{2} + 2\right)$

$= \dfrac{19}{3} + \dfrac{5}{2} + 1$

$= \dfrac{38 + 15}{6} + 1$

$= \dfrac{53}{6} + 1$

$= 8\dfrac{5}{6} + 1 = 9\dfrac{5}{6}$

6. **C** $\int_a^c f(x)dx = 0$

If $\int_a^b f(x)dx = A$, then $\int_b^c f(x)dx = -A$ and therefore $\int_a^c f(x)dx = A - A = 0$ because the graph is symmetrical.

7. **C** $f(x) = x^2$

$\int_0^3 x^2 dx = \dfrac{x^3}{3}\Big|_0^3$

$= \dfrac{27}{3} - 0 = 9$

8. **A** $y = 4 - x^2$

$\int_{-2}^0 (4 - x^2)dx = \left(4x - \dfrac{x^3}{3}\right)\Big|_{-2}^0$

$= 0 - \left(-8 + \dfrac{8}{3}\right)$

$= \dfrac{16}{3} = 5\dfrac{1}{3}$

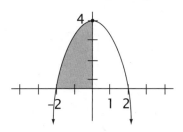

9. **B** $y = x^3 + 3x^2 - 4x$

$= x(x^2 + 3x - 4)$

$= x(x + 4)(x - 1)$

$x = 0, x = -4, x = 1$

CALCULUS

10. D $x^3 + 3x^2 - 4x = 0$
$x(x^2 + 3x - 4) = 0$
$x(x+4)(x-1) = 0$

$x = 0, 1, -4$

$\int_{-4}^{0} (x^3 + 3x^2 - 4x)dx$

$= \left(\frac{x^4}{4} + 3\frac{x^3}{3} - 4\frac{x^2}{2}\right)\Big|_{-4}^{0}$

$= 0 - (64 - 64 - 32) = 32$

$\int_{0}^{1} (x^3 + 3x^2 - 4x)dx$

$= \left(\frac{1}{4}x^4 + x^3 - 2x^2\right)\Big|_{0}^{1}$

$= \frac{1}{4} + 1 - 2 = -\frac{3}{4}$

Area $= 32 + \left|-\frac{3}{4}\right| = 32\frac{3}{4}$

Test 25
1. A

2. D   Let $u = x^2$; $\frac{du}{dx} = 2x$; $du = 2x\,dx$

$\frac{1}{2}\int_{0}^{\sqrt{\pi}/2} 2x\sin(x^2)dx$

$= \frac{1}{2}\int_{0}^{\sqrt{\pi}/2} \sin(u)du$

$= -\frac{1}{2}\cos(u)\Big|_{0}^{\sqrt{\pi}/2}$

$= -\frac{1}{2}\cos(x^2)\Big|_{0}^{\sqrt{\pi}/2}$

$= -\frac{1}{2}\left(\cos\left(\frac{\pi}{4}\right) - \cos(0)\right)$

$= -\frac{1}{2}\left(\frac{\sqrt{2}}{2} - 1\right)$

$= -\frac{\sqrt{2}}{4} + \frac{1}{2}$

3. D   Let $u = 2e^x - 1$; $\frac{du}{dx} = 2e^x$;
$du = 2e^x dx$

$\frac{1}{2}\int_{0}^{\ln(2)} \frac{2e^x dx}{2e^x - 1}$

$= \frac{1}{2}\int_{0}^{\ln(2)} \frac{du}{u}$

$= \frac{1}{2}\ln(u)\Big|_{0}^{\ln(2)}$

$= \frac{1}{2}\ln(2e^x - 1)\Big|_{0}^{\ln(2)}$

$= \frac{1}{2}\left[\ln(2e^{\ln 2} - 1) - \ln(2 - 1)\right]$

$= \frac{1}{2}\ln(2 \cdot 2 - 1)$

$= \frac{1}{2}\ln(3)$

$= \ln(\sqrt{3})$

4. **C** Let $u = \sin(2\theta)$; $\dfrac{du}{d\theta} = 2\cos(2\theta)$;
$du = 2\cos(2\theta)d\theta$

$$\int_{\pi/6}^{\pi/4} \cot(2\theta)\, d\theta = \frac{1}{2}\int_{\pi/6}^{\pi/4} \frac{\cos(2\theta)}{\sin(2\theta)}\, d\theta(2)$$

$$= \frac{1}{2}\int_{\pi/6}^{\pi/4} \frac{du}{u}$$

$$= \frac{1}{2}\ln(u)\Big|_{\pi/6}^{\pi/4}$$

$$= \frac{1}{2}\ln(\sin(2\theta))\Big|_{\pi/6}^{\pi/4}$$

$$= \frac{1}{2}\left(\ln\sin\left(\frac{\pi}{2}\right) - \ln\sin\left(\frac{\pi}{3}\right)\right)$$

$$= \frac{1}{2}\left(\ln 1 - \ln\frac{\sqrt{3}}{2}\right)$$

$$= \frac{1}{2}\left(0 - \ln\frac{\sqrt{3}}{2}\right)$$

$$\approx .07$$

5. **C** $\displaystyle\int_{\pi/3}^{2\pi/3} \csc(\theta)\cot(\theta)\, d\theta$

$$= -\csc(\theta)\Big|_{\pi/3}^{2\pi/3}$$

$$= -\left[\frac{2}{\sqrt{3}} - \frac{2}{\sqrt{3}}\right] = 0$$

6. **A** $\displaystyle\int_0^1 \frac{e^{3x}+2}{e^x}\, dx$

$$= \frac{1}{2}\int_0^1 e^{2x}dx \cdot 2 - 2\int_0^1 e^{-x}dx(-1)$$

$$= \frac{1}{2}e^{2x}\Big|_0^1 - 2e^{-x}\Big|_0^1$$

$$= \frac{1}{2}(e^2-1) - 2\left(\frac{1}{e}-1\right)$$

$$= \frac{1}{2}e^2 - \frac{1}{2} - \frac{2}{e} + 2$$

$$= \frac{3}{2} + \frac{e^2}{2} - \frac{2}{e}$$

7. **C** Let $u = e^x + 3$; $\dfrac{du}{dx} = e^x$; $du = e^x dx$

$$\int_0^1 e^x\sqrt{e^x+3}\, dx = \int_0^1 u^{1/2}du$$

$$= \frac{u^{3/2}}{3/2}\Big|_0^1$$

$$= \frac{2}{3}(e^x+3)^{3/2}\Big|_0^1$$

$$= \frac{2}{3}\left[(e+3)^{3/2} - 8\right]$$

8. **B** Let $u = x^3 - 3$; $\dfrac{du}{dx} = 3x^2$; $du = 3x^2 dx$

$$\frac{1}{3}\int_2^3 \frac{x^2 dx}{x^3-3}\cdot 3 = \frac{1}{3}\ln(u)\Big|_2^3$$

$$= \frac{1}{3}\ln(x^3-3)\Big|_2^3$$

$$= \frac{1}{3}(\ln(24) - \ln(5)) = \frac{1}{3}\ln\left(\frac{24}{5}\right)$$

9. **B** $\displaystyle\int_0^{\pi/4} \sec(\theta)\tan(\theta)\, d\theta = \sec(\theta)\Big|_0^{\pi/4}$

$$= \frac{2}{\sqrt{2}} - 1 = \sqrt{2} - 1$$

10. **A** $\displaystyle\int_0^1 e^{2x}(e^{3x}+1)\, dx$

$$= \int_0^1 e^{5x}dx + \int_0^1 e^{2x}dx$$

$$= \frac{1}{5}\int_0^1 e^{5x}dx\cdot 5 + \frac{1}{2}\int_0^1 e^{2x}\cdot 2 dx$$

$$= \frac{1}{5}e^{5x}\Big|_0^1 + \frac{1}{2}e^{2x}\Big|_0^1$$

$$= \frac{1}{5}(e^5-1) + \frac{1}{2}(e^2-1)$$

$$= \frac{1}{5}e^5 - \frac{1}{5} + \frac{1}{2}e^2 - \frac{1}{2}$$

$$= \frac{1}{5}e^5 + \frac{1}{2}e^2 - \frac{7}{10}$$

Test 26

1. **C** top curve − bottom curve in terms of x
2. **B** right curve − left curve in terms of y
3. **A** These curves intersect
    in the first quadrant when:
    $$-\frac{1}{3}x + \frac{4}{3} = x$$
    $$\frac{4}{3} = \frac{4}{3}x$$
    $$x = 1$$
    Intersection point $(1, 1)$

    They intersect
    in the second quadrant when:
    $$-\frac{1}{3}x + \frac{4}{3} = -x$$
    $$\frac{4}{3} = -\frac{2}{3}x$$
    $$x = -2$$
    Intersection point $(-2, 2)$

    Two integrals are required:
    $$\int_{-2}^{0} \left[\left(-\frac{1}{3}x + \frac{4}{3}\right) - (-x)\right]dx \text{ and}$$
    $$\int_{0}^{1} \left[\left(-\frac{1}{3}x + \frac{4}{3}\right) - x\right]dx \text{ which simplifies to:}$$
    $$\int_{-2}^{0} \left(\frac{2}{3}x + \frac{4}{3}\right)dx + \int_{0}^{1} \left(-\frac{4}{3}x + \frac{4}{3}\right)dx$$

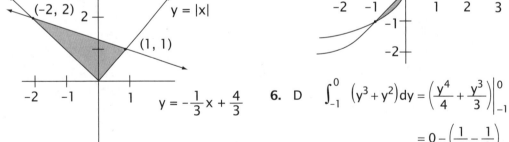

4. **C** $\int_{-2}^{0} \left(\frac{2}{3}x + \frac{4}{3}\right)dx + \int_{0}^{1} \left(-\frac{4}{3}x + \frac{4}{3}\right)dx$
    $$= \left(\frac{2}{3}\frac{x^2}{2} + \frac{4}{3}x\right)\Big|_{-2}^{0} + \left(-\frac{4}{3}\frac{x^2}{2} + \frac{4}{3}x\right)\Big|_{0}^{1}$$
    $$= 0 - \left(\frac{4}{3} - \frac{8}{3}\right) + \left(-\frac{2}{3} + \frac{4}{3}\right)$$
    $$= \frac{6}{3} = 2$$

5. **A** These graphs meet when:
    $$-y^2 = y^3$$
    $$y^3 + y^2 = 0$$
    $$y^2(y + 1) = 0$$
    $$y = 0, -1$$
    Intersection points $(0, 0)(-1, -1)$
    $$\int_{-1}^{0} (\text{right curve} - \text{left curve})dy =$$
    $$\int_{-1}^{0} (y^3 + y^2)dy$$

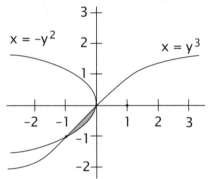

6. **D** $\int_{-1}^{0} (y^3 + y^2)dy = \left(\frac{y^4}{4} + \frac{y^3}{3}\right)\Big|_{-1}^{0}$
    $$= 0 - \left(\frac{1}{4} - \frac{1}{3}\right)$$
    $$= \frac{1}{12}$$

TEST 26 - TEST 27

7. C These curves intersect when:
$$\cos(\theta) + 2 = \sin(\theta) + 2$$
$$\cos(\theta) = \sin(\theta)$$
$$\theta = \frac{\pi}{4}$$

$$\int_0^{\frac{\pi}{4}} (\sin(\theta) + 2 - 2) d\theta + \int_{\frac{\pi}{4}}^{\frac{\pi}{2}} ((\cos(\theta) + 2) - 2) d\theta$$

$$= \int_0^{\frac{\pi}{4}} \sin(\theta) d\theta + \int_{\frac{\pi}{4}}^{\frac{\pi}{2}} \cos(\theta) d\theta$$

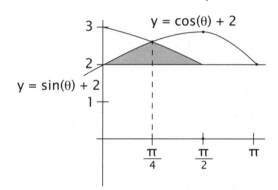

8. C $\quad -\int_0^{\frac{\pi}{4}} \sin(\theta) d\theta (-1) + \int_{\frac{\pi}{4}}^{\frac{\pi}{2}} \cos(\theta) d\theta$

$$= -\cos(\theta) \Big|_0^{\frac{\pi}{4}} + \sin(\theta) \Big|_{\frac{\pi}{4}}^{\frac{\pi}{2}}$$

$$= -\left(\frac{\sqrt{2}}{2} - 1\right) + \left(1 - \frac{\sqrt{2}}{2}\right)$$

$$= 2 - \sqrt{2}$$

9. A The intersection of $x = 1$ and $y = 2 + \ln(x)$ is the point $(1, 2)$

$$y = 2 + \ln(x)$$
$$y - 2 = \ln(x)$$
$$x = e^{y-2}$$

Solving for x and using the right curve minus the left curve.

$$\int_0^1 \left(1 - e^{y-2}\right) dy$$

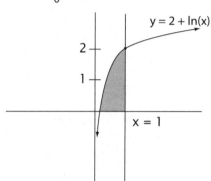

10. D $\quad \int_0^1 \left(1 - e^{y-2}\right) dy$

$$= \int_0^1 1 \, dy - \int_0^1 e^{y-2} \, dy$$

$$= y \Big|_0^1 - e^{y-2} \Big|_0^1$$

$$= 1 - \left(e^{-1} - e^{-2}\right)$$

$$= 1 - \frac{1}{e} + \frac{1}{e^2}$$

Test 27
1. D
2. A
3. A

4. B $y = \sin^{-1}(x^5)$
$$\frac{dy}{dx} = \frac{1}{\sqrt{1-(x^5)^2}} \cdot 5x^4$$
$$= \frac{5x^4}{\sqrt{1-x^{10}}}$$

5. D
$$\int_{\frac{1}{2}}^{\frac{\sqrt{2}}{2}} \frac{dx}{\sqrt{1-x^2}} = \sin^{-1}(x)\Big|_{\frac{1}{2}}^{\frac{\sqrt{2}}{2}}$$
$$= \sin^{-1}\left(\frac{\sqrt{2}}{2}\right) - \sin^{-1}\left(\frac{1}{2}\right)$$
$$= \frac{\pi}{4} - \frac{\pi}{6} = \frac{\pi}{12}$$

6. C $\lim_{x \to 0} \frac{\sin^{-1}(5x)}{7x}$

Has form $\frac{0}{0}$. Use L'Hôpital's Rule

$$= \lim_{x \to 0} \frac{\frac{1}{\sqrt{1-(5x)^2}} \cdot 5}{7}$$
$$= \lim_{x \to 0} \frac{5}{7\sqrt{1-25x^2}} = \frac{5}{7}$$

7. D $y = \cos^{-1}(\sqrt{x})$
The slope of the tangent line is the derivative.
$$\frac{dy}{dx} = \frac{-1}{\sqrt{1-(\sqrt{x})^2}} \cdot \frac{1}{2}x^{-\frac{1}{2}}$$
$$= \frac{-1}{2\sqrt{x}\sqrt{1-x}}$$
$$= \frac{-1}{2\sqrt{x-x^2}}$$

8. B

9. B   Let $u = x^2$; $\frac{dy}{dx} = 2x$; $du = 2x\,dx$
$$\int \frac{x}{1+x^4}dx \cdot 2 = \frac{1}{2}\int \frac{du}{1+u^2}$$
$$= \frac{1}{2}\tan^{-1}(u) + C$$
$$= \frac{1}{2}\tan^{-1}(x^2) + C$$

10. A   Let $u = \frac{x}{5}$; $\frac{du}{dx} = \frac{1}{5}$; $du = \frac{1}{5}dx$
$$\int \frac{dx}{\sqrt{25-x^2}} = \int \frac{dx}{\sqrt{25\left(1-\frac{x}{25}\right)}}$$
$$= \int \frac{dx}{5\sqrt{1-\frac{x^2}{25}}}$$
$$= \int \frac{du}{\sqrt{1-u^2}}$$
$$= \sin^{-1}(u) + C$$
$$= \sin^{-1}\left(\frac{x}{5}\right) + C$$

Test 28
1. D
2. B
3. A
4. B   Using #11: $a^2 = 4$ so $a = 2$.
5. D   Let $a = 3$; $b = -1$; $n = 4$
$$\int \left(\sqrt{3x-1}\right)^4 dx = \frac{2}{3}\frac{(3x-1)^6}{6} + C$$
$$= \frac{(3x-1)^6}{9} + C$$

6. A   Let $2a = \frac{1}{2}$; $a = \frac{1}{4}$; $u = x$; $du = dx$
$$\int \frac{dx}{x\sqrt{\frac{x}{2}-x^2}} = -4\sqrt{\frac{\frac{1}{2}-x}{x}} + C$$
$$= -4\sqrt{\frac{1-2x}{2x}} + C$$

7. C $\int_3^4 \frac{dx}{x^2\sqrt{x^2-4}}$

Let $a = 2$, $u = x$, $du = dx$.

$= \left(\frac{\sqrt{x^2-4}}{4x}\right)\Big|_3^4$

$= \frac{\sqrt{12}}{16} - \frac{\sqrt{5}}{12}$

$= \frac{2\sqrt{3}}{16} - \frac{\sqrt{5}}{12}$

$= \frac{\sqrt{3}}{8} - \frac{\sqrt{5}}{12} = \frac{3\sqrt{3} - 2\sqrt{5}}{24}$

8. B $\int_0^{\frac{\pi}{2}} \tan^2(2x)\, dx \quad a = 2$

$= \left(\frac{1}{2}\tan(2x) - x\right)\Big|_0^{\frac{\pi}{2}}$

$= \left(\frac{1}{2}\tan(\pi) - \frac{\pi}{2}\right) - \left(\frac{1}{2}\tan(0) - 0\right)$

$= -\frac{\pi}{2}$

9. D $\int_0^{\ln(2)} xe^{3x}\, dx \quad a = 3$

$= \left(\frac{e^{3x}}{9}(3x-1)\right)\Big|_0^{\ln(2)}$

$= \left[\frac{e^{3\ln(2)}}{9}(3\ln(2)-1)\right] - \frac{1}{9}(-1)$

$= \left[\frac{e^{\ln(2^3)}}{9}(3\ln(2)-1)\right] + \frac{1}{9}$

$= \frac{8}{9}(3\ln(2)-1) + \frac{1}{9}$

10. C $\int_2^3 \ln\left(\frac{x}{2}\right) dx = \left(x\ln\left(\frac{x}{2}\right) - x\right)\Big|_2^3 \quad a = \frac{1}{2}$

$= \left[3\ln\left(\frac{3}{2}\right) - 3\right] - \left[2\ln(1) - 2\right]$

$= \left[3\ln\left(\frac{3}{2}\right) - 3\right] + 2$

$= 3\ln\left(\frac{3}{2}\right) - 1$

Test 29

1. A $\frac{dy}{dx} + 2 = 0$

   For A, $\frac{dy}{dx} = -2$; plugging this into the differential equation gives the correct answer.

2. C $\frac{dy}{dx} = 4\sin(x) - \sin(x) \cdot e^y$

   $\frac{dy}{dx} = \sin(x)(4 - e^y)$

   $\frac{dy}{4 - e^y} = \sin(x)\, dx$

3. C The highest derivative present is a second derivative, so it has an order of 2.

4. D See Table of Selected Integrals in the instruction manual.

5. B $\frac{dy}{dx} = 2xy$

   $\frac{dy}{y} = 2x\, dx$

   $\ln(y) = x^2 + C$

   $y = e^{x^2 + C}$

6. D $\frac{dy}{dx}\cos(3x) = \sin(3x)$

   $\int dy = -\frac{1}{3}\int \frac{\sin(3x)}{\cos(3x)} dx(-3)$

   $y = -\frac{1}{3}\ln(\cos(3x)) + C$

7. C $\frac{dy}{dx} - 4x = 3$

   $\frac{dy}{dx} = 4x + 3$

   $\int dy = \int (4x + 3) dx$

   $y = 4\frac{x^2}{2} + 3x + C$

   $y = 2x^2 + 3x + C$

8. A $e^w \dfrac{dy}{dw} = e^{2w}$ $y(\ln(2)) = 5$

$dy = \dfrac{e^{2w}}{e^w} dw$

$\int dy = \int e^w dw$

$y = e^w + C$

$5 = e^{\ln(2)} + C$

$C = 3$

$y = e^w + 3$

$y = -\ln\left(\dfrac{2}{x} - 1\right)$

$y = \ln\left(\dfrac{1}{\frac{2}{x}-1}\right)$

$= \ln\left(\dfrac{\frac{1}{2-x}}{x}\right)$

$= \ln\left(\dfrac{x}{2-x}\right)$

9. B $\dfrac{dy}{dx} = \dfrac{2e^y}{x^2}$ $y(1) = 0$

$\int \dfrac{dy}{2e^y} = \int \dfrac{dx}{x^2}$

$-\dfrac{1}{2}\int e^{-y} dy(-1) = \int x^{-2} dx$

$-\dfrac{1}{2} e^{-y} = -\dfrac{1}{x} + C$

$e^{-y} = \dfrac{2}{x} - 2C$

$-y = \ln\left(\dfrac{2}{x} - 2C\right)$

$y = -\ln\left(\dfrac{2}{x} - 2C\right)$

$0 = -\ln(2 - 2C)$

$1 = (2 - 2C)^{-1}$

$1 = \dfrac{1}{2-2C}$

$2 - 2C = 1$

$-2C = -1$

$C = \dfrac{1}{2}$

10. B $(t^2 + 8)\dfrac{dR}{dt} = Rt$ $R(1) = 3e^2$

$\int \dfrac{dR}{R} = \dfrac{1}{2}\int \dfrac{t\,dt}{(t^2+8)} \cdot 2$

$\ln(R) = \dfrac{1}{2}\ln(t^2 + 8) + C$

$R = \sqrt{(t^2+8)} \cdot e^C$

$3e^2 = (\sqrt{1+8})(e^C)$

$3e^2 = 3e^C$

$C = 2$

$R = e^2\sqrt{t^2 + 8}$ is the particular solution.

Test 30

1. D
2. C
3. A $\int dy = y\,dt$

$\ln(y) = kt + C$

$y = e^{kt} + C$

$y = Ce^{kt}$

4. C $y = 1{,}000{,}000\, e^{kt}$

$400{,}000 = 1{,}000{,}000\, e^{5k}$

$\ln(.4) = 5k$

$k \approx -.1833$

$$500{,}000 = 1{,}000{,}000\, e^{-.1833t}$$
$$\ln\left(\frac{1}{2}\right) = -.1833t$$
$$t \approx 3.8 \text{ years}$$

5. D $50{,}000 = 1{,}000{,}000\, e^{-.1833t}$
$$\ln(.05) = -.1833t$$
$$t \approx 16.3 \text{ years}$$

6. B
$$T = (24-3)\, e^{kt} + 3$$
$$T = 21 e^{kt} + 3$$
$$15 = 21 e^{\frac{1}{2}k} + 3$$
$$12 = 21 e^{\frac{1}{2}k}$$
$$\ln\left(\frac{4}{7}\right) = \frac{1}{2}k$$
$$k \approx -1.119$$

7. B
$$T = 21 e^{-1.119t} + 3$$
$$7 = 21 e^{-1.119t} + 3$$
$$4 = 21 e^{-1.119t}$$
$$\ln\left(\frac{4}{21}\right) = -1.119t$$
$$t \approx 1.48 \text{ hours}$$

$1.48 \times 60 = 89$ minutes
Drink the ginger ale 89 minutes after 6pm, or 7:29pm.

8. C
$$\frac{dx}{dt} = -9.8t$$
$$\int dx = \int -9.8t$$
$$x = -9.8 \frac{t^2}{2} + C$$
$$x = -4.9t^2 + C$$

9. B When $t = 0$, $x = 500$
$$500 = -4.9(0)^2 + C$$
$$C = 500$$
$$x = -4.9t^2 + 500$$

The object will hit the ground when $x = 0$.

$$0 = -4.9t^2 + 500$$
$$t \approx 10.1 \text{ seconds}$$

10. A $\dfrac{dx}{dt} = -9.8t$ This is the velocity formula.

$$v = -9.8t$$
$$v(10.1) = -9.8(10.1)$$
$$v \approx -99 \,\frac{m}{\text{sec}}$$

Unit Test IV
I.
1. Let $u = 3 - x^2$; $du = -2x\,dx$
$$-\int 2x(3-x^2)^6 \, dx\,(-1) = -\int u^6\, du$$
$$= -\frac{u^7}{7} + C$$
$$= -\frac{1}{7}(3-x^2)^7 + C$$

2. Let $u = \sin(3\theta)$; $du = \cos(3\theta) \cdot d\theta \cdot 3$

$$\int \frac{2\cos(3\theta)d\theta}{(\sin(3\theta))^4} = \frac{2}{3}\int \frac{\cos(3\theta)d\theta \cdot 3}{(\sin(3\theta))^4}$$

$$= \frac{2}{3}\int \frac{du}{u^4}$$

$$= \frac{2}{3}\int u^{-4} du$$

$$= \frac{2}{3}(\sin(3\theta))^{-3} \cdot \frac{-1}{3} + C$$

$$= \frac{-2}{9 \cdot \sin^3(3\theta)} + C$$

3. $\int (\sec^2(2\theta) + 1) d\theta$

$$= \frac{1}{2}\int \sec^2(2\theta) d\theta \cdot 2 + \int d\theta$$

$$= \frac{1}{2}\int \sec^2(u) du + \int d\theta + C$$

$$= \frac{1}{2}\tan(2\theta) + \theta + C$$

4. $\int_1^3 (2x^2 - 3) dx = \int_1^3 2x^2 dx - \int_1^3 3 dx$

$$= 2\frac{x^3}{3}\Big|_1^3 - 3x\Big|_1^3$$

$$= \frac{2}{3}(27 - 1) - 3(3 - 1)$$

$$= \frac{2}{3}(26) - 6$$

$$= \frac{34}{3}$$

5. $\int_1^{e^2} \frac{dx}{x} = \ln(x)\Big|_1^{e^2}$

$$= \ln(e^2) - \ln(1) = 2$$

II. $\frac{dy}{dx} = x + \sin(x)$

$dy = (x + \sin(x)) dx$

$y = \frac{x^2}{2} - \cos(x) + C$

$5 = -1 + C$

$C = 6$

$y = \frac{x^2}{2} - \cos(x) + 6$

III. $y = \cos(x)$; $y = \sin(x)$; $x = 0$; $x = \frac{\pi}{2}$

$\cos(x) = \sin(x)$ when $x = \frac{\pi}{4}$

$$\int_0^{\frac{\pi}{4}} [\cos(x) - \sin(x)] dx + \int_{\frac{\pi}{4}}^{\frac{\pi}{2}} [\sin(x) - \cos(x)] dx$$

$$= [\sin(x) + \cos(x)]\Big|_0^{\frac{\pi}{4}} + [-\cos(x) - \sin(x)]\Big|_{\frac{\pi}{4}}^{\frac{\pi}{2}}$$

$$= \left(\frac{\sqrt{2}}{2} + \frac{\sqrt{2}}{2}\right) - (0 + 1) + (0 - 1) - \left(-\frac{\sqrt{2}}{2} - \frac{\sqrt{2}}{2}\right)$$

$$= \sqrt{2} - 2 + \sqrt{2} = 2\sqrt{2} - 2$$

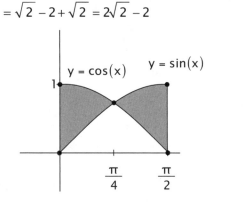

IV.
1. $\int_0^1 \frac{2dx}{1+x^2} = 2\int_0^1 \frac{dx}{1+x^2}$

$$= 2\tan^{-1}(x)\Big|_0^1$$

$$= 2(\tan^{-1}(1) - \tan^{-1}(0))$$

$$= 2\left(\frac{\pi}{4} - 0\right) = \frac{\pi}{2}$$

2. Let $u = x + 1$; $du = dx$; $n = 3$. Use #28.

$\int \sin^3(x+1) dx$

$$= -\frac{\sin^2(x+1)\cos(x+1)}{3} + \frac{2}{3}\int \sin(x+1) dx + C$$

$$= \frac{-\sin^2(x+1)\cos(x+1)}{3} - \frac{2}{3}\cos(x+1) + C$$

3. Let $u = x$; $a = 5$

$$\int \frac{dx}{x^2\sqrt{x^2-25}} = \frac{-\sqrt{x^2-25}}{-25x} + C$$
$$= \frac{\sqrt{x^2-25}}{25x} + C$$

V. $B = C_0 e^{kt}$
$= 100 e^{.02(3)}$
≈ 106 bacteria

$300 = 100 e^{.02t}$
$\ln(3) = .02t$
$t = 54.9$ hours

The bacteria will triple in approximately 54.9 hours.

VI. $y = x^2 - x - 2$

$$\int_0^2 0 - (x^2 - x - 2)dx + \int_2^3 (x^2 - x - 2)dx$$
$$= \int_0^2 (-x^2 + x + 2)dx + \frac{x^3}{3} - \frac{x^2}{2} - 2x \Big|_2^3$$
$$= \left(\frac{-x^3}{3} + \frac{x^2}{2} + 2x\right)\Big|_0^2 + \left(\frac{27}{3} - \frac{9}{2} - 6\right) - \left(\frac{8}{3} - 2 - 4\right)$$
$$= \left(\frac{-8}{3} + 2 + 4\right) + \frac{11}{6} = 5\frac{1}{6}$$

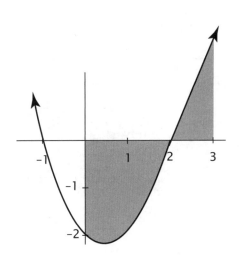

Final Test

I.

1. $f(g(x)) = f(3^x) = (3^x)^2 = 3^{2x}$
   $f(g(2)) = 3^{2(2)} = 3^4 = 81$

2. $f(x) + 2H(x) = x^2 + 2(1-x^2)$
   $= x^2 + 2 - 2x^2$
   $= 2 - x^2$

3. $\lim_{x \to 0} 3^x = 3^0 = 1$

4. $\lim_{x \to \infty} \frac{H(x)}{f(x)} \lim_{x \to \infty} \frac{1-x^2}{x^2} = \lim_{x \to \infty} \frac{\frac{1}{x^2} - 1}{1} = -1$

5. $\lim_{h \to 0} \frac{f(x+h) - f(x)}{h}$

   $= \lim_{h \to 0} \frac{2(x+h)^2 - 2x^2}{h}$

   $= \lim_{h \to 0} \frac{(2x^2 + 2xh + h^2) - 2x^2}{h}$

   $= \lim_{h \to 0} \frac{2x^2 + 4xh + 2h^2 - 2x^2}{h}$

   $= \lim_{h \to 0} \frac{4xh + 2h^2}{h}$

   $= \lim_{h \to 0} (4x + 2h) = 4x$

II.

1. $r = \sqrt{1 - 2\theta}$ $\frac{dr}{d\theta} = \frac{1}{2}(1 - 2\theta)^{-\frac{1}{2}}(-2)$
   $\frac{dr}{d\theta} = \frac{-1}{\sqrt{1-2\theta}}$

2. $w = xe^{2x}$ $\frac{dw}{dx} = x(2e^{2x}) + e^{2x}(1)$
   $= e^{2x}(2x+1)$

3. $y = \ln(\sin(\theta))$ $\frac{dy}{d\theta} = \frac{1}{\sin(\theta)} \cdot \cos(\theta)$
   $= \cot(\theta)$

III.

1. Let $u = \ln(x)$; $du = \frac{1}{x}dx$

$$\int_1^{e^2} \frac{\sqrt{\ln(x)}}{x}dx = \int_1^{e^2} u^{\frac{1}{2}}du$$

$$= \frac{u^{\frac{3}{2}}}{\frac{3}{2}}\Bigg|_1^{e^2}$$

$$= \frac{2}{3}(\ln(x))^{\frac{3}{2}}\Bigg|_1^{e^2}$$

$$= \frac{2}{3}\left[(\ln(e^2))^{\frac{3}{2}} - (\ln(1))^{\frac{3}{2}}\right]$$

$$= \frac{2}{3}(\sqrt{8})$$

$$= \frac{2}{3}(2\sqrt{2}) = \frac{4\sqrt{2}}{3}$$

2. Let $u = \sin(2x)$; $du = \cos(2x) \cdot dx \cdot 2$

$$\frac{1}{2}\int \sin(2x)\cos(2x)dx \cdot 2$$

$$= \frac{1}{2}\int u\,du$$

$$= \frac{1}{2}\cdot\frac{u^2}{2} + C$$

$$= \frac{1}{4}(\sin^2(2x)) + C$$

IV. $f(x) = -2x^3 + 6x^2 - 3$

$f'(x) = -6x^2 + 12x$
$-6x(x-2) = 0$
$x = 0$, $x = 2$ are critical points.

$f''(x) = -12x + 12$
$-12(x-1) = 0$
$x = 1$ possible inflection point

$f''(0) = 12$ so $x = 0$ is a min.
$f''(2) = -12$ so $x = 2$ is a max.

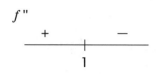

$x = 1$ is an inflection point

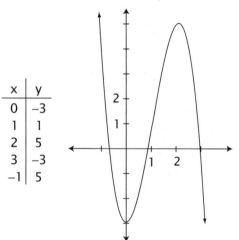

x	y
0	-3
1	1
2	5
3	-3
-1	5

f is concave up from $(-\infty, 1)$ and concave down from $(1, \infty)$.

V. $y = \sqrt{x-1}$
$y^2 = x - 1$; $x = y^2 + 1$

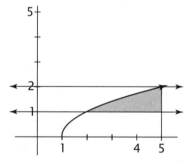

$$\int_1^2 (y^2 + 1)dy = \left(\frac{y^3}{3} + y\right)\Bigg|_1^2$$

$$= \left(\frac{8}{3} + 2\right) - \left(\frac{1}{3} + 1\right) = \frac{10}{3}$$

VI.
1. $\lim_{x\to 3}\dfrac{x^2-9}{x^2+x} = \lim_{x\to 3}\dfrac{(x-3)(x+3)}{x(x+1)} = 0$

2. $\lim_{x\to 3}\dfrac{x^2-9}{x-3} = \lim_{x\to 3}\dfrac{(x+3)(x-3)}{(x-3)} = 6$

3. $\lim_{x\to 0}\dfrac{\cos(x)-1}{\sin(x)} = \lim_{x\to 0}\dfrac{-\sin(x)}{\cos(x)} = 0$ using LR

4. $\lim_{x\to\infty}\left(1+\dfrac{5}{x}\right)^{2x}$

   $\dfrac{5}{x} \to 0$ so $\lim_{x\to\infty} 1^{2x} = 1$

5. $\lim_{x\to\infty}\dfrac{3x^2-4x+7}{x^2+3} = \lim_{x\to\infty}\dfrac{3-\dfrac{4}{x}+\dfrac{7}{x^2}}{1+\dfrac{3}{x^2}} = 3$

VII. $f(x) = x^2 - \ln(x)$

$f'(x) = 2x - \dfrac{1}{x}$

$f'(1) = 1$

$y = mx + b$
$1 = 1(1) + b$
$b = 0$

So the equation of the tangent line is $y = x$.

VIII. $y = x^2 + 1;\ y = x+3$

$x^2 + 1 = x + 3$
$x^2 - x - 2 = 0$
$(x+1)(x-2) = 0$
$x = -1, 2$

Points of intersection are $(-1, 2)$ and $(2, 5)$.

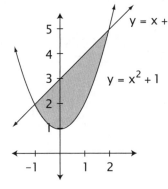

$\int_{-1}^{2}[(x+3)-(x^2+1)]dx$

$= \int_{-1}^{2}(-x^2+x+2)dx$

$= \left(-\dfrac{x^3}{3}+\dfrac{x^2}{2}+2x\right)\Big|_{-1}^{2}$

$= \left(-\dfrac{8}{3}+2+4\right)-\left(\dfrac{1}{3}+\dfrac{1}{2}-2\right)$

$= -\dfrac{8}{3}-\dfrac{5}{6}+8 = 4\dfrac{1}{2}$

IX. $2x + 4y = 16$

$y = 4 - \dfrac{1}{2}x$

$A(x) = xy$

$\quad = x\left(4-\dfrac{1}{2}x\right)$

$\quad = 4x - \dfrac{1}{2}x^2$

$A'(x) = 4 - x = 0$ when $x = 4$

$A''(x) = -1$

so $A''(4) = -1$ which yields a max.

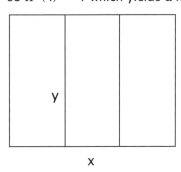

If $x = 4$, when $y = 4 - \dfrac{1}{2}(4) = 2$.

The dimensions of the largest possible area would be 4 ft by 2 ft.

Symbols, Formulas & Tables

Difference Formulas

$\sin(\alpha - \beta) = \sin\alpha \cos\beta - \cos\alpha \sin\beta$

$\cos(\alpha - \beta) = \cos\alpha \cos\beta + \sin\alpha \sin\beta$

$\tan(\alpha - \beta) = \dfrac{\tan\alpha - \tan\beta}{1 + \tan\alpha \tan\beta}$

Double angle formulas

$\sin 2\theta = 2\sin\theta \cos\theta$

$\cos 2\theta = 1 - 2\sin^2\theta$

$\tan 2\theta = \dfrac{2\tan\theta}{1 - \tan^2\theta}$

Half Angle formulas

$\sin\dfrac{\theta}{2} = \pm\sqrt{\dfrac{1 - \cos\theta}{2}}$

$\cos\dfrac{\theta}{2} = \pm\sqrt{\dfrac{1 + \cos\theta}{2}}$

$\tan\dfrac{\theta}{2} = \pm\sqrt{\dfrac{1 - \cos\theta}{1 + \cos\theta}}$

$\qquad = \dfrac{\sin\theta}{1 + \cos\theta}$

$\qquad = \dfrac{1 - \cos\theta}{\sin\theta}$

Negative Angles

$\sin -\alpha = -\sin\alpha$

$\cos -\alpha = \cos\alpha$

$\tan -\alpha = -\tan\alpha$

Law of Cosines

$a^2 = b^2 + c^2 - 2bc \cos(A)$

Sum Formulas

$\sin(\alpha + \beta) = \sin\alpha \cos\beta + \cos\alpha \sin\beta$

$\cos(\alpha + \beta) = \cos\alpha \cos\beta - \sin\alpha \sin\beta$

$\tan(\alpha + \beta) = \dfrac{\tan\alpha + \tan\beta}{1 - \tan\alpha \tan\beta}$

Pythagorean Identities

$\sin^2\theta + \cos^2\theta = 1$

$\tan^2\theta + 1 = \sec^2\theta$

$\cot^2\theta + 1 = \csc^2\theta$

Logarithmic Functions

I. $\ln(1) = 0$
II. $\ln(e) = 1$
III. $\ln(e^x) = x$
IV. $e^{\ln(x)} = x$
V. Product: $\ln(xy) = \ln(x) + \ln(y)$
VI. Quotient: $\ln(x/y) = \ln(x) - \ln(y)$
VII. Power: $\ln(x^a) = a \ln(x)$

Newton's Law of Cooling Model

$T = (T_0 - T_S)e^{Kt} + T_S$

Growth/Decay model

$y = C_0 e^{Kt}$

Chain Rule

$$\frac{dy}{dx} = \frac{dy}{du} \cdot \frac{du}{dt} \cdot \frac{dt}{dx}$$

Derivative Rules

1. $\frac{d}{dx}(C) = 0$

2. $\frac{d}{dx}(x) = 1$

3. $\frac{d}{dx}(u + v) = \frac{d}{dx}(u) + \frac{d}{dx}(v)$

4. $\frac{d}{dx}(Cv) = C\frac{d}{dx}(v)$

5. $\frac{d}{dx}(uv) = u\frac{d}{dx}(v) + v\frac{d}{dx}(u)$ **Product rule**

6. $\frac{d}{dx}v^n = nv^{n-1}\frac{d}{dx}(v)$ **Power rule**

7. $\frac{d}{dx}\left(\frac{u}{v}\right) = \frac{v\frac{d}{dx}(u) - u\frac{d}{dx}(v)}{v^2}$ **Quotient rule**

Derivatives of the Inverse Trigonometric Functions

1. $\frac{d}{dx}(\sin^{-1}(u)) = \frac{1}{\sqrt{1-u^2}} \frac{du}{dx}$ $-1 < u < 1$

2. $\frac{d}{dx}(\cos^{-1}(u)) = \frac{-1}{\sqrt{1-u^2}} \frac{du}{dx}$ $-1 < u < 1$

3. $\frac{d}{dx}(\tan^{-1}(u)) = \frac{1}{1+u^2} \frac{du}{dx}$

4. $\frac{d}{dx}(\cot^{-1}(u)) = \frac{-1}{1+u^2} \frac{du}{dx}$

5. $\frac{d}{dx}(\sec^{-1}(u)) = \frac{1}{|u|\sqrt{u^2-1}} \frac{du}{dx}$ $|u| > 1$

6. $\frac{d}{dx}(\csc^{-1}(u)) = \frac{-1}{|u|\sqrt{u^2-1}} \frac{du}{dx}$ $|u| > 1$

Fundamental Integration Formulas

1. $\int du = u + C$

2. $\int (du + dr + \ldots + dv) = \int du + \int dr + \int \ldots + \int dv$

3. $\int c\, du = c \int du$ where "c" is a constant

4. $\int u^n du = \dfrac{u^{n+1}}{n+1} + C \quad (n \neq -1)$

5. $\int \dfrac{du}{u} = \ln u + C$

6. $\int e^u du = e^u + C$

7. $\int \cos(u)\, du = \sin(u) + C$

8. $\int \sin(u)\, du = -\cos(u) + C$

9. $\int \sec^2(u)\, du = \tan(u) + C$

10. $\int \csc^2(u)\, du = -\cot(u) + C$

11. $\int \sec(u)\tan(u)\, du = \sec(u) + C$

12. $\int \csc(u)\cot(u)\, du = -\csc(u) + C$

Table of Selected Integrals

1. $\displaystyle\int \frac{u\,du}{(a+bu)^2} = \frac{1}{b^2}\left[\frac{a}{a+bu} + \ln(a+bu)\right] + C$

2. $\displaystyle\int \frac{u\,du}{(a+bu)^3} = \frac{1}{b^2}\left[-\frac{1}{a+bu} + \frac{a}{2(a+bu)^2}\right] + C$

3. $\displaystyle\int \frac{du}{u(a+bu)} = -\frac{1}{a}\ln\left(\frac{a+bu}{u}\right) + C$

4. $\displaystyle\int \frac{du}{u^2(a+bu)} = -\frac{1}{au} + \frac{b}{a^2}\ln\left(\frac{a+bu}{u}\right) + C$

5. $\displaystyle\int \frac{du}{a^2 + b^2 u^2} = \frac{1}{ab}\tan^{-1}\left(\frac{bu}{a}\right) + C$

6. $\displaystyle\int \frac{du}{a^2 - b^2 u^2} = -\frac{1}{2ab}\ln\left(\frac{a+bu}{a-bu}\right) + C$

7. $\displaystyle\int u(a^2 \pm b^2 u^2)^n\,du = \frac{(a^2 \pm b^2 u^2)^{n+1}}{\pm 2b^2(n+1)} + C$

8. $\displaystyle\int u\sqrt{a+bu}\,du = \frac{-2(2a-3bu)(a+bu)^{3/2}}{15b^2} + C$

9. $\displaystyle\int (u^2 \pm a^2)^{1/2}\,du = \frac{u}{2}\sqrt{u^2 \pm a^2} \pm \frac{a^2}{2}\ln\left(u + \sqrt{u^2 \pm a^2}\right) + C$

10. $\displaystyle\int \frac{du}{(u^2 \pm a^2)^{1/2}} = \ln\left(u + \sqrt{u^2 \pm a^2}\right) + C$

11. $\displaystyle\int \frac{du}{u(u^2 - a^2)^{1/2}} = \frac{1}{a}\sec^{-1}\left(\frac{u}{a}\right) + C$

12. $\displaystyle\int \frac{du}{u^2(u^2 \pm a^2)^{1/2}} = \frac{-\sqrt{u^2 \pm a^2}}{\pm a^2 u} + C$

13. $\displaystyle\int \frac{u^2\,du}{(a^2 - u^2)^{1/2}} = -\frac{u}{2}\sqrt{a^2 - u^2} + \frac{a^2}{2}\sin^{-1}\left(\frac{u}{a}\right) + C$

14. $\displaystyle\int \frac{du}{u^2(a^2 - u^2)^{1/2}} = -\frac{\sqrt{a^2 - u^2}}{a^2 u} + C$

15. $\displaystyle\int \sqrt{\frac{1+u}{1-u}}\,du = -\sqrt{1-u^2} + \sin^{-1}(u) + C$

16. $\int b^{au} du = \dfrac{b^{au}}{a \ln b} + C$

17. $\int u^n e^{au} du = \dfrac{u^n e^{au}}{a} - \dfrac{n}{a} \int u^{n-1} e^{au} du$

18. $\int \ln(u) du = u \ln(u) - u + C$

19. $\int u^n \ln(u) du = u^{n+1} \left[\dfrac{\ln(u)}{n+1} - \dfrac{1}{(n+1)^2} \right] + C$

20. $\int \dfrac{du}{u \ln(u)} = \ln[\ln(u)] + C$

21. $\int \sin^2(u) du = \dfrac{1}{2} u - \dfrac{1}{4} \sin(2u) + C$

22. $\int \sin(mu) \sin(nu) du = -\dfrac{\sin(m+n)u}{2(m+n)} + \dfrac{\sin(m-n)u}{2(m-n)} + C$

23. $\int u \sin(u) du = \sin(u) - u \cos(u) + C$

24. $\int \tan^n(u) du = \dfrac{\tan^{n-1}(u)}{n-1} - \int \tan^{n-2}(u) du$

25. $\int u(au+b)^{-1} du = \dfrac{u}{a} - \dfrac{b}{a^2} \ln(au+b) + C$

26. $\int (a^2 - u^2)^{\frac{1}{2}} du = \dfrac{u}{2} \sqrt{a^2 - u^2} + \dfrac{a^2}{2} \sin^{-1}\left(\dfrac{u}{a}\right) + C$

27. $\int \dfrac{du}{u(u^2 + a^2)^{\frac{1}{2}}} = -\dfrac{1}{a} \ln\left(\dfrac{a + \sqrt{u^2 + a^2}}{u}\right) + C$

28. $\int \sin^n(u) du = -\dfrac{\sin^{n-1}(u) \cos(u)}{n} + \dfrac{n-1}{n} \int \sin^{n-2}(u) du$

29. $\int \cos^n(u) du = \dfrac{\cos^{n-1}(u) \sin(u)}{n} + \dfrac{n-1}{n} \int \cos^{n-2}(u) du$

30. $\int e^{au} \ln(u) du = \dfrac{e^{au} \ln(u)}{a} - \dfrac{1}{a} \int \dfrac{e^{au}}{u} du$

Fundamental Theorem of Calculus (Part 1)

If f is continuous at every point on $[a, b]$ and F is an antiderivative of f on $[a, b]$ then:

$$\int_a^b f(x) dx = F(b) - F(a)$$

Glossary

A - B

Absolute (global) maximum or minimum - the highest or lowest point on the entire curve of a function

Absolute value - any value between the vertical symbols must be given as a positive number only

Acceleration - the rate of change of velocity

Amplitude - one-half the distance between the maximum and minimum values of a periodic function

Antiderivative - found by taking the derivative rules in reverse

Asymptotes - regions on a graph that are approached, but never touched, by a graphed function

C - E

Calculus - uses special techniques to deal with values that change at variable rates

Chain rule - used to find the derivative of composite functions

Circle - the curve defined by any quadratic expression $(x - h)^2 + (y - k)^2 = r^2$

Coefficient - a constant that is multiplied by a variable

Composite functions - two or more functions combined mathematically, the function of a function

Concavity - the curved area of a function

Constant - a quantity whose value is fixed

Constraint equation - used to solve for one variable in an optimization problem

Continuous - refers to a function which has no "gaps" in its graph

Cosecant (csc) - trig ratio formed by putting the hypotenuse over the opposite side; the reciprocal of the sine ratio

Cosine (cos) - trig ratio formed by putting the side adjacent to the given angle over the hypotenuse

Cotangent (cot) - trig ratio formed by putting the adjacent side over the opposite side; the reciprocal of the tangent ratio

Critical points - the places where it is possible for a function to have a maximum or minimum

Definite integral - used to find the area under a curvel

Dependent variable - its value depends on the value chosen for the independent variable

Derivative - denotes the rate of change at a given point of a function

Differentiation - the process of finding a derivative

Discontinuous - refers to a function which has "gaps" or "holes" in its graph

Domain - the set of all possible x values of a function

Ellipse - a "stretched" or "elongated" circle; formula is $(x - h)^2/a^2 + (y - k)^2/b^2 = 1$

Even functions - symmetric about the y-axis

Extrema - the maximum and minimum values of a function

F - H

Finite - having a countable number of elements

Frequency - the number of cycles a function completes in a normal period

Function - a relation where each value of x has one and only one value of y; symbol is $f(x)$

Fundamental Theorem of Calculus - relates derivatives to antiderivatives

Global (absolute) maximum or minimum - the highest or lowest point on the entire curve of a function

Greatest integer function - $f(x)$ is the greatest integer greater than or equal to x; $f(x) = [x]$

High order derivative - found by taking the derivative of a derivative as many times as necessary

Hyperbola - the curve defined by any quadratic expression $x^2/a^2 - y^2/b^2 = 1$ (when centered at the origin)

I - M

Implicit function - a function expressed by an equation, may not always be a true function

Implicit differentiation - a method used to find the derivative of an implicit function

Indefinite integral - notation used for the antiderivative

Independent variable - may be assigned any value desired

Inequality - a number sentence where the two sides are unequal for some or all values of the unknown

Infinite - continuing on indefinitely, cannot be counted

Inflection point - the point where the concavity of a graph changes from up to down or vice versa

Integral - in calculus, the inverse of the derivative; expresses the total change over a certain period

Integration - the process of finding an antiderivative

Interval - the area on a graph to which variables are restricted

Inverse - an operation which can "undo" another operation. The inverse of a function may or may not be a function itself.

Inverse trigonometric functions - used to find angle measurements when the side measurements of a triangle are known

L'Hôpital's Rule - connects limits with derivatives

Limit - a value that is approached but never reached by a mathematical function

Local maximum or minimum - the highest or lowest point on a particular part of the curve of a function

Logarithm - an alternative way to write an exponent; often written as log

Mean Value Theorem (MVT) - states that the average rate of change over the interval is equal to the instantaneous rate of change at one or more points

Multiplicity - refers to the number of times a given root of an equation occurs in the solution

N - P

Natural exponential function - the function described by $f(x) = e^x$

Natural log function - the function described by $f(x) = \ln x$. It is the inverse of the natural exponential function

Natural logarithm (natural log) - has a base of e (approximately 2.718)

Odd functions - symmetric about the origin

Optimization - the process of finding the maximum profit and minimum cost of producing a product

Parabola - the curve defined by any quadratic expression $y = a(x - h)^2 + k$

Period - the length of each repetition in a graph of a recurring function

Periodic function - a function which describes a recurring or cyclical behavior

Phase shift - determines the movement to the left or right on the graph of a function

Polar coordinates - values that determine a point on a circular, or polar, graph in terms of distance and direction

Polynomial - an algebraic expression with more than one term.

Pythagorean theorem - states that the sum of the squares of the legs of a right triangle equals the square of the hypotenuse.

Q - S

Quadrant - one of the four areas defined by the x- and y-axes of a Cartesian graph

Quadratic equation - an equation that has no variable with an exponential power higher than two

Radian - unit of angle measurement based on the relationship between the radius and circumference of a circle

Range - the set of all possible y values of a function

Rational roots test - a technique used to find possible solutions of higher degree polynomials

Related rate problems - use a stated rate of change for one variable to find the rate of change for another variable with a known relationship to the first.

Relation - a set of ordered pairs that may or may not be a function

Rolle's Theorem - a special case of the MTV, used in advanced algebra to prove the existence of a root of a polynomial

Roots - all values of x which make a polynomial equal to zero.

Secant (sec) - trig ratio formed by putting the hypotenuse over the adjacent side; the reciprocal of the cosine ratio

Secant line - a line that touches a curve in exactly two points

Sine (sin) - trig ratio formed by putting the side opposite the given angle over the hypotenuse

Step function - produces a graph with many discontinuities, example is greatest integer function

T - Z

Tangent (tan) - the trig ratio formed by putting the side opposite the given angle over the adjacent side

Tangent line - a line that touches a curve in exactly one point

Transformation - moving and modifying figures on a graph

Trigonometry - the study of triangles using known parts to find those that are unknown

Trigonometric functions - functions based on the relationships between the sides and angles of a triangle; sine, cosine, tangent, and their inverses

Variable - a quantity designated by a letter to which any value can be assigned

Velocity - the rate of change in the position of an object

Vertical line test - if any vertical line crosses a graph in more than one place, the graph does not represent a function

Vertical shift - the number of units a function is moved up or down on a graph

Secondary Levels Master Index

This index lists the levels at which main topics are presented in the instruction manuals for *Pre-Algebra* through *Calculus*. For more detail, see the description of each level at www.mathusee.com.

Absolute value .... Pre-Algebra, Algebra 1
Additive inverse .... Pre-Algebra
Age problems .... Algebra 2
Angles .... Geometry, PreCalculus
Angles of elevation & depression ..... PreCalc
Antiderivatives .... Calculus
Arc functions .... PreCalculus
Area.. Geometry
Associative property .... Pre-Algebra, Algebra 1
Axioms .... Geometry
Bases other than 10 .... Algebra 1
Binomial theorem......................... Algebra 2
Boat in current problems............. Algebra 2
Chemical mixtures .... Algebra 2
Circumference .... Geometry
Circumscribed figures .... Geometry
Cofunctions .... PreCalculus
Coin problems .... Algebra 1 & 2
Commutative property .... Pre-Algebra, Algebra 1
Completing the square................. Algebra 2
Complex numbers .... Algebra 2
Congruency .... Pre-Algebra, Geometry
Conic sections .... Algebra 1 & 2
Conjugate numbers .... Algebra 2
Consecutive integers.............. Algebra 1 & 2
Derivatives................................... Calculus
Determinants............................... Algebra 2
Difference of two squares ...... Algebra 1 & 2
Differential equations .... Calculus
Discriminants.............................. Algebra 2
Distance formula .... Algebra 2
Distance problems .... Algebra 2
Distributive property................ Pre-Algebra, Algebra 1
Expanded notation.................... Pre-Algebra
Exponents
    fractional............................Algebra 1 & 2
    multiply & divide ....Algebra 1 & 2
    negative .... Algebra 1 & 2
    notation .... Pre-Algebra
    raised to a power.............Algebra 1 & 2
Factoring polynomials............ Algebra 1 & 2
Functions.................. PreCalculus, Calculus

Graphs
    Cartesian coordinates .... Algebra 1
    circle .... Algebra 1 & 2, Calculus
    ellipse .... Algebra 1 & 2, Calculus
    hyperbola .... Algebra 1 & 2, Calculus
    inequality ....Algebra 1 & 2
    line................ Algebra 1 & 2, Calculus
    parabola........... Algebra 1 & 2, Calculus
    polar .... PreCalculus
    trig functions....... PreCalculus, Calculus
Greatest common factor............ Pre-Algebra
Identities .... PreCalculus
Imaginary numbers .... Algebra 2
Inequalities .... Algebra 1 & 2, PreCalculus
Inscribed figures .... Geometry
Integers .... Pre-Algebra
Integration .... Calculus
Interpolation .... PreCalculus
Irrational numbers .... Pre-Algebra
Latitude and longitude................. Geometry
Least common multiple............. Pre-Algebra
L'Hôpital's Rule .... Calculus
Limits .... PreCalculus, Calculus
Line
    graphing.........................Algebra 1 & 2
    properties of .... Geometry
Logarithms .... PreCalculus
Matrices....................................... Algebra 2
Maximum & minimum... Algebra 2, Calculus
Mean Value Theorem .... Calculus
Measurement, add & subtract ... Pre-Algebra
Metric-Imperial conversions ... Algebra 1 & 2
Midpoint formula .... Algebra 2
Motion problems......................... Algebra 2
Multiplicative inverse .... Pre-Algebra
Natural logarithms .... PreCalculus
Navigation .... PreCalculus
Negative numbers..................... Pre-Algebra
Number line............. Pre-Algebra, Algebra 1
Optimization................................ Calculus
Order of operations .... Pre-Algebra, Algebra 1
Parallel & perpendicular lines
    graphing.........................Algebra 1 & 2
    properties of .... Geometry

Pascal's triangle .... Algebra 2
Percent problems .... Algebra 2
Perimeter .... Geometry
Pi .... Pre-Algebra, Geometry
Place value .... Pre-Algebra
Plane .... Geometry
Points, lines, rays .... Geometry
Polar coordinates & graphs .... PreCalculus
Polygons
    area .... Geometry
    similar .... Pre-Algebra, Geometry
Polynomials
    add .... Pre-Algebra, Algebra 1
    divide .... Algebra 1
    factor .... Algebra 1 & 2
    multiply .... Pre-Algebra, Algebra 1 & 2
Postulates .... Geometry
Prime factorization................... Pre-Algebra
Proofs .... Geometry
Pythagorean theorem .... Pre-Algebra,
    Geometry, PreCalculus
Quadratic formula .... Algebra 2
Radians .... PreCalculus
Radicals .... Pre-Algebra,
    Algebra 1 & 2, Geometry
Ratio & proportion .... Pre-Algebra,
    Geometry, Algebra 2
Rational expressions .... Algebra 2
Real numbers .... Pre-Algebra
Reference angles .... PreCalculus
Related rates .... Calculus
Same difference theorem .... Pre-Algebra
Scientific notation .... Algebra 1 & 2
Sequences & series .... PreCalculus
Sets .... Geometry
Significant digits .... Algebra 1
Similar polygons ...... Pre-Algebra, Geometry
Simultaneous equations .... Algebra 1 & 2
Sine, cosine, & tangent, laws of. PreCalculus
Slope-intercept formula .... Algebra 1 & 2
Solve for unknown .... Pre-Algebra,
    Algebra 1 & 2
Special triangles .... Geometry, PreCalculus
Square roots ...... Pre-Algebra, Algebra 1 & 2
Surface area .... Pre-Algebra, Geometry
Temperature conversions .... Pre-Algebra
Time
    add & subtract .... Pre-Algebra
    military .... Pre-Algebra
Time & distance problems .... Algebra 2
Transformations .... Geometry
Transversals .... Geometry
Triangles .... Geometry, PreCalculus
Trig ratios .. Geometry, PreCalculus, Calculus
Unit multipliers .... Algebra 1 & 2
Vectors .... Algebra 2, PreCalculus
Volume .... Pre-Algebra, Geometry
Whole numbers .... Pre-Algebra

Calculus Index

Topic	Lesson
Absolute Value	1
Amplitude	5
Antiderivative	22, 24
Applications	18, 19, 20, 21, 30
Area	24, 26
Asymptotes	1, 3, 5, 8, 16
Axis of symmetry	2
Chain Rule	11
Circle	2
Coefficient	1
Completing the square	2
Composite functions	4, 11
Concavity	16
Conjugates	8
Constant	1
Constraint Equation	20
Continuity	1, 8
Cusp	11
Definite Integral	24, 25
Derivative	9, 10, 11, 12, 13, 14, 15, 16, 18, 19, 25, 27
Differential Equations	29, 30
Discontinuity	8, 11
Domain	1, 4
Economics Applications	19, 20
Ellipse	2
Equation of a line	2
Even Functions	17
Exponential Functions	6, 13, 30
Extrema	15
Fundamental Theorem of Calculus	24
Frequency	5
Function	1, 2, 4, 5, 6, 27
Function Notation	4
Graphs	1, 2, 3, 11, 15, 16
Greatest Integer Function	9
Half Life	30
High Order Derivatives	11, 16
Horizontal Line Test	4
Hyperbola	3
Implicit Differentiation	14
Inequalities	1, 3
Infinity	4, 8
Inflection	16
Integral	22, 23, 24, 25, 26, 28, 30
Integral Table	28
Integration Formulas	23, 28
Inverse Trigonometric Functions	27
Inverse Functions	4
Intervals	1, 4, 15, 24
L'Hôpital's Rule	17
Limits	7, 8, 17
Lines	2
Logarithmic Functions	6
Maximum & Minimum	15
Mean Value Theorem (MVT)	17
Natural Log	1, 4, 6, 13
Newton's Law of Cooling	30
Odd Functions	17
Optimization	20
Parabola	2, 15
Period	5
Phase Shift	5
Physics Applications	18
Polynomials	8
Pythagorean Theorem	9
Quadrant	5
Quadratic	2
Radian Measure	5
Range	1, 4
Rate of Change	9
Related Rates	21
Riemann Sums	24
Rolle's Theorem	17
Secant Lines	1
Semicircle	2
Step Function	9
Systems of equations	3
Tangent Lines	1, 11, 12, 15
Trigonometric Functions	5, 12, 27
Variables	1
Velocity	18
Vertical Line Test	4
Vertical Shift	5